T0281553

An Introduction to Analysis

Textbooks in Mathematics

Series editors:
Al Boggess, Kenneth H. Rosen

Functional Linear Algebra
Hannah Robbins

Introduction to Financial Mathematics
With Computer Applications
Donald R. Chambers, Qin Lu

Linear Algebra
An Inquiry-based Approach
Jeff Suzuki

The Geometry of Special Relativity
Tevian Dray

Mathematical Modeling in the Age of the Pandemic
William P. Fox

Games, Gambling, and Probability
An Introduction to Mathematics
David G. Taylor

Linear Algebra and Its Applications with R
Ruriko Yoshida

Maple™ Projects of Differential Equations
Robert P. Gilbert, George C. Hsiao, Robert J. Ronkese

Practical Linear Algebra
A Geometry Toolbox, Fourth Edition
Gerald Farin, Dianne Hansford

An Introduction to Analysis, Third Edition
James R. Kirkwood

Student Solutions Manual for Gallian's Contemporary Abstract Algebra, Tenth Edition
Joseph A. Gallian

Elementary Number Theory
Gove Effinger, Gary L. Mullen

Philosophy of Mathematics
Classic and Contemporary Studies
Ahmet Cevik

https://www.routledge.com/Textbooks-in-Mathematics/book-series/CANDHTEXBOOMTH

An Introduction to Analysis

Third Edition

James R. Kirkwood

CRC Press
Taylor & Francis Group
Boca Raton London New York

CRC Press is an imprint of the
Taylor & Francis Group, an **informa** business

A CHAPMAN & HALL BOOK

Third edition published 2021
by CRC Press
6000 Broken Sound Parkway NW, Suite 300, Boca Raton, FL 33487-2742

and by CRC Press
2 Park Square, Milton Park, Abingdon, Oxon, OX14 4RN

First edition published by Prindle, Weber, and Schmidt (PWS), 1989
Second edition published by Waveland Press, 2002

CRC Press is an imprint of Taylor & Francis Group, LLC

Library of Congress Cataloging-in-Publication Data

Names: Kirkwood, James R., author.
Title: An introduction to analysis / James R. Kirkwood.
Description: Third edition. | Boca Raton : Chapman & Hall/CRC Press, 2021. | Series: Textbooks in mathematics | Includes bibliographical references and index.
Identifiers: LCCN 2021000942 (print) | LCCN 2021000943 (ebook) | ISBN 9780367702359 (hardback) | ISBN 9781003145141 (ebook)
Subjects: LCSH: Mathematical analysis.
Classification: LCC QA300 .K523 2021 (print) | LCC QA300 (ebook) | DDC 515/.83--dc23
LC record available at https://lccn.loc.gov/2021000942
LC ebook record available at https://lccn.loc.gov/2021000943

ISBN: 978-0-367-70235-9 (hbk)
ISBN: 978-1-032-02186-7 (pbk)
ISBN: 978-1-003-14514-1 (ebk)

Contents

Preface

Preface to the First Edition

This book provides a mathematically rigorous introduction to analysis of real-valued functions of one variable. The only prerequisite is an elementary calculus sequence. A typical reader would be an average mathematics student for whom this is his or her first experience with an abstract mathematics course. Such students often find themselves floundering and discouraged after the first few weeks of the course, as it is usually impossible to make the transition from primarily computational to primarily theoretical mathematics without some frustration. This text is written in a manner that will help to ease this transition. I have followed the principle that the material should be as clear and intuitive as possible and still maintain mathematical integrity. In some instances this meant sacrificing elegance for clarity.

A major goal of this type of course is to teach students to understand mathematical proofs as well as to be able to formulate and write them. In constructing a proof, two processes are involved: (1) discovering the ideas necessary for the proof and (2) expressing these ideas in mathematical language. One of the pedagogical features of the text is frequently to supply the ideas of the proof and leave the write-up as an exercise. The text also states why a step in a proof is the reasonable thing to do and which techniques are recurrent.

Students often leave a first calculus sequence with little intuition about functions in an abstract setting. Examples, while no substitute for a proof, are a valuable tool in helping to develop intuition and are an important feature of this text. The text uses numerous examples to motivate what is to come, to highlight the need for a particular hypothesis in a theorem, and to make abstract ideas more concrete. Examples can also provide a vivid reminder that what one hopes might be true is not always true. Several of the exercises are devoted to the construction of counterexamples. Most of the exercises range in difficulty from the routine "getting your feet wet" types of problems to the moderately challenging problems.

The text provides ample material for a one-semester or a two-quarter course. The text begins with a discussion of the axioms of the real number system. The idea of a limit is introduced via sequences which seems to be the simplest setting in which limits can be rigorously discussed. A sufficient amount of the topology of the real number system is developed to obtain the familiar properties of continuous functions.

Chapters 1–6 would make a reasonable one-semester advanced calculus course and Chapters 4–6 contain material on continuous functions, differentiable functions, and the Riemann integral. Chapter 6 also contains an optional section on the Riemann-Stieltjes integral. In Chapter 6 the proof of the Riemann-Lebesgue Theorem is given. This result is often omitted in texts of this level. While this proof is substantially more difficult that the other results in the text, the statement of the theorem is not difficult to understand, and it is the fundamental existence theorem for Riemann integration. Having this result available enables the use of more natural and less tedious proofs for several other theorems. If time is limited, you might consider skipping the proof and concentrating on what the theorem says.

Chapter 7 covers series of real numbers and can be discussed immediately after Chapter 2 if you are willing to allow the integral test for convergence of series without a formal development of the Riemann integral. Chapter 8 deals with sequences and series of functions, and in Chapter 9 Fourier series are discussed. The chapter on Fourier series is intended to show an application of the earlier work, and the material is developed at a somewhat faster pace.

Obviously, the material in a text of this type is not new. All of the theorems and most of the exercises will be familiar to those who have worked in this area. I regret that it is only in rare instances that I am able to give credit to those who may have originated an idea. My intent is for the text to present the material in a way that will make the subject understandable and perhaps exciting to those who are beginning their study of abstract mathematics.

Additional comments for the Third Edition

The preface for the first edition also enunciates the goals and structure of this edition. The most significant change in this edition is the addition of a section on the Cantor set and the Cantor function and a small amount of material on connectedness.

The first edition of the text appeared around thirty years ago, and the results have been gratifying both from the reviews and informal comments of users and the number of adoptions. Personally, I have been most moved by the number of users who have said, "Finally, a math book I can understand." This was the major goal of the book.

Many people have been instrumental to the preparation of this work, but special thanks to my wife Bessie who did all the Latex work, without which the third edition would not have appeared. Her insights and cheerful attitude were immensely appreciated.

James Kirkwood

Introduction

In most elementary calculus courses, the emphasis is on learning the mechanics of calculus rather than on understanding the principles that justify the necessary techniques. Our work will focus on developing the theorems that one uses in calculus from the axioms of the real numbers. To do this, it is necessary to understand mathematical proofs. Before we begin our study, it is worthwhile to give a very brief introduction to "the rules of the game."

In a mathematical system such as the one that we shall develop, there are some objects and rules for manipulating and combining these objects. To develop any sort of theory, we need to have a common agreement about the terms we use and the rules that our system must obey. This agreement is described by a collection of definitions and axioms. There will necessarily be some undefined terms. A prime requirement (and one of the beauties) of mathematics is the lack of ambiguity of its language. It is incumbent upon us then to keep the undefined terms to an absolute minimum. As is common in modern mathematics, we shall take the word *set* as an undefined term.

Axioms and definitions are statements that all parties concerned agree to accept. They do not require proof. (In ancient times axioms were held to be self-evident truths.) An axiom is a statement that cannot be demonstrated in terms of simpler concepts. Axioms can be thought of as the fundamental building blocks of the theory. One characteristic that the axioms and definitions must have is consistency. This means that there is no statement that we can prove to be both true and false from our set of axioms and definitions. The axioms should also be independent. That is, no axiom should be derivable from the others. A third requirement for a set of axioms is that there should be sufficiently many of them to provide a meaningful theory. We want to derive as many results as we can about our system from the axioms and definitions. In other words, we seek to formulate and prove theorems. A theorem is a formula derivable from the axioms and definitions by the rules of inference.

The theorems that we shall prove are conditional. They have a hypothesis and a conclusion – sometimes called the antecedent and consequent. The hypothesis is a statement that is assumed to be true (the *if* part). The conclusion is a statement that can be deduced if the conditions of the hypothesis hold (the *then* part). We want to discuss some of the ways that theorems are stated and how we can prove them.

Suppose a theorem is of the form, "If A, then B." For example: "If f is a differentiable function, then f is continuous." One way to prove the theorem is directly. This means to show that if A is true, then B must necessarily follow. The statement, "If A, then B" can be alternatively worded, "A is sufficient for B."

There are also indirect methods of proof. The indirect method of proof that we shall most often use to prove a conditional theorem is to prove the *contrapositive* of the theorem. The contrapositive of "If A, then B is "If B is not true, then A is not true." The contrapositive of our preceding example is, "If f is a function that is not continuous then f is not differentiable." The statement, "If B is not true, then A is not true," can be alternatively worded, "B is necessary for A." One of the rules of inference is that a conditional theorem is true if the contrapositive is true and vice versa.

A second indirect method of proof of a conditional theorem is to show that it is impossible to have both the hypothesis true and the conclusion false. That is, one could prove, "If A, then B" by showing that "A and not B" is an impossibility. One example of where we shall use this technique is to prove that a convergent sequence can have only one limit point. In our proof we suppose that there exists a convergent sequence with more than one limit point and show that this is impossible. This method of proof is sometimes called *proof by contradiction.*

Another, sometimes useful, technique of proof is to consider various cases that may occur and to rule out all but one of them. For example, suppose x is a real number about which we know something and we want to show that $x > 0$. One of our axioms will be that if x is a real number, then exactly one of the following must be true: (i) $x < 0$, (ii) $x = 0$, (iii) $x > 0$. If we can rule out (i) and (ii), then (iii) must occur. Obviously, this is most useful when there are only a few cases to be considered.

If one interchanges the hypothesis and the conclusion of a theorem, the result is called the *converse* of the theorem. Thus the converse of "If A, then B" is "If B, then A." For example, the converse of "If x equals 4, then x is an integer" is "If x is an integer, then x equals 4." In this case the original statement is true, but the converse is not.

If a theorem, "If A, then B" and its converse, "If B, then A" are both true, then A and B are equivalent. To prove a theorem that is worded, "A is true if and only if B is true," we must prove two assertions: "If A, then B" and "If B, then A." Sometimes such a theorem is stated, "A is necessary and sufficient for B."

We shall often use examples to clarify explanations and to help make ideas more concrete. They are also useful in highlighting why a certain hypothesis is necessary in a theorem by showing that the theorem is not true if the hypothesis is omitted. Sometimes a problem will be stated, "Is it true that..." In such a situation you must prove that the statement is true or find an example where the assertion does not hold. Such an example is called a *counterexample.* Obviously, showing that something is true for any number of examples does not constitute a proof that a theorem is true.

Next, a word about definitions. Definitions are to be interpreted in the biconditional (if and only if) form, even though it has become customary not to state them in that fashion. For example, suppose we were to make the following definition: "A triangle with three sides of equal length is an equilateral triangle." We interpret the definition to mean that every triangle with three equal sides is an equilateral triangle and any equilateral triangle has three equal sides.

Notice the use of the words *every* and *any* in the previous sentence. These words could have been interchanged and the sentence would have exactly the same implications. Other terms that will typically have the same meaning as these two are *each* and *all.* Thus, for example, we could equally well have said, "Each triangle with three equal sides is an equilateral triangle."

This should be enough to get us started. We shall discuss other points of logic as we encounter them.

1

The Real Number System

The pattern of mathematics is to propose a set of axioms that a collection of objects and operations on the objects must satisfy and then to derive as many conclusions as possible about the system under consideration. The system is then the objects and operations together with the axioms. While courses in plane geometry emphasize this pattern, this is often not the case in algebra. Thus, one is often adept at algebraic manipulations without realizing why these operations are legitimate. Why, for example, is $3+2 = 2+3$ or $5 \times 0 = 0$? Throughout the text we shall deal with the real number system. In this first chapter, we present the preliminary material that provides the foundation for our work, including the axioms of the real numbers, and some of the consequences of these axioms.

1.1 Sets and Functions

Sets and their Operations

Perhaps the two most fundamental concepts in mathematics are sets and functions. In this section we describe some of their properties. The term *set* is left undefined and is synonymous with *collection* or *aggregate* or any of several other terms. A set may contain nothing or it may contain some objects called *elements* or *members*. If the set contains nothing, it is called the *null set* or the *empty set* and is denoted by $\emptyset$. One property that a set must have is that it must be well-defined. This means that given a set and an object, it is possible to say whether or not the object is a member of the set. Usually, sets will be denoted with upper case letters and elements by lower case letters. To express the idea that "x is an element of the set A," we write

$$x \in A$$

and to express "x is not an element of the set A," we write

$$x \notin A.$$

It may seem that anything can be a set. This is not the case. Later in this chapter, we shall show that there can be no *largest* set. Thus there is no set that contains *everything*. There are other restrictions that a set must satisfy in order to avoid some paradoxes, but these will not affect our work.[1]

Definition: Let A be a set. The *complement of A*, denoted A^c is the set of elements not in A.

We have just said that there is no largest set, and if A^c is everything not in A, then there is a difficulty. In particular, what should $\emptyset^c$ be? This problem is overcome by having a *universal* set in the background that contains just the objects in which we are interested. What this universal set actually is depends on the situation under consideration and is usually clear from the context of the discussion. For us, the universal set will usually be the real numbers. Thus, if A is the set whose elements are the numbers 1 and 2, then A^c is the set whose elements are all real numbers except 1 and 2.

There are two ways in which we shall most often describe a set. One is by simply listing the elements in the set. The other is by describing a property that those elements in the set, and only those elements, satisfy. For example, if A is the set made up of the elements 1 and 2, then we could write

$$A = \{1, 2\}$$

or

$$A = \{x | x^2 - 3x + 2 = 0\}.$$

The latter way of describing the elements in a set is called *set builder* notation. If we write

$$A = \{x | P\}$$

it means that A consists of all elements (in the universal set) that satisfy property P. In our example property P is the condition that the number be a solution to the equation $x^2 - 3x + 2 = 0$.

Definition: Let A and B be sets. We say that A is a *subset* of B if every element of A is also an element of B.

If A is a subset of B, then we write $A \subset B$ or $B \supset A$.

Definition: Two sets are *equal* if they contain exactly the same elements.

That is, a set is determined by its elements and the order of the elements is immaterial. The usual way to prove that two sets A and B are equal is to prove that $A \subset B$ and $B \subset A$.

Given two sets, we can construct new sets using these sets as building blocks. We now describe the most common ways of doing this.

---

[1] One of the most famous paradoxes arises if we try to construct a set that has itself as an element; that is, a set A with $A \in A$. This paradox is called Russell's paradox after the twentieth-century philosopher and mathematician, Bertrand Russell.

Definition: Let A and B be sets. Then

(i) the *union* of A and B, denoted $A \cup B$, is the set of elements that are in A or B.

(ii) The *intersection* of A and B, denoted $A \cap B$, is the set of elements that are in both A and B.

(iii) The *complement* of B relative to A (or A minus B), denoted $A \setminus B$, is the set of elements that are in A but not in B.

Thus

$$A \cup B = \{x | x \in A \text{ or } x \in B\}$$

$$A \cap B = \{x | x \in Ax \in B\}$$

$$A \setminus B = \{x | x \in A \text{ and } x \notin B\} = A \cap B^c.$$

When we use the word *or*, we mean it in the inclusive sense. That is, $A \cup B$ consists of the elements that are in A or B, or both A and B.

Example
Let $A = \{2, 3, 4\}$ and $B = \{3, 4, 5\}$. Then $A \cup B = \{2, 3, 4, 5\}$, $A \cap B = \{3, 4\}$, $A \setminus B = \{2\}$, $B \setminus A = \{5\}$.

The ideas of intersection and union are extendable to larger collections of sets. If J represents a set such that A_j is a set for each $j \in J$ (such a set J is called an *indexing* set) then

$$\bigcup_{j \in J} A_j = \{x | x \in A_j \text{ for } some \ j \in J\}$$

and

$$\bigcap_{j \in J} A_j = \{x | x \in A_j \text{ for } every \ j \in J\}$$

Definition: Let A and B be sets. If $A \cap B = \emptyset$, then A and B are said to be *disjoint*. If $\{A_j | j \in J\}$ is a collection of sets, the collection is said to be *pairwise disjoint* if

$$A_i \cap A_j = \emptyset \text{ if } i, j \in J \text{ and } i \neq j.$$

The last type of set construction that we consider is the Cartesian product.
Definition: Let $A_1, A_2, \ldots, A_n$ be a finite collection of sets. The *Cartesian product* of $A_1, A_2, \ldots, A_n$, denoted $A_1 \times A_2 \times \cdots \times A_n$, is defined by

$$A_1 \times A_2 \times \cdots \times A_n = \{(a_1, a_2, \ldots, a_n) | a_i \in A_i, i = 1, \ldots, n\}.$$

Thus the Cartesian product of two sets A and B is the set of all ordered pairs where the first coordinate comes from A and the second coordinate comes from B. We show in Exercise 9, Section 1.1, that if A and B are nonempty sets, then $A \times B$ is different from $B \times A$ unless $A = B$.

Set theory is a mathematical subject in itself, and we shall discuss only those results needed to prove some later theorems. To illustrate a technique that is often used to prove elementary set theoretical properties, we present the following result.

Theorem 1.1: If A and B are sets, then

(a) $A \setminus (B \cup C) = (A \setminus B) \cap (A \setminus C)$.

(b) $A \setminus (B \cap C) = (A \setminus B) \cup (A \setminus C)$.

Proof of (a): To accomplish the proof, we show

(i) $A \setminus (B \cup C) \subset (A \setminus B) \cup (A \setminus C)$, and

(ii) $(A \setminus B) \cap (A \setminus C) \subset A \setminus (B \cup C)$.

To prove (i), we choose $x \in A \setminus (B \cup C)$ arbitrarily and must show $x \in (A \setminus B) \cap (A \setminus C)$. Since $x \in A \setminus (B \cup C)$, it must be that $x \in A$ and $x \notin (B \cup C)$. Since $x \notin (B \cup C)$, then $x \notin B$ and $x \notin C$. Thus $x \in (A \setminus B)$ and $x \in (A \setminus C)$, so $x \in (A \setminus B) \cap (A \setminus C)$, which proves (i).

To prove (ii), we choose $x \in (A \setminus B) \cap (A \setminus C)$ arbitrarily and must show $x \in (A \setminus (B \cup C)$. Since $x \in (A \setminus B)$, $x \in A$ and $x \notin B$; and since $x \in (A \setminus C)$, $x \in A$ and $x \notin C$. Thus $x \notin (B \cup C)$ and $x \in A$, so $x \in A \setminus (B \cup C)$, which proves (ii).

The proof of (b) is left for Exercise 7, Section 1.1. ∎

Corollary 1.1 (De Morgan's Laws): If B and C are sets, then

(a) $(B \cup C)^c = B^c \cap C^c$.

(b) $(B \cap C)^c = B^c \cup C^c$.

Proof of (a): Let A be the universal set in the theorem. Then

$$A \setminus (B \cup C) = A \cap (B \cup C)^c = (B \cup C)^c$$

(since A is the universal set). But

$$A \setminus (B \cup C) = (A \setminus B) \cap (A \setminus C) = B^c \cap C^c.$$

Thus $(B \cup C)^c = B^c \cap C^c$. ∎

Functions

The second fundamental concept we discuss in this section is that of a function.

Definition: Let A and B be sets. A *function f* from the set A to the set B is a rule that associates with each element x in the set A a unique element $f(x)$ in the set B.

If f is a function from A to B, then we write $f : A \to B$, and if $f(a) = b$, then we say that f maps a to b.

Notes:

1. While every element in A must be mapped to some element in B by f if f is a function from A to B, it is *not* necessary that every element in B have some element in A mapped into it by f. For example, if $A = \{2, 3\}$, $B = \{4, 5, 6\}$, then $f : A \to B$ defined by $f(2) = 4$ and $f(3) = 6$ satisfies the definition of a function even though no element in A is mapped to 5.

2. It is possible that different elements of A may be mapped to the same element of B. If $A = \{0, 1\}$ and $B = \{3, 4\}$ and if $f(0) = 3$ and $f(1) = 3$ then f is a function from A to B.

Definition: Let f be a function from A to B. The set of elements in B that have some point of A mapped into them by f is called the *range* of f or the *image* of A under f.The set A is called the *domain* of f.

The range of f will be denoted by either $\mathcal{R}(f)$ or $f(A)$. The domain of f will be denoted by $\mathcal{D}(f)$. Thus, if $f : A \to B$, then

$$\mathcal{R}(f) = f(A) = \{y \in B|\ \text{there is an } x \in A \text{ with } f(x) = y\}.$$

Definition: Two functions f and g are *equal* if $\mathcal{D}(f) = \mathcal{D}(g)$ and $f(x) = g(x)$ for each x in their common domain.

Most often our functions will map from a subset of the real numbers into the real numbers and will be defined by an equation. We assume that the domain is the set of all real numbers for which $f(x)$ is defined unless it is otherwise stated. For example, if

$$f(x) = \frac{x + 1}{x - 3},$$

then the domain of f is assumed to be all real numbers except $x = 3$. In situations like this, it is common to be slightly abusive of the definition and say that f is a function from the real numbers to the real numbers, even though the domain of f is not all the real numbers.

Definition: The function $f : A \to B$ is said to be a *one-to-one* (1-1) *function* function if different elements of A are mapped by f to different elements of B. The function f is said to be an *onto function* if for each $y \in B$ there is an $x \in A$ with $f(x) = y$.

Thus in a 1-1 function, if $x_1, x_2 \in A$ with $x_1 \neq x_2$, then $f(x_1) \neq f(x_2)$. An alternative way to state the definition of a 1-1 function is to say that if $x_1, x_2 \in A$ with $f(x_1) = f(x_2)$, then it must be that $x_1 = x_2$.

Example 1.2:
Let $A = \{1, 3, 5\}$ $B = \{2, 10, 12, 15\}$, and $f : A \to B$ be defined by

$$f(1) = 10, f(3) = 2, f(5) = 12.$$

Then f is 1-1 but not onto, since no element of A is mapped to 15.

We can represent a function as a collection of ordered pairs with the first coordinate in the pair representing a point in the domain and the second coordinate representing the point in the range to which the first coordinate is mapped. In Example 1.2 we would write

$$f = \{(1, 10), (3, 2), (5, 12)\}$$

Conversely, a set of ordered pairs *may* represent a function under this convention. It will represent a function if no two different ordered pairs have the same *first* coordinate. This is required because a function cannot send the same point in the domain to two different points in the range. If we represent a function as a collection of ordered pairs, then the function is 1-1 if no two different ordered pairs have the same *second* coordinate.

Example 1.3:
Let f be a function from the real numbers to the real numbers defined by $f(x) = 3x - 7$. We shall show that f is 1-1 and onto. To show f is 1-1, we suppose that $f(x_1) = f(x_2)$ and show that it must be true that $x_1 = x_2$. Thus if

$$3x_1 - 7 = 3x_2 - 7,$$

then

$$3x_1 = 3x_2,$$

so

$$x_1 = x_2$$

as required. To show f is onto, we must choose an arbitrary real number y and find a real number x for which $f(x) = y$. So if

$$3x - 7 = y,$$

then

$$x = \frac{y + 7}{3}.$$

That is,

$$f\left(\frac{y+7}{3}\right) = y.$$

Notice that if y is a real number, so is $(y+7)/3$.

Example 1.4:

Let be a function from the real numbers to the real numbers defined by $f(x) = (x^2 - 4)/(x - 2)$. Notice that $\mathcal{D}(f) = \{x | x \neq 2\}$. We shall show that f is 1-1 but not onto. Suppose $x_1, x_2 \in \mathcal{D}(f)$ with $f(x_1) = f(x_2)$. Then

$$\frac{x_1^2 - 4}{x_1 - 2} = \frac{x_2^2 - 4}{x_2 - 2};$$

or, since $x_1, x_2 \neq 2$, we can write this as

$$x_1 + 2 = x_2 + 2,$$

so that $x_1 = x_2$ and f is 1-1.

Next we show that f is not onto. To do this, we must find a number y for which there is no $x \in \mathcal{D}(f)$ with $f(x) = y$. The clue to finding this number y is the fact that $f(x) = x + 2$ if $x \neq 2$ and 2 is not in $\mathcal{D}(f)$. Thus $2 + 2 = 4$ is a reasonable candidate for y. Suppose $f(x) = 4$. Then

$$\frac{x^2 - 4}{x - 2} = 4,$$

so

$$x^2 - 4 = 4x - 8 \text{ or } x^2 - 4x + 4 = 0.$$

But this equation is true only for $x = 2$ and $2 \notin \mathcal{D}(f)$. Thus there is no $2 \in \mathcal{D}(f)$ with $f(x) = 4$.

Definition: Suppose f and g are functions from the real numbers to the real numbers. Then:

(i) $f + g$ is the function whose domain is $\mathcal{D}(f) \cap \mathcal{D}(g)$ and $(f + g)(x) = f(x) + g(x)$ for $x \in \mathcal{D}(f) \cap \mathcal{D}(g)$.

(ii) $f \cdot g$ is the function whose domain is $\mathcal{D}(f) \cap \mathcal{D}(g)$ and $f \cdot g(x) = f(x) \cdot g(x)$ for $x \in \mathcal{D}(f) \cap \mathcal{D}(g)$.

(iii) f/g is the function whose domain is

$$\mathcal{D}(f/g) = \mathcal{D}(f) \cap \mathcal{D}(g) \cap \{x \in \mathcal{D}(g) | g(x) \neq 0\}$$

and $(f/g)(x) = f(x)/g(x)$ for all $x \in \mathcal{D}(f/g)$.

(iv) If α is a real number, then αf is a function whose domain is $\mathcal{D}(f)$ and $(\alpha f)(x) = \alpha \cdot f(x)$ for $x \in \mathcal{D}(f)$.

Notice that (iv) is actually a special case of (ii).

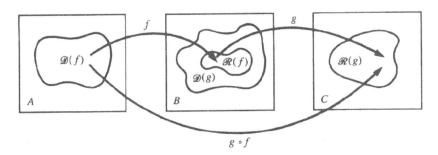

Figure 1.1
Function composition

There is really no reason to require that our functions f and g must have domain and range in the real numbers, but we do need to be able to add, multiply, and divide $f(x)$ and $g(x)$ for the definition to make sense.

There is another way to combine functions that will also be important to us.

Definition: Let A, B, and C be sets and f and g be functions with $\mathcal{D}(f) \subset A$, $\mathcal{R}(f) \subset B$, $\mathcal{D}(g) \subset B$ with $\mathcal{R}(f) \subset \mathcal{D}(g)$ and $\mathcal{R}(g) \subset C$. Then the *composition* of g with f, denoted $g \circ f$, is the function from $\mathcal{D}(f)$ to C defined by $(g \circ f)(x) = g(f(x))$.

Figure 1.1 illustrates this definition.

Example 1.5:
Let $f(x) = x^2 + 3$ and $g(x) = \sqrt{x-2}$. Then

$$\mathcal{D}(f) \supset \mathcal{R}(f) = \{x | x \geq 3\}$$

and $g \circ f(x) = g(f(x)) = g(x^2 + 3) = \sqrt{(x^2 + 3) - 2} = \sqrt{x^2 + 1}$.

Theorem 1.2: Suppose $f : A \to B$ and $g : B \to C$.

(a) If f and g are 1-1 functions, then $g \circ f$ is a 1-1 function.

(b) If f and g are onto functions, then $g \circ f$ is an onto function.

Proof: We give the proof of part (a) and leave the proof of part (b) for Exercise 1.7, Section 1.1.

To prove part (a), we choose $x_1, x_2 \in A$ with $x_1 \neq x_2$ and must show $(g \circ f)(x_1) \neq (g \circ f)(x_2)$. Now if $x_1 \neq x_2$, then $f(x_1) \neq f(x_2)$, since f is 1-1.

If $f(x_1) \neq f(x_2)$, then $g(f(x_1)) \neq g(f(x_2))$, since g is 1-1. Thus if $x_1 \neq x_2$, then $g(f(x_1)) \neq g(f(x_2))$; that is, $(g \circ f)(x_1) \neq (g \circ f)(x_2)$. ∎

Suppose that $f : X \to Y$ is a 1-1 function. Then we can define a function $f^{-1} : \mathcal{R}(f) \to X$ by $f^{-1}(y) = x$ if and only if $y = f(x)$. The reason f needs to be 1-1 for f^{-1} to exist as a function can be seen when we consider what would happen if there were $x_1, x_2 \in X$ with $f(x_1) = f(x_2) = y$. What should we do with $f^{-1}(y)$? Should it go to x_1 or x_2? This is a fatal predicament and is the reason f must be 1-1. Also notice that if f and f^{-1} are as we said, then $f^{-1}(f(x)) = x$ for $x \in \mathcal{D}(f)$ and $f(f^{-1})(y) = y$ for $y \in \mathcal{D}(f^{-1})$. Thus $f \circ f^{-1}$ is the *identity* function on $\mathcal{D}(f^{-1})$ (that is, $(f \circ f^{-1})(y) = y$), and $f^{-1} \circ f$ is the identity function on $\mathcal{D}(f)$. With this in mind, we make the following definition.

Definition: Let $f : X \to Y$ be a 1-1 function. Then the function $f^{-1} : \mathcal{R}(f) \to X$ defined by $f^{-1}(y) = x$ if $y = f(x)$ is called the *inverse* of the function f.

Example 1.6:
Let $f(x) = x^3 + 1$. We shall later establish methods to show this is a 1-1 function. For now, assume that it is. We want to find f^{-1}. We know $f^{-1}(f(x)) = x$; so if we let $y = f(x)$, this condition becomes $f^{-1}(y) = x$. Thus if we can solve for x in terms of y, we shall be done. So let

$$y = x^3 + 1 \text{ or } x = \sqrt[3]{y - 1}$$

Then

$$f^{-1}(y) = x = \sqrt[3]{y - 1}.$$

If we visualize a function f as a set of ordered pairs, then f^{-1} is the function that reverses the coordinates. So if

$$f = \{(x_1, y_1), (x_2, y_2), \dots, (x_n, y_n)\},$$

then

$$f^{-1} = \{(y_1, x_1), (y_2, x_2), \dots, (y_n, x_n)\}.$$

This gives another way of seeing why f needs to be 1-1. For f^{-1} to be a function, all the first coordinates of its ordered pairs must be different. Thus $y_i \neq y_j$ if $i \neq j$, which means f is 1-1.

An important observation: If f is a 1-1 function, then $\mathcal{D}(f) = \mathcal{R}(f^{-1})$ and $\mathcal{D}(f^{-1}) = \mathcal{R}(f)$.

The notation f^{-1} is used another way. Associated with *every* function $f : X \to Y$ (whether f is 1-1 or not) is a mapping from *subsets of* Y to *subsets of* X, denoted by f^{-1} and defined by

$$f^{-1}(B) = \{x \in X | f(x) \in B\}$$

for $B \subset Y$. Thus $f^{-1}(B) \subset X$ for $B \subset Y$. The set $f^{-1}(B)$ is called the *inverse image of B*. To say that $f(x) \in B$ is equivalent to saying $x \in f^{-1}(B)$. The use of f^{-1} to denote two different kinds of functions is unfortunate but universal. However, the way in which f^{-1} is to be interpreted is usually clear from the context of the problem and rarely leads to confusion.

Example 1.7:
Let $X = \{1, 2, 3, 4\}$, $Y = \{2, 4, 5\}$, and $f : X \to Y$ be defined by $f(1) = 5$, $f(2) = 4$, $f(3) = 5$, $f(4) = 4$. Then

$$f^{-1}(\{4, 5\}) = \{1, 2, 3, 4\},$$

$$f^{-1}(\{5\}) = \{1, 3\},$$

$$f^{-1}(\{2\}) = \emptyset.$$

Example 1.8:
Let $f : X \to Y$ with $A_1, A_2 \subset X$. We show that $f(A_1 \cup A_2) \subset f(A_1) \cup f(A_2)$. Let $y \in f(A_1 \cup A_2)$. Then there is an $x \in A_1 \cup A_2$ with $f(x) = y$. Now $x \in A_1$ or $x \in A_2$. If $x \in A_1$, then $y = f(x) \in f(A_1)$, and if $x \in A_2$, then $y = f(x) \in f(A_2)$. Thus $y = f(x) \in f(A_1)$ or $y = f(x) \in f(A_2)$, so $y \in f(A_1) \cup f(A_2)$.

Example 1.9:
Let $f : X \to Y$ with $B_1, B_2 \subset X$. We show that $f^{-1}(B_1) \cap f^{-1}(B_2) \subset f^{-1}(B_1 \cap B_2)$.
Let $x \in f^{-1}(B_1) \cap f^{-1}(B_2)$. Then $x \in f^{-1}(B_1)$ so $f(x) \in B_1$, and $x \in f^{-1}(B_2)$ so $f(x) \in B_2$. Thus $f(x) \in B_1$ and $f(x) \in B_2$, so $f(x) \in B_1 \cap B_2$, and therefore $x \in f^{-1}(B_1 \cap B_2)$.

Exercises 1.1

1. Find $A \cup B$, $A \cap B$, $A \setminus B$, $B \setminus A$, $A \times B$, and $B \times A$ if A and B are

   (a) $A = \{2, 3, 6\}$, $B = \{1, 3, 5\}$.
   (b) $A = \{0, 1\}$, $B = \emptyset$.
   (c) $A = \{1, 2, 3, 4\}$, $B = \{2, 3\}$.

2. Find $\mathcal{D}(f)$ and $\mathcal{R}(f)$ if $f(x)$ is given by

   (a) $f(x) = |x|/x$.
   (b) $f(x) = 2/(x^2 + 1)$.
   (c) $f(x) = 1/\sqrt{x - 3}$.

3. Suppose $f(x) = 3x^2 + 7$. Find $f^{-1}(A)$ if

   (a) $A = \{9, 10, 11\}$.
   (b) $A = \{1, 2, 3\}$.

4. Find $(f \circ g)(1)$ and $(g \circ f)(1)$ if f and g are defined by

   (a) $f(x) = \sin x$, $g(x) = \ln x$.
   (b) $f(x) = \sqrt{x-1}$, $g(x) = x^3 + 4$.
   (c) $f(x) = \pi$, $g(x) = \tan x$.

5. Let $f(x) = x$ and $g(x) = 1/x$. Is $f = g \circ g$?

6. Find the domain of $f \circ g$ and $g \circ f$ if $f(x) = \sqrt{x+1}$ and $g(x) = 1/(x-6)$.

7. (a) Prove that $A \setminus (B \cap C) = (A \setminus B) \cup (A \setminus C)$ for any sets A, B, and C.
   (b) Prove that $(B \cap C)^c = B^c \cup C^c$ for any sets A, B, and C.

8. Let $\{A_i | i \in I\}$ be a collection of sets. Show that:

   (a)
   $$\left(\bigcap_{i \in I} A_i \right)^c = \bigcup_{i \in I} (A_i^c).$$

   (b)
   $$\left(\bigcup_{i \in I} A_i \right)^c = \bigcap_{i \in I} (A_i^c).$$

9. Show that $A \times B = B \times A$ for nonempty sets A and B if and only if $A = B$.

10. Suppose $B = B_1 \cup B_2$. Show that $A \times B = (A \times B_1) \cup (A \times B_2)$. Is it always true that

    (a) $(A_1 \cap A_2) \times (B_1 \cap B_2) = (A_1 \times B_1) \cap (A_2 \times B_2)$?
    (b) $(A_1 \cup A_2) \times (B_1 \cup B_2) = (A_1 \times B_1) \cup (A_2 \times B_2)$?

11. Show that for any sets A, B, and C

    (a) $A \cap B \subset A$.
    (b) $A \cup B \supset A$.
    (c) $A \cup B = A$ if and only if $B \subset A$.
    (d) $A \cap B = A$ if and only if $A \subset B$.
    (e) $A \cup (B \cap C) = (A \cup B) \cap (A \cup C)$.
    (f) $A \cap (B \cup C) = (A \cap B) \cup (A \cap C)$.
    (g) $A \setminus B = A$ if and only if $A \cap B = \emptyset$.
    (h) $A \setminus B = \emptyset$ if and only if $B \supset A$.

Note: One can generalize parts e and f to show that

$$A \cup \left(\bigcap_{i \in I} B_i \right) = \bigcap_{i \in I} (A \cup B_i)$$

and

$$A \cap \left(\bigcup_{i \in I} B_i \right) = \bigcup_{i \in I} (A \cap B_i)$$

for any collection of sets $\{B_i | i \in I\}$. We shall use this generalization in later chapters.

12. Show that for any sets A and B

    (a) $A = (A \cap B) \cup (A \setminus B)$.

    (b) $A \cup B = (A \setminus B) \cup (B \setminus A) \cup (A \cap B)$ and that the sets $A \setminus B$, $B \setminus A$, and $A \cap B$ are pairwise disjoint.

13. Let $f : X \to Y$ with $A_1, A_2 \subset X$ and $B_1, B_2 \subset Y$. Show that:

    (a) $f(A_1 \cup A_2) = f(A_1) \cup f(A_2)$.

    (b) $f(A_1 \cap A_2) \subset f(A_1) \cap f(A_2)$.

    (c) $f^{-1}(B_1 \cup B_2) = f^{-1}(B_1) \cup f^{-1}(B_2)$.

    (d) $f^{-1}(B_1 \cap B_2) = f^{-1}(B_1) \cap f^{-1}(B_2)$.

    (e) $f^{-1}(Y \setminus B_1) = X \setminus f^{-1}(B_1)$.

    (f) Give an example where $f(A_1 \cap A_2) \neq f(A_1) \cap f(A_2)$.

Note: One can generalize parts a–d to arbitrary collections of sets. For example, it can be shown that

$$f^{-1}\left(\bigcup_{i \in I} B_i \right) = \bigcup_{i \in I} f^{-1}(B_i)$$

for any collection of subsets Y, $\{B_i | i \in I\}$. We shall use this generalization in later chapters.

14. Prove that if $f : X \to Y$ and for every pair of sets $A, B \subset X$, $f(A \cap B) = f(A) \cap f(B)$, then f is a 1-1 function.

15. Prove that if $f : X \to Y$ is a 1-1 function, then for any sets $A, B \subset X$, $f(A \cap B) = f(A) \cap f(B)$.

16. Suppose $f : X \to Y$ and $A \subset X$, $B \subset Y$. Show that

    (a) $f(f^{-1}(B)) \subset B$. Give an example where $f(f^{-1}(B)) \neq B$.

    (b) $A \subset f^{-1}(f(A))$. Give an example where $A \neq f^{-1}(f(A))$.

    (c) f is a 1-1 function if and only if $f^{-1}(f(A)) = A$ for every $A \subset X$.

17. Suppose $f : X \to Y$ and $g : Y \to Z$ are onto functions. Show that $f \circ g$ maps X onto Z.

18. Show that if f and g are 1-1 functions such that $f \circ g$ is defined, then $(f \circ g)^{-1} = g^{-1} \circ f^{-1}$.

19. Show that $f : X \to Y$ is 1-1 and onto if and only if for each set $A \subset X$, $f(A^c) = [f(A)]^c$.

20. Suppose that A_k is a set for each positive integer k.

    (a) Show that $x \in \cap_{n=1}^{\infty} (\cup_{k=n}^{\infty} A_k)$ if and only if $x \in A_k$ for infinitely many sets A_k.

    (b) Show that $x \in \cup_{n=1}^{\infty} (\cap_{k=n}^{\infty} A_k)$ if and only if $x \in A_k$ for all but finitely many of the sets A_k.

1.2 Properties of the Real Numbers as an Ordered Field

The underlying structure for the topics that we shall study is the real number system. The properties of the real numbers can be broadly grouped into two classes: algebraic and geometric, or topological. In this section we begin by stating some of the axioms that govern the algebraic properties of the real number system and deriving some familiar consequences.

Axioms of a Field

The real number system, which will be denoted from now on by $\mathbb{R}$, is an example of a mathematical structure known as a *complete ordered field*.

Definition: A *field* F is a nonempty set together with two operations $+$ and $\cdot$ called addition and multiplication, which satisfy the following axioms:

(A1) The operations $+$ and $\cdot$ are *binary* operations; that is, if $a, b \in F$, then $a + b$ and $a \cdot b$ are uniquely determined elements of F.

(A2) The operations $+$ and $\cdot$ are associative. That is, for $a, b, c \in F$

$$(a + b) + c = a + (b + c) \text{ and } (a \cdot b) \cdot c = a \cdot (b \cdot c).$$

(A3) The operations $+$ and $\cdot$ are commutative. That is, for $a, b \in F$

$$a + b = b + a \text{ and } a \cdot b = b \cdot a.$$

(A4) The distributive property holds. That is, for $a, b, c \in F$

$$a \cdot (b + c) = (a \cdot b) + (a \cdot c).$$

(A5) There exist additive and multiplicative identities. That is, there are elements 0 and 1 in F for which

$$0 + a = a \text{ and } 1 \cdot a = a$$

for every $a \in F$.

(A6) There exists an additive inverse for each $a \in F$. That is, if $a \in F$, there is an element in F denoted $-a$ for which

$$a + (-a) = 0.$$

(A7) For each $a \in F$ for which $a \neq 0$, there is a multiplicative inverse. That is, if $a \in F$ and $a \neq 0$, there is an element in F denoted a^{-1} or $1/a$ for which

$$a \cdot a^{-1} = 1.$$

We shall prove a few propositions to show how these axioms can be used to derive some of the other well-known properties of the real numbers. This is not a complete list, and a thorough study of these properties is better left to a course in abstract algebra.

Theorems 1.3 and 1.4 provide the key to the proofs of these properties.

Theorem 1.3: Let F be a field. Then the additive identity and the multiplicative identity are unique.

Proof: We show that the additive identity is unique and leave it for Exercise 1, Section 1.2, to show that the multiplicative identity is unique. The usual way to prove uniqueness of an object is to suppose that there are two elements that possess the property that defines the object and then to show that the elements are equal.

Suppose F is a field with two additive identities, call them 0 and $\bar{0}$. Since 0 is an additive identity

$$\bar{0} + 0 = \bar{0},$$

and since $\bar{0}$ is an additive identity

$$0 + \bar{0} = 0.$$

Since addition is commutative, we have

$$\bar{0} = \bar{0} + 0 = 0 + \bar{0} = 0.$$

So $\bar{0}$ and 0 are the same. Thus there is only one additive identity. ■

Theorem 1.4: Let F be a field. Then the additive inverse and multiplicative inverse of an element are unique.

Proof: We do the proof for the multiplicative inverse. Let $a \in F$, $a \neq 0$. Suppose b and $\bar{b}$ are multiplicative inverses of a. That is, suppose

$$a \cdot b = b \cdot a = 1 \text{ and } a \cdot \bar{b} = \bar{b} \cdot 1 = 1.$$

Then

$$\bar{b} = \bar{b} \cdot 1 = \bar{b} \cdot (a \cdot b) = (\bar{b} \cdot a) \cdot b = 1 \cdot b = b$$

where we have used the associative property in the third equality. Thus $\bar{b} = b$, and the multiplicative inverse is unique.

It is left for Exercise 1, Section 1.2, to show that the additive inverse is unique. ∎

Theorem 1.5: Let F be a field. Then $a \cdot 0 = 0$ for every $a \in F$.

Proof: For $a \in F$, $a + 0 = a$. Then $a \cdot a = a \cdot (a + 0) = a \cdot a + a \cdot 0$. Adding $-(a \cdot a)$ to both sides of the equation gives $0 = a \cdot 0$. ∎

Theorem 1.6: Let F be a field and $a, b \in F$. Then

((a)) $a \cdot (-b) = (-a) \cdot b = -(a \cdot b)$.

((b)) $-(-a) = a$.

((c)) $(-a) \cdot (-b) = a \cdot b$.

Proof: The main tools that we use to construct the proof are Theorem 1.5 and the fact that additive inverses are unique.

(a) By Theorem 1.5 and Axiom A4

$$0 = a \cdot 0 = a \cdot (b + (-b)) = a \cdot b + a \cdot (-b).$$

Thus $-(a \cdot b)$ and $a \cdot (-b)$ are additive inverses of $a \cdot b$. Since additive inverses are unique, $-(a \cdot b) = a \cdot (-b)$. Showing that $-(a \cdot b) = (-a) \cdot b$ is similar and is left as Exercise 2, Section 1.2.

(b) By definition $-(-a)$ is the additive inverse of $(-a)$. But $a + (-a) = 0$, so a is also the additive inverse of $(-a)$. Using the fact that additive inverses are unique, we conclude $a = -(-a)$.

(c) By Theorem 1.5

$$0 = (-a) \cdot 0 = (-a) \cdot (b + (-b)) = (-a) \cdot b + (-a) \cdot (-b).$$

Thus $(-a) \cdot (-b)$ is the additive inverse of $(-a) \cdot b$. By part a of this theorem, $(-a) \cdot b = -(a \cdot b)$, so $(-a) \cdot (-b)$ and $a \cdot b$ are additive inverses of $-(a \cdot b)$. Thus $a \cdot b = (-a) \cdot (-b)$, since additive inverses are unique. ∎

Hereafter $a \cdot b$ will usually be written as ab and $a + (-b)$ will be written as $a - b$.

The Order Axiom

Definition: Let F be a field. Then F is an *ordered field* if it satisfies the additional axiom:

(A8) There is a nonempty subset P of F (called the positive subset) for which

(i) If $a, b \in P$, then $a + b \in P$ (closure under addition).

(ii) If $a, b \in P$, then $ab \in P$ (closure under multiplication).

(iii) For any $a \in F$ exactly one of the following holds: $a \in P$, $-a \in P$, or $a = 0$ (law of trichotomy).

For us, P will be the set of positive real numbers. We use this axiom to define an order relation on the real numbers.

Definition: Let F be an ordered field, and let P be the positive subset of F. Let $a, b \in F$. We say $a < b$ if $b - a \in P$. We say $a \leq b$ if $a < b$ or $a = b$. The statements $a < b$ and $b > a$ are equivalent.

Remark: This ordering also induces an ordering on any subset of the real numbers.

Notice that if $a, b \in F$, then either $b - a \in P$ $(a < b)$ or $-(b - a) = a - b \in P$ $(b < a)$ or $b - a = 0$ $(a = b)$ by the law of trichotomy. Thus we can *compare* any two elements of an ordered field. Also $a > 0$ if and only if $a \in P$, and if $a > 0$, then $-a < 0$.

Theorem 1.7: Let F be an ordered field. For $a, b, c \in F$ the following hold:

(a) If $a < b$, then $a + c < b + c$.

(b) If $a < b$ and $b < c$, then $a < c$. (This says the relation $<$ is *transitive*.)

(c) If $a < b$ and $c > 0$, then $ac < bc$.

(d) If $a < b$ and $c < 0$, then $bc < ac$.

(e) If $a \neq 0$, then $a^2 > 0$.

Proof: The main device that is used in proving the theorem is Axiom A8 of the definition of an ordered field.

(a) By definition of the relation $<$, $a < b$ means $(b - a) \in P$. Then

$$b - a = (b - a) + 0 = (b - a) + (c - c) = (b + c) - (a + c) \in P.$$

Thus $(b + c) > (a + c)$. The observant reader will protest that we have not shown that $-a - c = -(a + c)$. This is left as Exercise 2, Section 1.2.

(b) Since $a < b$ and $b < c$, then $(b - a) \in P$ and $(c - b) \in P$. Since P is closed under addition,

$$(c - b) + (b - a) = c + ((-b) + b) + (-a) = c - a \in P.$$

Thus $a < c$.

(c) The hypotheses imply that $(b - a) \in P$ and $c \in P$. Since P is closed under multiplication, $(b - a)c = (bc - ac) \in P$. That is, $ac < bc$.

(d) The proof is similar to part c with one modification and is left as Exercise 3, Section 1.2.

(e) Since $a \neq 0$, either $a > 0$ (i.e., $a \in P$) or $a < 0$ (i.e., $0 - a = -a \in P$) by the trichotomy property of P. Since P is closed under multiplication, in the first case we have $a \cdot a = a^2 \in P$. In the second case $(-a) \cdot (-a) = a^2 \in P$. So in either case, $a^2 > 0$. ∎

Definition: An *interval* of real numbers is a set A containing at least two numbers such that if $r, s \in A$ with $r < s$ and if t is a number such that $r < t < s$, then $t \in A$.

A set consisting of a single point is sometimes called a degenerate interval. An interval is one of the following sets:

$$\mathbb{R} = (-\infty, \infty)$$

(i) $(a, b) = \{x | a < x < b\}$

(ii) $[a, b] = \{x | a \leq x \leq b\}$

(iii) $[a, b) = \{x | a \leq x < b\}$

(iv) $(a, b] = \{x | a < x \leq b\}$

(v) $(-\infty, a) = \{x | x < a\}$

(vi) $(-\infty, a] = \{x | x \leq a\}$

(vii) $(a, \infty) = \{x | x > a\}$

(viii) $[a, \infty) = \{x | x \geq a\}$,

where $a < b$.

Intervals of types (i), (v), and (vii) are called *open intervals*, and intervals of types (ii), (vi), and (viii) are called *closed intervals* for reasons that we shall see later. The interval $(-\infty, \infty)$ is both open and closed.

The real numbers have several subsets that we shall often use. Since we use them so often, we establish some notation for then, including

$\mathbb{N}$ = natural numbers = $\{1, 2, 3, \dots\}$

$\mathbb{Z}$ = integers = $\{\dots, -2, -1, 0, 1, 2, \dots\}$

$\mathbb{Q}$ = rational numbers = $\{x \mid x \in \mathbb{R}, x = p/q \text{ where } p, q \in \mathbb{Z}, \text{ and } q \neq 0\}$.

Real numbers that are not rational are called *irrational numbers.*

We assume that the usual laws of exponents are valid (e.g., $x^{a+b} = x^a \cdot x^b$, $x^{a-b} = x^a/x^b$, etc.). There are a couple of things about which we want to be careful, however. One is the use of rational and irrational exponents.

We cannot prove the next theorem now, because it depends on an axiom of the real numbers we have not yet stated – the completeness axiom. We shall prove this theorem later, but it is convenient to state it now because we use it to discuss some order properties.

Theorem 1.8: Let x be a positive real number, and let n be a positive integer. Then there is a unique positive number y such that $y^n = x$. ■

Using Theorem 1.8, we can make the following definition.

Definition: Let x be a positive real number and n a positive integer. Then $x^{1/n} = \sqrt[n]{x}$ is that unique positive real number for which $(x^{1/n})^n = x$.

Thus if p and q are positive integers, then $x^{p/q} = (x^{1/q})^p$.

Theorem 1.9: Let x be a positive real number, and let s_1 and s_2 be positive rational numbers with $s_1 < s_2$. Then

(a) $x^{s_1} < x^{s_2}$ if $x > 1$.

(b) $x^{s_1} > x^{s_2}$ if $0 < x < 1$.

We leave the proof as Exercise 10, Section 1.2. ■

Since $x^{-a} = 1/x^a$ we can make analogous conclusions for negative rational numbers.

These theorems are important because later they will enable us to define positive numbers raised to irrational powers. For example, we need them to know what $2^{\sqrt{3}}$ means if we are going to discuss the function $f(x) = 2^x$ where x can be any real number.

Mathematical Induction

Some of the steps involved in the proofs of Theorems 1.8 and 1.9 and some other theorems we shall prove involve a method of proof called *proof by induction.* We digress for a bit to discuss this technique.

We state the principles involved and give some examples of how the technique is used but do not elaborate on the reasons for its validity. There is more than one form of mathematical induction, but the one that we use is as follows:

Induction Principle: Suppose that for each positive integer n there is a statement $P(n)$. Suppose that

(i) $P(1)$ is true.

(ii) For any positive integer k, if $P(k)$ is true, then $P(k+1)$ must be true.

Then $P(n)$ is true for every positive integer n.

Example 1.10:
We show that for each positive integer n

$$\sum_{j=1}^{n} j = \frac{n(n+1)}{2}.$$

According to the Induction Principle, we must first check that $P(1)$ is true. $P(1)$ is the statement

$$\sum_{j=1}^{1} j = \frac{1(1+1)}{2}$$

which is true. Next we assume that $P(k)$ is true. That is, we assume

$$\sum_{j=1}^{k} j = \frac{k(k+1)}{2}. \tag{1.1}$$

We must next show that $P(k+1)$ is true; that is,

$$\sum_{j=1}^{k+1} j = \frac{(k+1)[(k+1)+1]}{2}.$$

The verification that $P(k+1)$ is true will almost always use the assumption that $P(k)$ is true. We exhibit that here. Now

$$\sum_{j=1}^{k+1} j = \sum_{j=1}^{k} j + (k+1) = \frac{k(k+1)}{2} + (k+1)$$

where the second equality follows from (1.1). Then

$$\frac{k(k+1)}{2} + (k+1) = \frac{k(k+1) + 2(k+1)}{2} = \frac{(k+1)(k+2)}{2}$$

$$= \frac{(k+1)[(k+1)+1]}{2}$$

which shows that $P(k + 1)$ is true. Thus by the Induction Principle, we can conclude that

$$\sum_{j=1}^{n} j = \frac{n(n+1)}{2}$$

for every positive integer n.

As the second example, we prove the Binomial Theorem. To do this, we need to establish some notation. For a positive integer k we define $k!$ (read "k factorial") to be the product of all the positive integers up to and including k. That is, $k! = 1 \cdot 2 \cdot 3 \cdots k$. Also 0! is defined to be 1. For $m \geq k$ and m and k non negative integers, let

$$\binom{m}{k} = \frac{m!}{k!(m-k)!}.$$

To facilitate the proof of the Binomial theorem, we first prove the following theorem.

Theorem 1.11: For m and j positive integers with $j \leq m$

$$\binom{m}{j} + \binom{m}{j-1} = \binom{m+1}{j}.$$

Proof: We have

$$\binom{m}{j} + \binom{m}{j-1} = \frac{m!}{j!(m-j)!} + \frac{m!}{[m-(j-1)]!(j-1)!}$$

$$= \frac{m!(m+1-j)}{j!(m-j)!(m+1-j)} + \frac{m!j}{(m+1-j)!(j-1)!j}$$

$$= \frac{m![(m+1-j)+j]}{(m+1-j)!j!} = \frac{(m+1)!}{(m+1-j)!j!} = \binom{m+1}{j}. \quad \blacksquare$$

Theorem 1.12 (Binomial Theorem): Let a and b be real numbers, and let m be a positive integer. Then

$$(a+b)^m = \sum_{j=0}^{m} \binom{m}{j} a^j b^{m-j}.$$

Proof: We do the proof by induction. For $m = 1$ we have

$$\sum_{j=0}^{1} \binom{1}{j} a^j b^{1-j} = \binom{1}{0} b + \binom{1}{1} a = b + a = (a+b)^1.$$

Assume the theorem is true for $m = k$. That is, assume that

$$(a + b)^k = \sum_{j=0}^{k} \binom{k}{j} a^j b^{k-j}. \tag{1.2}$$

We must show the theorem is true for $m = k + 1$. That is, we must show that

$$(a + b)^{k+1} = \sum_{j=0}^{k+1} \binom{k+1}{j} a^j b^{k+1-j}. \tag{1.3}$$

Once this is done, the Induction Principle says that we have accomplished the proof.

To prove (1.3):

$$(a + b)^{k+1} = (a + b)^k (a + b) = \left(\sum_{j=0}^{k+1} \binom{k}{j} a^j b^{k-j} \right) (a + b).$$

by (1.2). Expanding this product gives

$$\sum_{j=0}^{k} \binom{k}{j} a^{j+1} b^{k-j} + \sum_{j=0}^{k} \binom{k}{j} a^j b^{k+1-j}.$$

We eventually want to add these two summands. We begin by letting $j + 1 = t$ in the first summand. Using this substitution

$$\sum_{j=0}^{k} \binom{k}{j} a^{j+1} b^{k-j} = \sum_{t=1}^{k+1} \binom{k}{t-1} a^t b^{k-(t-1)} \tag{1.4}$$

$$= \sum_{t=1}^{k} \binom{k}{t-1} a^t b^{k+1-t} + a^{k+1}.$$

In the second summand, we split the first term off so that

$$\sum_{j=0}^{k} 6k \binom{k}{j} a^j b^{k+1-j} = \sum_{j=1}^{k} \binom{k}{j} a^t b^{k+1-j} + b^{k+1}. \tag{1.5}$$

In 1.4 we rename the index of summation j. Then from 1.4 and 1.5, we have

$$(a + b)^{k+1} = a^{k+1} + \sum_{j=1}^{k} \binom{k}{j-1} a^j b^{k+1-j} + \sum_{j=1}^{k} \binom{k}{j} a^j b^{k+1-j} + b^{k+1}$$

$$= a^{k+1} + \sum_{j=1}^{k} \left[\binom{k}{j-1} + \binom{k}{j} \right] a^j b^{k+1-j} + b^{k+1}. \tag{1.6}$$

By Theorem 1.11, $\binom{k}{j-1} + \binom{k}{j} = \binom{k+1}{j}$, so 1.6 can be written

$$a^{k+1} + \sum_{j=1}^{k} \binom{k+1}{j} a^j b^{k+1-j} + b^{k+1} = \sum_{j=0}^{k+1} \binom{k+1}{j} a^j b^{k+1-j},$$

since $a^{k+1} = \binom{k+1}{k+1} a^{k+1} b^0$ and $b^{k+1} = \binom{k+1}{0} a^0 b^{k+1}$. ∎

The Absolute Value Function

An extremely important function on the real numbers is the *absolute value function*. It is particularly important because it provides a basis for the geometrical or topological properties of the real numbers, and these are the properties with which we shall be primarily concerned.

Definition: For $a \in \mathbb{R}$, the *absolute value* of a, denoted $|a|$, is given by

$$|a| = \begin{cases} a, & \text{if } a \geq 0 \\ -a, & \text{if } a < 0. \end{cases}$$

The following theorem gives some of the properties of the absolute value function that we shall often use.

Theorem 1.13: For $a, b \in \mathbb{R}$ the following hold:

(a) $|a| \geq 0$ with equality if and only if $a = 0$.

(b) $|a| = |-a|$.

(c) $-|a| \leq a \leq |a|$.

(d) $|ab| = |a| \cdot |b|$.

(e) $1/|b| = |1/b|$ if $b \neq 0$.

(f) $|a/b| = |a|/|b|$ if $b \neq 0$.

(g) $|a| < b$ if and only if $-b < a < b$.

(h) $|a + b| \leq |a| + |b|$ (triangle inequality).

(i) $||a| - |b|| \leq |a - b|$.

Proof: We give the proof of part (h) and leave the proof of the remaining parts as Exercises 13 and 14, Section 1.2. To prove $|a+b| \leq |a| + |b|$, note that by part (c)

$$-|a| \leq a \leq |a| \text{ and } -|b| \leq b \leq |b|$$

so that by adding the inequalities (which we show is valid in Exercise 4, Section 1.2),

$$-|a| - |b| \leq a + b \leq |a| + |b| \text{ or } -(|a| + |b|) \leq a + b \leq |a| + |b|$$

so

$$|a + b| \le |a| + |b|$$

by part (g). ∎

The absolute value function and the triangle inequality will be used extensively. Suppose we define a function $d : \mathbb{R} \times \mathbb{R} \to \mathbb{R}$ by $d(a, b) = |a - b|$. Then

(i) $d(a, b) \ge 0$ and $d(a, b) = 0$ only if $a = b$, since $|a - b| \ge 0$, and $|a - b| = 0$ only if $a = b$.

(ii) $d(a, b) = d(b, a)$ since $d(a, b) = |a - b| = |-(a - b)| = |b - a| = d(b, a)$.

(iii) $d(a, c) \le d(a, b) + d(b, c)$, since

$$d(a, c) = |a - c|$$
$$= |(a - b) + (b - c)| \le |a - b| + |b - c| \text{ (by the triangle inequality)}$$
$$= d(a, b) + d(b, c).$$

Definition: If X is a set and d is a function $d : X \times X \to \mathbb{R}$ satisfying the preceding properties (i), (ii), and (iii), then d is called a *metric* on X.

A metric may be thought of as a function that measures the distance between points, and it will often be advantageous to think of $|a - b|$ as the distance between a and b.

Exercises 1.2

1. (a) Show that the multiplicative identity in a field is unique.

   (b) Show that the additive inverse in a field is unique.

2. If F is a field and $a, b \in F$, show that

   (a) $-(a \cdot b) = (-a) \cdot b$.

   (b) $-a - b = -(a + b)$.

3. Show that if F is an ordered field and $a, b, c \in F$ with $c < 0$ and $a < b$, then $ac > bc$.

4. Show that if F is an ordered field and $a, b, c, d \in F$ with $a < b$ an $c < d$, then

   (a) Show that $a + c < b + d$.

   (b) Give an example to show that it is not necessarily true that $ac < bd$.

5. Show that if F is an ordered field and $a, b \in F$ with $a \le b$ and $b \le a$, then $a = b$.

6. Show that for any positive integer n

   (a) $\sum_{j=1}^{n} j^2 = n(n+1)(2n+1)/6$.

   (b) $\sum_{j=1}^{n} j^3 = [n(n+1)/2]^2$.

   (c) For any real number $x \ne 1$

   $$1 + x + x^2 + \cdots + x^n = \frac{1 - x^{n+1}}{1 - x}.$$

   The following exercises refer to the field of real numbers.

7. Show that

   (a) $1 > 0$.

   (b) If $0 < a < b$, then $0 < 1/b < 1/a$.

   (c) If $0 < a < b$, then $a^n < b^n$ for any positive integer n.

   (d) Show that if a and b are positive umbers and $a^n < b^n$ for some positive integer n, then $a < b$.

   (e) Show for any real numbers a and b that $|a| \le |b|$ if and only if $a^2 \le b^2$.

   (f) Prove Theorem 1.10.

8. Use induction to prove the following:

   (a) If $0 < x < 1$ and n is a positive integer, then $x^{n+1} < x^n < 1$. If m and n are positive integers with $m > n$, then $x^m < x^n$.

   (b) If $x > 1$ and n is a positive integer, then $x^{n+1} > x^n > 1$. If m and n are positive integers with $m > n$, then $x^m > x^n$.

   (c) If $x > 0$ and n is a positive integer, then $0 < x < (x+1)^n$.

9. (a) Show that if $0 < x < 1$ and n is a positive integer, then $x^{1/n} < 1$. If m and n are positive integers with $m > n$, then $x^{1/m} > x^{1/n}$.

   (b) Show that if $x > 1$ and n is a positive integer, then $x^{1/n} > 1$. If m and n are positive integers with $m > n$, then $x^{1/m} < x^{1/n}$.

10. (a) Show that if $0 < x < 1$ and s_1 and s_2 are positive rational numbers with $s_1 < s_2$, then $x^{s_1} > x^{s_2}$. [Hint: Write $s_1 = a/n$, $s_2 = b/n$ where a, b, and n are positive integers with $a < b$.]

    (b) Show that if $x > 1$ and s_1 and s_2 are positive rational numbers with $s_1 < s_2$, then $x^{s_1} < x^{s_2}$.

11. If $a, b \geq 0$, show that $\sqrt{ab} \leq (a+b)/2$. This says that the arithmetic mean of two non negative numbers dominates the geometric mean.

12. Use induction to show that $(1+a)^n \geq 1 + na$ for n a positive integer and $a > -1$.

13. Prove parts (a)-(g) of Theorem 1.13.

14. (a) Show that $|a| - |b| \leq |a - b|$.
    (b) Show that $||a| - |b|| \leq |a - b|$.

15. Prove by induction that if $a_1, a_2, \ldots, a_n$ are real numbers, then
$$|a_1 + \cdots + a_n| \leq |a_1| + \cdots + |a_n|.$$

16. Suppose a_n, L, and ϵ are real numbers. Show that $|a_n - L| < \epsilon$ if and only if $a_n \in (L - \epsilon, L + \epsilon)$.

17. Show that the function d is a metric where
$$d(x, y) = \frac{|x - y|}{1 + |x + y|}.$$

18. (a) Let $a_1, \ldots, a_n$ and $b_1, \ldots, b_n$ be real numbers. Prove the Schwarz inequality. That is,
$$\sum_{i=1}^{n} a_i b_i \leq \left(\sum_{i=1}^{n} a_i^2 \right)^{1/2} \left(\sum_{i=1}^{n} b_i^2 \right)^{1/2}.$$

    [*Hint:* For any real numbers α and β, $\sum_{i=1}^{n}(\alpha a_i - \beta b_i)^2 \geq 0$. Expand this and choose α and β judiciously.]

    (b) When does equality hold in the Schwarz inequality?

19. (a) Use the Schwarz inequality to prove the Minkowski inequality. That is,
$$\left(\sum_{i=1}^{n}(a_i + b_i)^2 \right)^{1/2} \leq \left(\sum_{i=1}^{n} a_i^2 \right)^{1/2} + \left(\sum_{i=1}^{n} b_i^2 \right)^{1/2}$$

    for real numbers $a_1, \ldots, a_n$ and $b_1, \ldots, b_n$.

    (b) When does equality hold in Minkowski's inequality?

20. Show that if $a_1, \ldots, a_n > 0$, then $(\sum_{k=1}^{n} a_k)(\sum_{k=1}^{n} 1/a_k) \geq n^2$. When does equality occur?

21. Let $f(x) = a_n x^n + a_{n-1} x^{n-1} + \cdots + a_1 x + a_0$ ($a_n \neq 0$.) Let $A = \max\{|a_0|, |a_1|, \ldots, |a_n|\}$, and let $B = nA/|a_n|$. Show that $f(x) \neq 0$ if $|x| > B$.

22. Show that if $a < b + \epsilon$ for every $\epsilon > 0$, then $a \leq b$.

1.3   The Completeness Axiom

We begin this section by discussing the final axiom of the real numbers–that
of completeness. This completeness axiom may be the least familiar of the
axioms of the real numbers, but it is of fundamental importance anytime that
limits are discussed. It is interesting to notice how this axiom is needed when
we discuss the Archimedean property of real numbers, Cauchy sequences, and
the existence of the Riemann integral, to name a few examples. There is
more than one way to state the completeness axiom of the real numbers. Our
approach uses the idea of a bounded set.

Definition: Let A be a set of real numbers. If there is a real number b for
which $x \le b$ for every $x \in A$, then b is said to be an *upper bound* for A. A set
that has an upper bound is said to be *bounded above*. If there is a number c
such that $c \le x$ for every $x \in A$, then c is said to be a *lower bound* for A. A
set that has a lower bound is said to be *bounded below*. A set that is bounded
above and below is said to be *bounded*. A set that is not bounded is said to
be *unbounded*.

Definition: Let A be a set of real numbers that is bounded above. The
number b is called the *least upper bound* (or *supremum*) of the set A (denoted
l.u.b. A or sup A) if

(i) b is an upper bound of A, and

(ii) If c is an upper bound of A, then $b \le c$.

Note that (ii) is equivalent to saying that if $d < b$, then d is not an upper
bound of A.

Definition: Let A be a set of real numbers that is bounded below. The
number b is called the *greatest lower bound* (or *infimum*) of the set A (denoted
g.l.b. A or inf A) if

(i) b is a lower bound of A, and

(ii) If c is a lower bound of A, then $c \le b$.

Definition: Let S be an ordered field. Then S is said to be *complete* if for
any nonempty subset A of S that is bounded above, the least upper bound of
A is in S.

The Axiom of Completeness: The final axiom of the real numbers is:

(A9) The real numbers are complete.

The definition of completeness could have been stated in terms of lower bounds. The definition we gave is equivalent to the following, as we show in Exercise 7, Section 1.3: Let S be an ordered field. Then S is complete if for any nonempty subset A of S that is bounded below, the greatest lower bound of A is in S.

An example of an ordered field that is *not* complete is the rational numbers. For example, if we let

$$A = \{x \in \mathbb{Q} | x < \sqrt{2}\},$$

then A is bounded above, but it has no least upper bound *in the rational numbers.*

Theorem 1.14: If the least upper bound and greatest lower bound of a set of real numbers exist, they are unique.

Proof: We show that the least upper bound of a set is unique. Let A be a set of real numbers that is bounded above. Suppose α and $\bar{\alpha}$ are both least upper bounds of A. Thus α and $\bar{\alpha}$ are both upper bounds of A. Since α is a least upper bound and $\bar{\alpha}$ is an upper bound, then

$$\alpha \leq \bar{\alpha}.$$

Likewise, since $\bar{\alpha}$ is a least upper bound and α is an upper bound,

$$\bar{\alpha} \leq \alpha.$$

Thus $\bar{\alpha} = \alpha$, so the least upper bound is unique.

We leave the proof of the uniqueness of the greatest lower bound as Exercise 5, Section 1.3. ■

Using the axiom of completeness and uniqueness of the least upper bound, we can define what it means to raise a positive number to an irrational power.

Definition: Let $x > 1$ be a positive real number and r a positive irrational number. Then the number x^r is the least upper bound of the set A where

$$A = \{x^p | p \text{ is a positive rational number less than } r\}.$$

If $0 < x < 1$ and r is a positive irrational number, then x^r is defined to be $1/y^r$ where $y = 1/x > 1$.

Since $x^{-r} = 1/x^r$, this also gives a way of defining a positive number raised to a negative irrational exponent.

Theorem 1.15 gives an alternative way of defining the least upper bound and greatest lower bound. This theorem will provide the characterization of these numbers that we shall use most often.

Theorem 1.15:

(a) A number α is the least upper bound of a set of real numbers A if and only if

  (i) α is an upper bound of A, and

  (ii) Given any $\epsilon > 0$, there is a number $x(\epsilon)$ in A for which $x(\epsilon) > \alpha - \epsilon$.

(b) A number β is the greatest lower bound of a set of real numbers A if and only if

  (i) β is a lower bound of A, and

  (ii) Given any $\epsilon > 0$, there is a number $x(\epsilon)$ in A for which $x(\epsilon) < \beta + \epsilon$.

Before proving Theorem 1.15, it is worthwhile to give a detailed discussion of condition (ii) of part (a), since this will be a recurrent theme throughout the text. One thinks of ϵ as being a very small positive number. Suppose $\alpha =$ l.u.b. A. What (ii) is saying is that if *first we choose* ϵ, *then* we can find a number $x(\epsilon)$ in A that is larger than $\alpha - \epsilon$. This means $\alpha = \epsilon$ is not an upper bound for A. Thus for any $\epsilon > 0$, $\alpha - \epsilon$ is not an upper bound for A. The order of doing things here is *crucial. First,* ϵ is chosen, and then $x(\epsilon)$ is found. The number $x(\epsilon)$ in A will often depend on ϵ as the notation is supposed to emphasize.

Proof: We give the proof of part (a) and leave the proof of part (b) as Exercise 3, Section 1.3.

First, suppose $\alpha =$ l.u.b. A. We show that (i) and (ii) must hold. By definition, (i) holds. To show that (ii) holds, let $\epsilon > 0$ be given. Then $\alpha - \epsilon < \alpha$, and so $\alpha - \epsilon$ cannot be an upper bound for A, since any upper bound of A is greater than or equal to the least upper bound of A, that is, α. Since $\alpha - \epsilon$ is not an upper bound of A, there is some element of A, call it $x(\epsilon)$ larger than $\alpha - \epsilon$.

Conversely, suppose (i) and (ii) hold. Then, by (i), α is an upper bound of A. Suppose $\bar{\alpha}$ is also an upper bound of A. We must show $\alpha \leq \bar{\alpha}$. Suppose this is not the case; that is, suppose $\bar{\alpha} < \alpha$. Then let

$$\epsilon = \alpha - \bar{\alpha} > 0,$$

so that

$$\bar{\alpha} = \alpha - \epsilon.$$

By (ii) there is a number $x(\epsilon) \in A$ with

$$x(\epsilon) > \alpha - \epsilon = \bar{\alpha},$$

so $\bar{\alpha}$ is not an upper bound for A. ∎

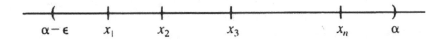

Figure 1.2
Least upper bound illustration

Example 1.11:
The purpose of this example is to clarify the comments preceding the proof of Theorem 1.15. Let

$$A = \left\{ \frac{1}{2}, \frac{2}{3}, \frac{3}{4}, \frac{4}{5}, \ldots \right\} = \left\{ \frac{n}{n+1} \middle| n \in \mathbb{N} \right\}.$$

We claim 1 is the least upper bound of A. Clearly,

$$\frac{n}{n+1} < 1 \text{ for } n \in \mathbb{N},$$

so 1 is an upper bound for A. Now choose $\epsilon = .01 > 0$.

Then we claim there is an element in A larger than $1 - \epsilon = .99$. In fact, $999/1000 = .999 > .99$ and $999/1000$ is in A.

Suppose that we had chosen $\epsilon = .0001$. Then $1 - \epsilon = .9999$, and the number we chose before, $999/1000$, no longer works. We have to choose another number in A, say $99,999/100,000 = .99999$, to exceed $1 - \epsilon$.

The point is that in this example no number in A exceeds $1 - \epsilon$ for every $\epsilon > 0$. However, it seems plausible (we need Theorem 1.18 to actually prove it) that if we choose $\epsilon > 0$ *first, then* we can find an element of A that exceeds $1 - \epsilon$.

Example 1.1 shows that the least upper bound of a set need not be an element of the set. On the other hand, it might be. If we modify our example by adding 1 to the set, so that if $B = A \cup \{1\} = \{1, 1/2, 2/3, 3/4, \ldots\}$, then 1 is the least upper bound of B and $1 \in B$.

Notice that if the set A is a finite set, then l.u.b. A is just the largest number in the set. As we have seen, if A is infinite, then l.u.b. A may or not be in A. The following result may provide some additional insight into the concept of the least upper bound of a set.

Theorem 1.16: Let $\alpha = $ l.u.b. A, and suppose $\alpha \notin A$. Then for any $\epsilon > 0$, the interval $(\alpha - \epsilon, \alpha)$ contains an infinite number of points of A.

Proof: Choose $\epsilon > 0$. By definition of l.u.b. A and since $\alpha \notin A$, the interval $(\alpha - \epsilon, \alpha)$ contains at least one point of A. Suppose that interval contains only a finite number of points of A (call them $x_1, x_2, \ldots, x_n$), and suppose they occur as in Figure 1.2. Since there are only a finite number of these points, we can select the largest, which in this case is x_n. Now x_n is larger than any other number in A but is smaller than α. Thus α is not the least upper bound of A.

We proved this result by proving the *contrapositive* of the proposition. That is, we assumed the conclusion was false and showed the hypothesis could not hold. This is a common method of proof.

Definition: Let A be a set of real numbers, and suppose c is a real number. Then cA is the set of real numbers given by

$$cA = \{cx | x \in A\}.$$

Example 1.12:
    Let $A = \{-1, 2, 4\}$. Then $3A = \{-3, 6, 12\}$ and $-2A = \{2, -4, -8\}$. Notice that

$$\sup A = 4, \quad \sup 3A = 12 = 3 \sup A,$$
$$\inf A = -1, \quad \inf 3A = -3 = 3 \inf A,$$

and

$$\inf(-2A) = -8 = -2 \sup A,$$
$$\sup(-2A) = 2 = -2 \inf A.$$

This is an example of an important principle, which is given by Theorem 1.17.

Theorem 1.17: Let A be a bounded set of real numbers, and suppose c is a real number. Then

(a) If $c > 0$

    (i) l.u.b. $(cA) = c$ l.u.b. A.
    (ii) g.l.b. $(cA) = c$ g.l.b. A.

(b) If $c < 0$

    (i) l.u.b. $(cA) = c$ g.l.b. A.
    (ii) g.l.b. $(cA) = c$ l.u.b. A.

Proof: We give the proof of part (ii) of (b) and leave the proof of the remaining parts as an exercise.
    Let A be a set of real numbers with $\alpha = $ l.u.b. A and suppose $c < 0$. For any $x \in A$, $x \leq \alpha$. Thus, since $c < 0$,

$$cx \geq c\alpha.$$

Therefore $c\alpha \leq cx$ for any $x \in A$, so $c\alpha$ is a lower bound for cA. Next we show cA is the greatest lower bound.
    Let $\epsilon > 0$ be given. We must show that there is a number $cx(\epsilon) \in cA$ with $cx(\epsilon) < c\alpha + \epsilon$. Now $\epsilon/(-c) > 0$, and since $\alpha = \sup A$, there is a number $x(\epsilon) \in A$ with $x(\epsilon) > \alpha - \epsilon/(-c)$. Then

$$cx(\epsilon) < c\alpha - c\left(\frac{\epsilon}{-c}\right) = c\alpha + \epsilon. \quad \blacksquare$$

Our intuition tells us that $1/N$ can be made arbitrarily small by making N sufficiently large. This is an important fact. Theorem 1.18 is the basis for this result.

Theorem 1.18 (The Archimedean Principle): If a and b are real numbers with $a > 0$, then there is a positive integer n such that $na > b$.

Proof: Let $A = \{ka | k \text{ is a positive integer}\}$. We shall show that A is *not* bounded above. If A is not bounded above, then b is not an upper bound. Thus there is some positive integer n for which $na > b$. To show that A is not bounded above, we suppose that A is bounded above and show that this is impossible.

If A is bounded above, then by the completeness property, A has a least upper bound, which we call α. Since $\alpha > 0$, there is a number in A, call it Na, such that

$$\alpha - a < Na.$$

But then

$$\alpha < Na + a = (N+1)a$$

and $(N+1)a \in A$ so that α is not an upper bound for A. ∎

Corollary 1.18: Let a be a positive real number and b be any real number. Then there is a positive integer n such that

$$\frac{b}{n} < a. \qquad ∎$$

With the Archimedean Principle in mind, we can return to the example where $A = \{n/(n+1) | n \in \mathbb{N}\}$ to give a formal argument regarding why 1 is the least upper bound of A. We already know 1 is an upper bound of A. Now let $\epsilon > 0$ be given. Taking $a = \epsilon$ and $b = 1$ in the Archimedean Principle, we know there is a positive integer N with $N\epsilon > 1$ so that

$$\epsilon > \frac{1}{N}.$$

Clearly, N depends on ϵ, and we emphasize this by writing $N(\epsilon)$ for N. Now

$$\frac{N(\epsilon) - 1}{N(\epsilon)} \in A \text{ and } 1 - \epsilon < 1 - \frac{1}{N(\epsilon)} = \frac{N(\epsilon) - 1}{N(\epsilon)},$$

which proves that 1 is the least upper bound of A.

Cardinality

We can classify sets by the number of elements they contain or their *cardinality*. Intuitively, sets have the same cardinality if they have the same number of elements.

Definition: Two sets A and B are said to have the *same cardinality* if there is a 1-1 onto function from the set A to the set B. In this case we say that A is *equivalent* to B.

Definition: A set is said to be *finite with cardinality n* if S is equivalent to $\{1, 2, \ldots, n\}$ where n is some positive integer. The empty set is also finite with cardinality 0. Sets that do not have finite cardinality are *infinite* sets. A set S is called *countable* if it has the same cardinality as some subset of the positive integers. Sets that are not countable are said to be *uncountable*.

Saying that a set S is countable is equivalent to saying there is a 1-1 (not necessarily onto) function $f : S \to \mathbb{N}$. In particular, finite sets are countable.

It is often convenient to think of countable sets as those sets whose elements can be enumerated. Thus if A is a finite set with cardinality N, we can write

$$A = \{x_1, x_2, \ldots, x_N\}$$

and if B is an infinite countable set, we write

$$B = \{x_1, x_2, \ldots\}.$$

Example 1.13:

The set of integers is countable. To show this, we must construct a 1-1 function from $\mathbb{Z}$ to $\mathbb{N}$. We do this by mapping

$$0 \to 1, 1 \to 3, 2 \to 5, 3 \to 7, \ldots$$

(that is, by sending the nonnegative integers to the odd positive integers) and mapping

$$-1 \to 2, -2 \to 4, -3 \to 6, \ldots$$

(the negative integers to the even positive integers). In equation form, f can be written

$$f(n) = \begin{cases} 2n + 1 & \text{if } n \geq 0 \\ -2n & \text{if } n < 0. \end{cases}$$

It is left for Exercise 16, Section 1.3, to show that f is 1-1.

Theorem 1.19: The union of a countable collection of countable sets is countable.

Remark: The hypotheses say that we have a collection of sets and that the number of sets in our collection is countable. Also, each set of this collection has a countable number of elements. We must show that the total number of elements is countable.

Proof: Since the number of sets is countable, we may enumerate the sets by naming them $E_1, E_2, E_3, \ldots$. Each set within the collection has countably many elements, so we may enumerate the elements of E_1 by

$$x_{11}, x_{12}, x_{13}, \ldots$$

and similarly for every other set in the collection. The first subscript indicates the set from which the element comes; the second subscript indicates the number of the element within that set. Thus x_{mn} is the nth element from the set E_m. What we must prove is that

$$\cup_i E_i = \cup_i \left(\cup_j \{x_{ij}\} \right)$$

is countable. Let

$$f : \cup_i \cup_j \{x_{ij}\} \to \mathbb{N}$$

be defined by

$$f(x_{ij}) = 2^i 3^j.$$

This is a 1-1 mapping by the Fundamental Theorem of Arithmetic; that is, since 2 and 3 are prime numbers,

$$2^p 3^q = 2^r 3^s,$$

if and only if $p = r$ and $q = s$.∎

Using this theorem, it is not difficult to prove that the rational numbers are countable, which we do in Exercise 11, Section 1.3. At this point, one may begin to expect that every set is countable. This is not the case.

Theorem 1.20: The real numbers in $(0,1)$ form an uncountable set.

Proof: We use the fact that any real number in $(0,1)$ can be represented as an infinite decimal. For a proof of this fact, see *Advanced Calculus* by Hans Sagan, pp. 36–40 (Boston: Houghton Mifflin, 1974.) Suppose that this set is countable. Thus the set can have its elements enumerated $x_1, x_2, x_3, \ldots$ So

$$x_1 = .a_{11}a_{12}a_{13} \cdots$$

$$x_2 = .a_{21}a_{22}a_{23} \cdots$$

$$x_3 = .a_{31}a_{32}a_{33} \cdots$$

$$\vdots$$

We shall construct a number in $(0,1)$ different from every x_i. Let $b = .b_1 b_2 b_3 \cdots$ where

$$b_1 = \begin{cases} 1 & \text{if } a_{11} \neq 1 \\ 2 & \text{if } a_{11} = 1, \end{cases}$$

$$b_2 = \begin{cases} 1 & \text{if } a_{22} \neq 1 \\ 2 & \text{if } a_{22} = 1, \end{cases}$$

and in general

$$b_n = \begin{cases} 1 & \text{if } a_{nn} \neq 1 \\ 2 & \text{if } a_{nn} = 1 \end{cases}.$$

Thus b is different from every x_i, since they must differ in the ith decimal place. Thus any countable collection is not large enough to include all the numbers in $(0,1)$.■

Note: While it is true that every number has a decimal representation, this representation is not always unique. For example, $.5 = .499....$ The number b that we have constructed does have a unique representation.

Cantor's Theorem

This section is optional and may be omitted without loss of continuity.

A lengthy discussion of cardinality would take us too far from our main objectives. For those who are interested in further reading on the subject, the book *Naive Set Theory* by Paul Halmos (New York: Van Nostrand, 1961) is an excellent source. As a final example, we show why there can be no *largest* set. We don't really have sufficient background to do this, but if we accept a few intuitively plausible propositions, which can be rigorously developed, then we can proceed.

Associated with each set is a number called a cardinal number. For finite sets the cardinal number of a set is the cardinality of the set. Cardinal numbers have an order on them that is used to order sets by the number of elements. We assume

(i) If X and Y are sets and there is a 1-1 function from X *into* Y, then the cardinal number of X is no larger than the cardinal number of Y.

(ii) If (i) holds, and if there is no *onto* function from X to Y, then the cardinal number of Y is strictly larger than the cardinal number of X.

(iii) We interpret the idea of the cardinal number of Y being larger than the cardinal number of X as meaning that Y has more elements than X.

Keeping these propositions in mind, we prove the following theorem.

Theorem 1.21 (Cantor's Theorem): For X any set, let $\mathcal{P}(X)$ denote the set of subsets of X. Then the cardinal number of $\mathcal{P}(X)$ is strictly larger than the cardinal number of X.

Proof: First note that it is possible to find a 1-1 map of X *into* $\mathcal{P}(X)$, namely, the function that sends $x \in X$ to $\{x\} \in \mathcal{P}(X)$. We now show that no function from X to $\mathcal{P}(X)$ can be onto.

Suppose that $f : X \to \mathcal{P}(X)$ and that f is onto. We shall show that this is impossible by finding a set $A \in \mathcal{P}(X)$ (i.e., A is a subset of X) that has no element from X sent onto it by f. We construct the set A as follows: For each element $x \in X$, $f(x) \in \mathcal{P}(X)$. That is, $f(x)$ is a subset of X, so either $x \in f(x)$ or $x \notin f(x)$. Define

$$A = \cup_{x \in X}\{x | x \notin f(x)\}.$$

Notice that A depends on f. Now, if f is onto, there is an element $a \in X$ such that $f(a) = A$. Either $a \in A$ or $a \notin A$. If $a \in A$, then $a \notin f(a)$ (this is the way elements get to be in A), so $a \notin f(a) = A$. Thus $a \in A$ leads to a contradiction. Suppose $a \notin A = f(a)$. Then $a \notin f(a)$, so $a \in A$, which is another contradiction. We are thus forced into concluding that there is no element from X that is mapped to the set A, so f cannot be onto. ∎

Exercises 1.3

1. Let $A = \{1, 1/2, 1/4, 1/8, \dots\}$.

   (a) Find the least upper bound and greatest lower bound of A.

   (b) Let $\epsilon = 10^{-6}$.

      (i) Find a number in A that exceeds (l.u.b. A) $- \epsilon$.

      (ii) Find a number in A that is smaller than (g.l.b. A) $+ \epsilon$.

2. Find the least upper bound and the greatest lower bound of the following sets. For a given $\epsilon > 0$, find a number in the set that exceeds (l.u.b. A) $- \epsilon$ and a number in the set that is smaller than (g.l.b. A) $+ \epsilon$.

   (a) $A = \{(4 + x)/x \mid x \geq 1\}$.

   (b) $A = \{\sqrt{x + 1}/x \mid x \geq 2\}$.

   (c) $A = \{x \mid x^2 - x < 6\}$.

3. Prove part (b) of Theorem 1.15.

4. (a) Prove that between any two real numbers there is a rational number.

   (b) Prove that between any two real numbers there is an irrational number. You may use the fact that $\sqrt{2}$ is irrational.

   (c) Prove that any interval contains infinitely many rational numbers and infinitely many irrational numbers.

5. If A is a bounded set, prove that the greatest lower bound of A is unique.

6. (a) Show that the set $A = \{x | a < x < b\}$ contains neither its least upper bound nor its greatest lower bound.

(b) Show that the set $B = \{x | a \leq x \leq b\}$ contains its least upper bound and its greatest lower bound.

7. Show that the following statement is equivalent to saying, "S is complete."

If S is an ordered field and A is a nonempty subset of S that is bounded below, then A has a greatest lower bound that is an element of S.

8. Let A and B be sets of real numbers. Define a set $A + B$ by

$$A + B = \{a + b | a \in A \text{ and } b \in B\}.$$

(a) Let $A = \{-1, 0, 3, 5\}$ and $B = \{-2, 12, 20\}$. Find $A + B$.

(b) Show that if A and B are bounded sets, then

$$\text{l.u.b. } (A + B) = (\text{l.u.b. } A) + (\text{ l.u.b. } B)$$

and

$$\text{g.l.b. } (A + B) = (\text{ g.l.b. } A) + (\text{ g.l.b. } B)$$

(c) Define a set $AB = \{ab | a \in A \text{ and } b \in B\}$. Give an example where

$$\text{g.l.b. } (AB) \neq (\text{ g.l.b. } A)(\text{ g.l.b. } B).$$

9. Let y be a positive real number, and let n be a positive integer. Let

$$A = \{x | x^n < y\}.$$

(a) Show that A is nonempty and bounded above.

(b) Let $z = \text{l.u.b. } A$. Show that $z^n = y$. Thus $z = y^{1/n}$, and we have proven Theorem 1.8. [*Hint:* Suppose $z^n \neq y$. If $z^n < y$, then $y - z^n = \epsilon > 0$. Use the Binomial Theorem to find a number δ such that $(z + \delta)^n < y$. This shows that z was not l.u.b. A. A similar argument works if one assumes $y < z^n$.]

(c) Let

$$B = \{x | x^n > y \text{ and } x > 0\}.$$

Show that B is nonempty and bounded below.

(d) Show that l.u.b. $A = $ g.l.b. B.

10. Let A be an uncountable set and B a countable set. Show that $A \setminus B$ is uncountable.

11. (a) Prove that the rational numbers are countable.

(b) Prove that the irrational numbers are uncountable.

12. Show that the set of open intervals with rational endpoints is countable.

13. (a) Let A and B be countable sets. Show that the Cartesian product $A \times B$ is countable.

(b) Let $A_1, \ldots, A_n$ be countable sets. Show that

$$A_1 \times \cdots \times A_n = \{(a_1, \ldots, a_n) | a_i \in A_i\}$$

is countable.

14. Show that the set of polynomials with rational coefficients is countable.

15. Construct a 1-1 onto function from $\{0, 1, 2, \ldots\}$ to $\{1, 2, 3, \ldots\}$. Prove that the function you have constructed is 1-1 and onto.

16. Show that the function $f : \mathbb{Z} \to \mathbb{N}$ given by

$$f(n) = \begin{cases} 2n + 1 & \text{if } n \geq 0 \\ -2n & \text{if } n < 0 \end{cases}$$

is 1-1 and onto.

17. Let a be greater than 1 and r a positive irrational number. Let

$$A = \{a^x | x \text{ is a positive rational number} < r\}$$

and

$$B = \{a^x | x \text{ is a positive rational number} > r\}.$$

Show that:

(a) The set A is bounded above, and the set B is bounded below.

(b) If $s \in A$ and $t \in B$, then $s < t$.

(c) What assumptions, if any, must be made to show l.u.b. A = g.l.b. B?

18. (a) Show that if $a > 1$ and x and y are real numbers with $x < y$, then $a^x < a^y$. Recall that we have proven this in the case that x and y are rational numbers.

(b) Show that if $0 < a < 1$ and if x and y are real numbers with $x < y$, then $a^x > a^y$.

2

Sequences of Real Numbers

2.1 Sequences of Real Numbers

One of the great advantages of calculus is that it enables us to solve problems of a dynamic nature, that is, problems in which a change in the variables occurs. The technique that calculus uses to deal with this type of problem is the limiting process. In this chapter we introduce sequences of real numbers. Sequences provide perhaps the simplest setting for the rigorous study of limits, and sequences will also be indispensable in studying more complex topics.

Definition: A *sequence* of real numbers is a function from the positive integers into the real numbers.

The function concept is not the most convenient way to visualize a sequence, and we shall establish a more intuitive viewpoint. If f is the function in the definition, then the range of f is the set

$$\{f(1), f(2), \dots\}.$$

The numbers $f(1), f(2), \dots$ are called the *terms* of the sequence and $f(n)$ is called the nth term of the sequence. The domain of a sequence is always the positive integers. Therefore if we list only the range or terms of the sequence in their natural order of appearance, then the sequence will be completely described. It is customary to further simplify the notation and write f_n for $f(n)$. So now our sequence is written as a subscripted variable within braces $\{f_1, f_2, \dots\}$ or more often as $\{f_n\}$. Thus $\{x_n\}$ will represent a sequence whose first term is x_1, whose second term is x_2, and so on.

Sequences are often viewed as an infinite string of numbers. Two sequences are equal if and only if they are equal *term by term*. That is, not only must the numbers in the sequences be the same, but they must also appear in the same order. Thus the sequence $\{1, 2, 3, 4, 5, \dots\}$ is not equal to the sequence $\{2, 1, 3, 4, 5, \dots\}$ even though they consist of the same numbers.

Often a sequence is represented by a function within braces, which describes the nth term of the sequence. For example, we might represent the sequence $\{1, 1/2, 1/3, 1/4, \dots\}$ as $\{1/n\}$, since the nth term of the sequence is equal to $1/n$.

Example 2.1:
The sequence $1, 1, 1, \ldots$ or $\{1, 1, 1, \ldots\}$ may be represented as $\{1\}$ or $\{1\}_{n=1}^{\infty}$ where the indices are used to emphasize that the object under consideration is a sequence and not merely a set. (Remember, a set is defined by the elements within the set and is independent of the order in which the elements appear, whereas a sequence is determined by the elements *and* the order in which they appear.)

Example 2.2:
The sequence $2, 4, 6, 8, \ldots$ or $\{2, 4, 6, 8, \ldots\}$ may be represented as $\{2n\}$ or $\{2n\}_{n=1}^{\infty}$.

Definition: We say that the sequence of real numbers $\{x_n\}$ *converges to the number L* if, for any $\epsilon > 0$, there is a positive integer $N(\epsilon)$ such that if n is a positive integer larger than $N(\epsilon)$, then $|x_n - L| < \epsilon$. In this case we say that L is the *limit* of the sequence $\{x_n\}$, and we write $\lim x_n = L$ or $\{x_n\} \to L$. If a sequence does not converge, then it is said to *diverge*.

The intuition of this definition is that x_n can be made arbitrarily close to L by taking n sufficiently large.

The definition of convergence is extremely important. The ϵ in the definition should be thought of as a very small number. Then $|x_n - L|$ measures the distance between the nth term of the sequence and the number L. The definition says that if we go far enough out in the sequence (to the $N(\epsilon)$ term), then *every* term in the sequence past that point is within the distance ϵ of L. The order of selection of ϵ and $N(\epsilon)$ is crucial. First ϵ (our small number) is chosen, and then $N(\epsilon)$ is determined. $N(\epsilon)$ will usually depend on ϵ, as the notation $N(\epsilon)$ is supposed to emphasize. Figure 2.1 demonstrates the definition using the graph of a sequence. Since the domain of a sequence is the positive integers, the graph will be a set of points rather than a curve. Notice that some of the terms that precede the $N(\epsilon)$ term in the definition may lie within ϵ of L. Every point in the graph of the sequence whose x coordinate (which represents the number of the term) exceeds $N(\epsilon)$ must lie between $L - \epsilon$ and $L + \epsilon$.

Example 2.3:
The sequence $\{a_n\} = \{1 + 1/(2n)\}$ converges to 1. As n becomes larger, $1/(2n)$ becomes nearer to 0, and so intuitively $1 + 1/(2n)$ should become nearer to 1. To prove this using the definition, we

(i) let $\epsilon > 0$ be given, and then

(ii) find how large n must be to ensure that $|a_n - L| < \epsilon$.

In this example $a_n = 1 + 1/(2n)$ and $L = 1$. So

$$|a_n - L| = \left|\left(1 + \frac{1}{2n}\right) - 1\right| = \frac{1}{2n} < \epsilon \text{ if } n > \frac{1}{2\epsilon}.$$

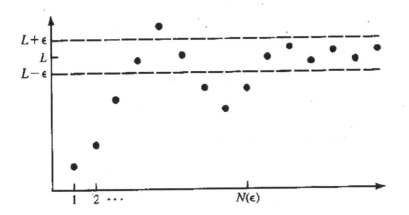

Figure 2.1
A convergent sequence

Thus if $N(\epsilon)$ is at least as large as $1/(2\epsilon)$ (which exists by the Archimedean Principle) and if $n > N(\epsilon)$, then $|a_n - L| < \epsilon$.

Notice that $N(\epsilon)$ depends on ϵ; in fact, no number N can work for every ϵ. That is, *first ϵ is given and then $N(\epsilon)$ is found.*

Example 2.4:
Let $\{x_n\} = \{2 - 1/n^2\}$. Intuitively this sequence appears to converge to 2. To prove that this is the case, we follow the pattern of the previous example. So we

(i) Let $\epsilon > 0$ be given, and then

(ii) find how large n must be to ensure that

$$|x_n - L| = \left|\left(2 - \frac{1}{n^2}\right) - 2\right| < \epsilon.$$

Thus we need $|-1/n^2| < \epsilon$. This will be true if $1/n^2 < \epsilon$ or $n^2 > 1/\epsilon$, so $n > 1/\sqrt{\epsilon}$. In this case, we could take $N(\epsilon)$ to be any positive integer as large as $1/\sqrt{\epsilon}$. So if $n > N(\epsilon) \geq 1/\sqrt{\epsilon}$, then $|(2-1/n^2)-2| < \epsilon$, and we have proved that $\{2 - (1/n^2)\} \to 2$ as our intuition suggested.

Example 2.5:
Let $\{x_n\}$ be the sequence

$$\{1, 2, 1, 4, 1, 6, 1, 8, \dots\}.$$

This sequence diverges for the following reason: Suppose that we choose $\epsilon = \frac{1}{2}$.

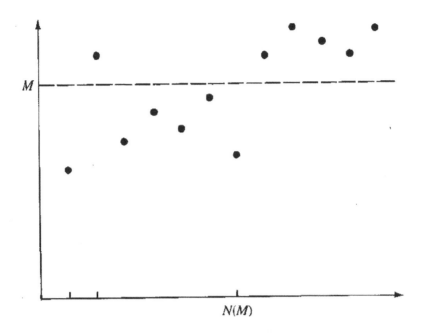

$N(M)$

Figure 2.2
A sequence that diverges to infinity

Then no matter what values we take for L and $N(\epsilon)$, there are numbers in the sequence beyond the $N(\epsilon)$ term such that $|x_n - L| > \frac{1}{2}$.

There are two ways in which a sequence can diverge that will be of special interest to us. We now define these.

Definition: A sequence of real numbers $\{x_n\}$ is said to *diverge to infinity* if, given any number M, there is a positive integer $N(M)$ such that if $n > N(M)$, then $x_n > M$. Figure 2.2 illustrates this idea.

Definition: A sequence of real numbers $\{x_n\}$ is said to *diverge to negative infinity* if, given any number K, there is a positive integer $N(K)$ such that if $n > N(K)$, then $x_n < K$. In this case we write $\lim x_n = -\infty$ or $\{x_n\} \to -\infty$.

In this definition one should think of K as being a very large negative number.

Example 2.6:
Let $\{x_n\} = \{n^2\}$. We show that $\{x_n\}$ diverges to ∞. Choose a positive number M. If we let N(M) be the smallest integer larger than $\sqrt{M}$ (e.g. if

$M = 101$, then $N(M) = 11$), and if $n > N(M)$, then

$$x_n = n^2 > N(M)^2 > (\sqrt{M})^2 = M.$$

Example 2.7:
Let $\{x_n\} = \{1, 2, 1, 3, 1, 4, \ldots\}$. Then $\{x_n\}$ diverges because there is no number L for which *all* the terms beyond some point may be made arbitrarily close to L. The sequence does not diverge to ∞ because there is no point in the sequence beyond which the terms are arbitrarily large. (The 1's are always there.)

We now prove some theorems that describe the behavior of convergent sequences.

Theorem 2.1: A sequence of real numbers can converge to at most one number.
Writing mathematical proofs involves two processes: (1) formulating the ideas of the proof and (2) writing these ideas in mathematical language appealing only to the definitions and previously proven results. Here we give the ideas involved in the proof and leave it for Exercise 5, Section 2.1, to express these ideas in mathematical language.
Proof: We shall suppose that there is a sequence $\{x_n\}$, that converges to two different numbers, L and M, and show this is impossible. Picture L and M on the number line in Figure 2.3. Draw an interval about L and an interval about M that do not intersect each other, as shown in Figure 2.4. In writing your proof, you should specify how large these intervals are.

Figure 2.3
Two possible limits of a sequence

Figure 2.4
Intervals about L and M

If $\{x_n\}$ converges to L, then all the terms beyond some point, say N_1, in the sequence must be in $(L - \epsilon, L + \epsilon)$. If $\{x_n\}$ converges to M, then all the terms in the sequence beyond some point, say N_2, must lie in $(M - \epsilon, M + \epsilon)$. Then if n is larger than both N_1 and N_2, x_n must be in both $(L - \epsilon, L + \epsilon)$ and $(M-\epsilon, M+\epsilon)$. But this is impossible, since the intervals do not intersect. ∎

Theorem 2.2: The sequence of real numbers $\{a_n\}$ converges to L if and only if for every $\epsilon > 0$, all but a finite number of terms of $\{a_n\}$ lie in the interval $(L - \epsilon, L + \epsilon)$.

Proof: Suppose that the sequence $\{a_n\}$ converges to L. Let $\epsilon > 0$ be given. By definition, there is a number $N(\epsilon)$ such that if $n > N(\epsilon)$, then $|a_n - L| < \epsilon$. Then $-\epsilon < a_n - L < \epsilon$ or $L - \epsilon < a_n < L + \epsilon$; that is, $a_n \in (L - \epsilon, L + \epsilon)$ if $n > N(\epsilon)$. Thus all of the terms of $\{a_n\}$ except possibly $a_1, a_2, \ldots, a_{N(\epsilon)}$ lie in $(L - \epsilon, L + \epsilon)$.

Conversely, suppose that given $\epsilon > 0$ all but a finite number of terms of $\{a_n\}$ lie in $(L - \epsilon, L + \epsilon)$. We shall show that $\{a_n\}$ converges to L. Let $a_{n_1}, a_{n_2}, \ldots, a_{n_k}$ denote those terms of $\{a_n\}$ that are *not* in $(L-\epsilon, L+\epsilon)$. Let $N(\epsilon) = \max\{n_1, n_2, \ldots, n_k\}$. If $n > N(\epsilon)$, then $a_n \in (L - \epsilon, L + \epsilon)$ so that $|a_n - L| < \epsilon$. ∎

Definition: A sequence is *bounded* if the terms of the sequence form a bounded set.

Theorem 2.3: If $\{a_n\}$ is a convergent sequence of real numbers, then the sequence $\{a_n\}$ is bounded.

Proof: Suppose the sequence $\{a_n\}$ converges to L. Then there is a number N such that all terms of the sequence $\{a_n\}$ beyond the Nth term lie in the interval $(L - 1, L + 1)$. The set of numbers $\{a_1, a_2, \ldots, a_N, L - 1, L + 1\}$ is a finite set and so is bounded. But the upper and lower bounds for this set provide bounds for the set of terms of the sequence $\{a_n\}$. ∎

Given a sequence, we can form a new sequence by multiplying each term in the given sequence by some fixed number. For example, if

$$\{a_n\} = \{1/n\} = \{1, \frac{1}{2}, \frac{1}{3}, \frac{1}{4}, \ldots\}$$

we can multiply each number in the sequence by 5 to obtain

$$\{5, \frac{5}{2}, \frac{5}{3}, \frac{5}{4}, \ldots\}$$

which could be written $\{5a_n\}$. Likewise, if we have two given sequences, we can form new sequences by adding, multiplying or dividing (provided none of the divisor terms are 0) the terms of one of the sequences with the terms of the other, term by term. Thus if

$$\{a_n\} = \left\{\frac{2}{n^2}\right\} = \left\{2, \frac{2}{4}, \frac{2}{9}, \frac{2}{16}, \ldots\right\},$$

and

$$\{b_n\} = \left\{\frac{1}{n}\right\} = \left\{1, \frac{1}{2}, \frac{1}{3}, \frac{1}{4}, \dots\right\},$$

then

$$\{a_n + b_n\} = \left\{2 + 1, \frac{2}{4} + \frac{1}{2}, \frac{2}{9} + \frac{1}{3}, \frac{2}{16} + \frac{1}{4}, \dots\right\},$$

$$\{a_n b_n\} = \left\{2, \frac{1}{4}, \frac{2}{27}, \frac{1}{32}, \dots\right\},$$

and

$$\left\{\frac{a_n}{b_n}\right\} = \left\{2, 1, \frac{2}{3}, \frac{1}{2}, \dots\right\}.$$

Theorem 2.4: Suppose that $\{a_n\}$ and $\{b_n\}$ are sequences of real numbers such that $\{a_n\} \to a$ and $\{b_n\} \to b$. Then

(a) $\{a_n + b_n\} \to a + b$.

(b) $\{ca_n\} \to ca$ for any number c.

(c) $\{a_n b_n\} \to ab$.

(d) If $b \neq 0$ and $b_n \neq 0$ for any n, then $\{a_n/b_n\} \to a/b$.

Proof of (a): We give the ideas of the proof and then give a formal argument. We are given that $\{a_n\} \to a$ and $\{b_n\} \to b$ and must show that $\{a_n + b_n\} \to a + b$. That is, we must show that given $\epsilon > 0$, there is a number $N(\epsilon)$ such that if $n > N(\epsilon)$, then $|(a_n + b_n) - (a + b)| < \epsilon$. The hypotheses say that $|a_n - a|$ and $|b_n - b|$ will be small if n is large. Thus we rewrite

$$|(a_n + b_n) - (a + b)| = |(a_n - a) + (b_n - b)| \leq |a_n - a| + |b_n - b|$$

by the triangle inequality. We have now split the quantity that we want to show is small into two parts, each of which can be made small. This type of application of the triangle inequality is very common and extremely important.

Now if $|a_n - a| < \epsilon/2$ and $|b_n - b| < \epsilon/2$, then

$$|(a_n + b_n) - (a + b)| < \epsilon,$$

and we are done.

We now give a formal proof. Since $\{a_n\} \to a$, there is a number $N_1(\epsilon/2)$ such that if $n > N_1(\epsilon/2)$, then $|a_n - a| < \epsilon/2$; since $\{b_n\} \to b$, there is a number $N_2(\epsilon/2)$ such that if $n > N_2(\epsilon/2)$, then $|b_n - b| < \epsilon/2$. (The notation N_1 and N_2 is used to emphasize that we may have to go out different distances to acquire the same degree of closeness for the different sequences.)

We want both conditions to hold, so let

$$N(\epsilon) = \max\left\{N_1\left(\frac{\epsilon}{2}\right),\ N_2\left(\frac{\epsilon}{2}\right)\right\}.$$

Then if $n > N(\epsilon)$,

$$|(a_n + b_n) - (a + b)| \leq |a_n - a| + |b_n - b| < \frac{\epsilon}{2} + \frac{\epsilon}{2} = \epsilon.$$

The proof of (b) is left for Exercise 7, Section 2.1.

Proof of (c): Here we want to show that if $\epsilon > 0$ is given, there is a number $N(\epsilon)$ such that if $n > N(\epsilon)$, then $|a_n b_n - ab| < \epsilon$.

We shall again use the triangle inequality and an additional common device – adding and subtracting the same number. Now,

$$|a_n b_n - ab| = |a_n b_n - a_n b + a_n b - ab| \leq |a_n b_n - a_n b| + |a_n b - ab|$$

$$= |a_n||b_n - b| + |b||a_n - a|.$$

This was done because we know $|b_n - b|$ and $|a_n - a|$ can be made small. We shall choose n large enough to make $|a_n||b_n - b| < \epsilon/2$ and $|b||a_n - a| < \epsilon/2$. Since $\{a_n\}$ is convergent, it is bounded by Theorem 2.3. Thus there is an $M > 0$ with $|a_n| < M$ for every n. Since $\{a_n\} \to a$, there is a number $N_1(\epsilon)$ such that if $n > N_1(\epsilon)$, then

$$|a_n - a| < \frac{\epsilon}{2|b| + 1}.$$

(We use $2|b| + 1$ in case $b = 0$.) Since $\{b_n\} \to b$, there is a number $N_2(\epsilon)$ such that if $n > N_2(\epsilon)$, then

$$|b_n - b| < \frac{\epsilon}{2M}.$$

Then if $n > \max\{N_1(\epsilon), N_2(\epsilon)\}$, both conditions hold, so

$$|a_n b_n - ab| \leq |a_n||b_n - b| + |b||a_n - a|$$

$$< M|b_n - b| + |b||a_n - a|$$

$$< M\left(\frac{\epsilon}{2M}\right) + \frac{|b|\epsilon}{2|b| + 1} < \epsilon.$$

Proof of (d): We must show that given $\epsilon > 0$, there is a number $N(\epsilon)$ such that if $n > N(\epsilon)$, then $|(a_n/b_n) - (a/b)| < \epsilon$.

The idea is the same as in part (c). We try to write $|(a_n/b_n) - (a/b)|$ as the sum of two terms, where one term has $|a_n - a|$ as a factor and the other has $|b_n - b|$ as a factor. Thus

$$\left|\frac{a_n}{b_n} - \left(\frac{a}{b}\right)\right| = \left|\frac{a_n b - b_n a}{b_n b}\right|$$

$$= \frac{1}{|b_n b|} |a_n b - ab + ab - b_n a|$$

$$\leq \frac{|b||a_n - a|}{|b_n b|} + \frac{|a||b - b_n|}{|b_n b|}$$

$$= \frac{1}{|b_n|} |a_n - a| + \frac{|a|}{|b_n b|} |b - b_n|. \tag{2.1}$$

Now we have our two terms in the desired form. We want to make each term on the right-hand side of (2.1) smaller than $\epsilon/2$. The one problem that could possibly occur would be if $1/|b_n|$ could grow indefinitely large. Exercise 10, Section 2.1 shows that this cannot happen under our hypotheses. That is, there is a number $M > 0$ with $(1/|b_n|) < M$ for every n.

Since $\{a_n\} \to a$, there is a number $N_1(\epsilon)$ such that if $n > N_1(\epsilon)$, then $|a_n - a| < \epsilon/(2M)$; and since $\{b_n\} \to b$, there is a number $N_2(\epsilon)$ such that if $n > N_2(\epsilon)$, then

$$|b_n - b| < \frac{\epsilon |b|}{2M|a| + 1}.$$

So if $N(\epsilon) = \max\{N_1(\epsilon), N_2(\epsilon)\}$, and if $n > N(\epsilon)$, then

$$\left| \left(\frac{a_n}{b_n}\right) - \left(\frac{a}{b}\right) \right| \leq \frac{1}{|b_n|} |a_n - a| + \frac{|a|}{|b_n b|} |b - b_n| < M|a_n - a| + \frac{|a|M}{|b|} |b_n - b|$$

$$< \frac{M\epsilon}{2M} + \frac{|a|M}{|b|} \left(\frac{\epsilon |b|}{2M|a| + 1} \right) < \epsilon. \quad \blacksquare$$

We prove the following important result in Exercises 8 and 9, Section 2.1.

Theorem 2.5:

(a) Suppose that $\{a_n\}$ converges to L and that $a_n \leq K$ for every n. Then $L \leq K$.

(b) Suppose that $\{a_n\}$ and $\{b_n\}$ are sequences with $a_n \leq b_n$ for every n. Also suppose that $\{a_n\} \to L$ and $\{b_n\} \to K$. Then $L \leq K$.

(c) If $\{a_n\}$ and $\{b_n\}$ are sequences with $0 \leq a_n \leq b_n$ for every n and if $\{b_n\} \to 0$, then $\{a_n\} \to 0$.

(d) If $\{a_n\}$, $\{b_n\}$, and $\{c_n\}$ are sequences with $a_n \leq b_n \leq c_n$ for every n and $\{a_n\} \to L$ and $\{c_n\} \to L$, then $\{b_n\} \to L$.

Definition: Let $\{a_n\}$ be a sequence of real numbers. If $a_{n+1} \geq a_n$ ($a_n \geq a_{n+1}$) for every n, the sequence is *monotone increasing (decreasing)*.

If $a_{n+1} > a_n$ ($a_n > a_{n+1}$) for every n, the sequence is *strictly monotone increasing (decreasing)*.

Any of these types of sequences is called a monotone sequence.

Monotone sequences are nice because the converse of Theorem 2.3 is true. That is, bounded monotone sequences converge, as we show in Theorem 2.6. This is not true for arbitrary bounded sequences, as the sequence $\{(-1)^n\}$ shows.

Theorem 2.6: A bounded monotone sequence converges.

Proof: We give the proof only for bounded monotone increasing sequences. Let $\{a_n\}$ be such a sequence. To show that the sequence converges, we need a candidate for the number to which it converges. There is really only one – the least upper bound of the set of terms of $\{a_n\}$. Call this number L. By definition of least upper bound, L has two properties:

(i) $a_n \leq L$ for every n.

(ii) For any given $\epsilon > 0$, there is a positive integer $N(\epsilon)$ for which $a_{N(\epsilon)} > L - \epsilon$.

Since $\{a_n\}$ is monotone increasing, if $n > N(\epsilon)$, then $a_n \geq a_{N(\epsilon)} > L - \epsilon$. But $a_n \leq L$. Thus $L - \epsilon < a_n \leq L$ for $n > N(\epsilon)$; that is,

$$|a_n - L| < \epsilon \text{ for } n > N(\epsilon). \qquad \blacksquare$$

Corollary 2.6:

(a) A monotone increasing sequence either converges or it diverges to ∞.

(b) A monotone decreasing sequence either converges or it diverges to $-\infty$.

Example 2.8
Let $\{x_n\}$ be the sequence of numbers defined by

$$x_1 = \sin 1$$
$$x_2 = \text{ larger of } \{\sin 1, \sin 2\}$$
$$\vdots$$

$x_k = $ largest of $\{\sin 1, \sin 2, \ldots, \sin k\}$
where the angle is given in radian measure. Then $\{x_n\}$ is monotone increasing and bounded above by 1, since $\sin x \leq 1$. Thus by Theorem 2.6, $\{x_n\}$ converges. Notice that we have not exhibited the number to which $\{x_n\}$ converges.

Example 2.9:
Later we shall define the natural logarithm function, but you are probably

already familiar with its importance and the frequency with which its base, the number e, occurs in mathematics. We shall not use sequences to define e, but one way that it can be defined is by

$$e = \lim_{n \to \infty} \left(1 + \frac{1}{n}\right)^n.$$

Here we shall show that the sequence $\{x_n\} = \{(1 + 1/n)^n\}$ is an increasing sequence.

To show the sequence is increasing, we use the Binomial Theorem. We have

$$x_n = \left(1 + \frac{1}{n}\right)^n = \sum_{k=0}^{n} \binom{n}{k} \left(\frac{1}{n}\right)^k (1)^{n-k}$$

$$= \sum_{k=0}^{n} \frac{n(n-1)\cdots(n-k+1)}{k!} \left(\frac{1}{n}\right)^k$$

$$= \sum_{k=0}^{n} \frac{1}{k!}(1)\left(1 - \frac{1}{n}\right)\left(1 - \frac{2}{n}\right)\cdots\left(1 - \frac{k-1}{n}\right) \leq \sum_{k=1}^{n} \frac{1}{k!},$$

and

$$x_{n+1} = \sum_{k=0}^{n+1} \frac{1}{k!}(1)\left(1 - \frac{1}{n+1}\right)\left(1 - \frac{2}{n+1}\right)\cdots\left(1 - \frac{k-1}{n+1}\right).$$

Now,

$$1 - \frac{j}{n+1} > 1 - \frac{j}{n}$$

so that the kth term for x_{n+1} is larger than the kth term for x_n. In addition, x_{n+1} has one additional term so that $x_{n+1} > x_n$.

Later we shall show that this sequence is bounded above so that $\{(1 + 1/n)^n\}$ is a convergent sequence.

The next theorem is a special case of a more general result that we shall prove later.

Theorem 2.7: Let $A_n = [a_n, b_n]$ be a sequence of closed intervals such that $A_n \supset A_{n+1}$ for $n = 1, 2, \ldots$. Suppose that $\lim_{n \to \infty}(b_n - a_n) = 0$. Then there is a real number p for which

$$\cap_{n=1}^{\infty} A_n = \{p\}.$$

That is, there is exactly one point common to every A_n. ■

We leave the proof for Exercise 12, Section 2.1.

Definition: A sequence of sets $\{A_n\}$ such that $A_n \supset A_{n+1}$ is called a *nested* sequence of sets.

In Chapter 1 we defined the least upper bound and greatest lower bound of a set of real numbers. Theorem 2.8 gives a fact that will be important later.

Theorem 2.8: Let A be a nonempty set of real numbers that is bounded above. Then there is a sequence of numbers $\{x_n\}$ such that

(i) $x_n \in A$, $n = 1, 2, \ldots$

(ii) $\lim x_n = $ l.u.b. A.

An analogous result holds for g.l.b. A.

Proof: Let $\alpha = $ l.u.b. A. If $\alpha \in A$, then, by definition of l.u.b. A, for each positive integer n there is a number $x_n \in A$ with

$$\alpha - \frac{1}{n} < x_n < \alpha.$$

We claim that $\lim x_n = \alpha$. Let $\epsilon > 0$ be given. By the Archimedean Principle, there is a positive integer $N(\epsilon)$ such that $1/N(\epsilon) < \epsilon$. Thus if $n > N(\epsilon)$, then

$$|x_n - \alpha| = \alpha - x_n < \frac{1}{n} < \frac{1}{N(\epsilon)} < \epsilon,$$

so that $\lim x_n = \alpha$. ∎

From now on, we shall refer to a sequence of real numbers as merely a sequence.

Definition: A sequence $\{a_n\}$ is called a *Cauchy sequence* if, given any $\epsilon > 0$, there is a positive integer $N(\epsilon)$ such that if $n, m > N(\epsilon)$, then

$$|a_n - a_m| < \epsilon.$$

In Exercises 13 and 14, Section 2.3, we prove the following very important result.

Theorem 2.9 A sequence converges if and only if it is a Cauchy sequence.

This theorem turns out to be surprisingly powerful and we shall use it extensively. Part of its power comes from the fact that, to show a sequence converges using this theorem, we do not need to have a candidate for the number to which the sequence converges.

Exercises 2.1

1. Each of the following sequences converges.

(a) Find the number to which the sequence converges.

(b) If L is the number to which the sequence converges, find how large n must be so that $|a_n - L| < 0.01$.

(c) For $\epsilon > 0$ find $N(\epsilon)$ such that $|a_n - L| < \epsilon$ if $n > N(\epsilon)$.

(i) $\{a_n\} = \{3 - 1/(4n)\}$.
(ii) $\{a_n\} = \{(2n + 3)/(3n - 1)\}$.
(iii) $\{a_n\} = \{(n^2 + 1)/n^2\}$.
(iv) $\{a_n\} = \{1 + 1/2^n\}$.

2. Construct a sequence $\{a_n\}$ that converges to 0, but $a_n \neq 0$ for any n.

3. Assume that $\pi = 3.1415926535\ldots$ is irrational. Describe how you would construct a sequence $\{a_n\}$ that has all of the following properties: (i) Each term of $\{a_n\}$ is rational, (ii) $\{a_n\}$ is monotone increasing, and (iii) $\{a_n\}$ converges to π. This shows that the rational numbers are not complete.

4. Let r be a rational number.

(a) Construct a strictly increasing sequence of rational numbers that converges to r.

(b) Construct a strictly increasing sequence of irrational numbers that converges to r. (You may use the fact that π is irrational.)

5. Prove Theorem 2.1.

6. Give an example of a bounded sequence that does not converge.

7. Prove part (b) of Theorem 2.4.

8. (a) Suppose that $\{a_n\} \to a$ and that $a_n \geq 0$ for every n. Show that $a \geq 0$. [*Hint:* Suppose that $a < 0$ and take $\epsilon = -a/2$. Show that $\{a_n\}$ cannot converge to a.]

(b) Show that if $\{a_n\} \to L$ and $a_n \leq K$ for every n, then $L \leq K$.

(c) Suppose that $\{a_n\}$ and $\{b_n\}$ are sequences with $a_n \leq b_n$ for every n, and suppose that $\{a_n\} \to a$ and $\{b_n\} \to b$. Show that $a \leq b$.

(d) If $\{a_n\}$ and $\{b_n\}$ are sequences with $0 \leq a_n \leq b_n$ for every n and $\{b_n\} \to 0$, show that $\{a_n\} \to 0$.

9. (a) Suppose that $\{a_n\}$ and $\{b_n\}$ are sequences with $\{a_n\} \to L$. Show that if $\{a_n - b_n\} \to 0$, then $\{b_n\} \to L$.

(b) Show that if $\{a_n\}$, $\{b_n\}$, and $\{c_n\}$ are sequences with $a_n \leq b_n \leq c_n$ for every n and if $\{a_n\} \to L$ and $\{c_n\} \to L$, then $\{b_n\} \to L$.

10. Suppose that $\{b_n\} \to b$ and $b \neq 0$ and $b_n \neq 0$ for any n. Show that there is a number $K > 0$ such that $|b_n| > K$ for every n. You may assume $b_n > 0$ to simplify the write-up.

11. (a) If $\{a_n\} \to a$, show that $\{|a_n|\} \to |a|$.

(b) Find an example of a sequence $\{a_n\}$ such that $\{|a_n|\} \to |a|$ but $\{a_n\}$ does not converge.

(c) If $\{|a_n|\} \to 0$, show that $\{a_n\} \to 0$.

12. Let I_n be an interval $[a_n, b_n]$. Suppose that $\{I_n\}$ is a sequence of nested intervals; that is, $I_n \supseteq I_{n+1}$ for each n so that $a_n \le a_{n+1}$ and $b_n \ge b_{n+1}$ for each n.

(a) Show that $\{a_n\}$ and $\{b_n\}$ are convergent sequences.

(b) If $\{a_n\} \to a$ and $\{b_n\} \to b$, show that $a \le b$.

(c) If $\{b_n - a_n\} \to 0$, show that $a = b$.

(d) If $\{I_n\}$ is a nested sequence of closed bounded intervals whose lengths go to 0, show that $\cap_{n=1}^{\infty} I_n$ consists of exactly one point.

(e) Is the result of part (d) true if the intervals are not required to be closed?

(f) Is the result of part (d) true if the intervals are not required to be bounded?

13. Complete the proof of Theorem 2.6 by showing that a bounded monotone decreasing sequence converges.

14. Give an example of two divergent sequences whose sum converges.

15. (a) Prove that the sequence $\{(1 - n^2)/n\}$ diverges to $-\infty$.

(b) Prove that $\{n^2 - 4n + 7\}$ diverges to ∞.

16. Let $\{a_n\}$ and $\{b_n\}$ be sequences that diverge to ∞.

(a) Show that $\{a_n b_n\}$ diverges to ∞.

(b) Show that $\{a_n + b_n\}$ diverges to ∞.

17. Give example of sequences $\{a_n\}$ and $\{b_n\}$ that diverge to ∞ for which

(a) $\{a_n - b_n\}$ diverges to ∞.

(b) $\{a_n - b_n\}$ converges to 5.

18. Prove that if $\{a_n\}$ is a sequence of nonnegative numbers and $\{a_n\} \to a$, then $\{\sqrt{a_n}\} \to \sqrt{a}$.

19. Show that the sequence $\{a_n\}$ converges where

(a) $a_n = 1 \cdot 3 \cdot 4 \cdots (2n - 1)/(2 \cdot 4 \cdot 6 \cdots (2n))$.

(b) $a_1 = .1, a_2 = .12, \ldots, a_n = .12 \ldots n$.

20. (a) Suppose that $\{a_n\}$ is bounded and that $\{b_n\} \to 0$. Prove that $\{a_n b_n\}$ converges to 0.

(b) Find an example where $\{b_n\} \to 0$ but $\{a_n b_n\}$ does not converge to 0.

21. Show that a monotone increasing sequence that does not converge must diverge to ∞.

22. (a) Suppose that $\{a_n\} \to a$. Let $\{b_n\}$ be the sequence defined by

$$b_n = \frac{a_1 + \cdots + a_n}{n}.$$

Show that $\{b_n\} \to a$.

(b) Let $\{a_n\} = \{(-1)^n\}$. Show that $\{a_n\}$ diverges but the sequence $\{b_n\}$ defined by

$$b_n = \frac{a_1 + \cdots + a_n}{n}$$

converges.

23. (a) Show that if $n \geq 3$, then $\{n^{1/n}\}$ is a monotone decreasing sequence.

(b) Show that $\lim n^{1/n} = 1$.

24. Suppose that $\{x_n\}$ is a sequence of positive numbers and

$$\lim \left(\frac{x_{n+1}}{x_n} \right) = L.$$

(a) Show that if $L > 1$, then $\lim x_n = \infty$, and if $L < 1$, $\lim x_n = 0$.

(b) Construct a sequence of positive numbers $\{x_n\}$ such that

$$\lim \frac{x_{n+1}}{x_n} = 1$$

and the sequence $\{x_n\}$ diverges.

(c) Show that $\lim n^k / \alpha^n = 0$ for $\alpha > 1$ and k a positive integer.

25. Give a condition equivalent to the statement, "The sequence $\{x_n\}$ does not converge to L," that does not use negatives.

2.2 Subsequences

Definition: Let $\{a_n\}$ be a sequence of real numbers, and let

$$n_1 < n_2 < \cdots < n_k < \cdots$$

be a strictly increasing sequence of positive integers. Then

$$\{a_{n_1}, a_{n_2}, \ldots, a_{n_k}, \ldots\}$$

is called a *subsequence* of $\{a_n\}$ and is denoted by $\{a_{n_k}\}$.

Notice that the second subscript gives the number of the term in the subsequence. For example, a_{n_5} is the fifth term in the subsequence.

Thus a subsequence is obtained from a sequence by choosing certain terms from the original sequence but not altering the relative order in which the terms occur. If

$$n_1 < n_2 < \cdots < n_k < \cdots$$

is a strictly increasing sequence of positive integers, then $n_k \geq k$ for every k. Also notice that a sequence is a subsequence of itself by choosing $n_k = k$ for every k.

Convergence of the subsequence $\{a_{n_k}\}$ of $\{a_n\}$ is simply convergence of the sequence $\{b_n\}$ where $b_k = a_{n_k}$. Thus the subsequence $\{a_{n_k}\}$ converges to L if $|a_{n_k} - L|$ may be made arbitrarily small by taking k sufficiently large.

Given a sequence that does not converge, it is sometimes possible to extract a subsequence that does converge. For example, the sequence

$$\{-1, 1, -1, 1, -1, 1, \dots\} = \{(-1)^n\}$$

does not converge, but the subsequence

$$\{(-1)^{2n}\} = \{1, 1, 1, \dots\}$$

converges to 1. The numbers $2, 4, \dots, 2k, \dots$ are the $n_1, n_2, \dots, n_k, \dots$ in the definition.

Definition: Let $\{a_n\}$ be a sequence of real numbers. We say that the number L is a *subsequential limit* of $\{a_n\}$ if there is a subsequence of $\{a_n\}$ that converges to L.

We shall adopt the *convention* that ∞ or $-\infty$ are subsequential limit points of a sequence if there are subsequences that diverge to those values. We shall also adopt the convention that ∞ is the least upper bound of a set that is not bounded above and $-\infty$ is the greatest lower bound of a set that is not bounded below. With these conventions, every sequence has a subsequential limit point and every set has a least upper bound and a greatest lower bound. Whenever a letter such as L is used to denote a subsequential limit point, it will be understood to represent a real number. We shall often refer to a subsequential limit of a sequence as merely a limit point of the sequence.

Theorem 2.10: A sequence $\{a_n\}$ converges to L if and only if every subsequence of $\{a_n\}$ converges to L.

Proof: We first show that if $\{a_n\}$ converges to L, then any subsequence of $\{a_n\}$ converges to L. Let $\{a_{n_k}\}$ be a subsequence of $\{a_n\}$, and let $\epsilon > 0$ be given. Since $\{a_n\}$ converges to L, there is a number $N(\epsilon)$ such that if $n > N(\epsilon)$, then $|a_n - L| < \epsilon$. We must show that there is a number $K(\epsilon)$ such that if $k > K(\epsilon)$, then $|a_{n_k} - L| < \epsilon$. (Remember, the k counts the terms of the subsequence.) Since $n_1 < n_2 < \cdots$ is an increasing sequence of positive integers, $n_k \geq k$ for every positive integer k. Therefore, if $k > N(\epsilon)$, then a_{n_k} is a term of $\{a_n\}$ past the $N(\epsilon)$ term, and so $|a_{n_k} - L| < \epsilon$. (Thus we can take $K(\epsilon) = N(\epsilon)$.)

Conversely, let $n_k = k$, $k = 1, 2, \ldots$ so that the subsequence $\{a_{n_k}\}$ is the original sequence $\{a_n\}$. Then when we say this subsequence converges to L, we are also saying that $\{a_n\}$ converges to L. ■

We next give a characterization of subsequential limit points.

Theorem 2.11: The number L is a subsequential limit point of the sequence $\{a_n\}$ if and only if for any $\epsilon > 0$, the interval $(L - \epsilon, L + \epsilon)$ contains infinitely many terms of $\{a_n\}$.

Proof: First suppose that L is a subsequential limit point of $\{a_n\}$. Then there is a subsequence $\{a_{n_k}\}$ of $\{a_n\}$ that converges to L. Let $\epsilon > 0$ be given. Then there is a number $K(\epsilon)$ such that if $k > K(\epsilon)$, then $|a_{n_k} - L| < \epsilon$; that is, $a_{n_k} \in (L - \epsilon, L + \epsilon)$, and so $(L - \epsilon, L + \epsilon)$ contains infinitely many terms of $\{a_n\}$.

Conversely, suppose that for any $\epsilon > 0$ the interval $(L - \epsilon, L + \epsilon)$ contains infinitely many terms of $\{a_n\}$. We shall show that L is a subsequential limit point of $\{a_n\}$ by constructing a subsequence that converges to L. The construction proceeds inductively.

Step 1. Let $\epsilon = 1$. There is some term of $\{a_n\}$ in $(L - 1, L + 1)$, say a_{n_1}.
Step 2. Let $\epsilon = 1/2$. There are infinitely many terms of $\{a_n\}$ in $(L - 1/2, L + 1/2)$, so we can choose one beyond the n_1 term. Call it a_{n_2}. That is, it is possible to choose $n_2 > n_1$ with $a_{n_2} \in (L - 1/2, L + 1/2)$.

Induction Step: Assume that for $n = k$ (a positive integer) it is possible to find a term of $\{a_n\}$ beyond the n_{k-1} term in $(L - 1/k, L + 1/k)$. Choose one of these terms, and let it be a_{n_k}.

Let $n = k + 1$. Letting $\epsilon = 1/(k+1)$ in the hypothesis, there are infinitely many terms of $\{a_n\}$ in $(L - 1/(k+1), L + 1/(k+1))$. Choose one beyond the n_kth term, and let it be $a_{n_{k+1}}$.

Thus we shown by induction that it is possible to extract a subsequence $\{a_{n_k}\}$ of $\{a_n\}$ such that $a_{n_k} \in (L - 1/k, L + 1/k)$ for $k = 1, 2, \ldots$. We show that $\{a_{n_k}\}$ converges to L.

Let $\epsilon > 0$ be given. By the Archimedean Principle, there is a positive integer $K(\epsilon)$ with $1/K(\epsilon) < \epsilon$. Thus if $k > K(\epsilon)$, then

$$a_{n_k} \in \left(L - \frac{1}{k}, L + \frac{1}{k}\right) \subset (L - \epsilon, L + \epsilon),$$

so that $\{a_{n_k}\} \to L$. ■

Corollary 2.11: Let $\{a_n\}$ be a sequence of real numbers. Then L is a subsequential limit of $\{a_n\}$ if and only if given any $\epsilon > 0$ and any positive integer N, there is a positive integer $n(\epsilon, N) > N$ for which $|a_{n(\epsilon,N)} - L| < \epsilon$.

Remark: Following our usual notation, one should think of ϵ as being a very small number and N as being very large. As an example of what the corollary says, consider the sequence

$$\{a_n\} = \{1, \frac{1}{2}, 2, \frac{1}{4}, 3, \frac{1}{6}, 4, \dots\}.$$

The odd terms of the sequence are becoming larger, whereas the even terms are approaching 0. Thus 0 is a limit point of $\{a_n\}$ (as is ∞, but that does not concern us here). In the corollary if we choose $\epsilon = 10^{-3}$ and $N = 1000$, we must find a term $n(10^{-3}, 1000)$ in the sequence beyond the 1000th term for which $|a_{n(10^{-3},1000)} - 0| < 10^{-3}$. The 2002nd term in $\{a_n\}$ has value less than 10^{-3} and is beyond the 1000th term. Thus one possibility for $n(10^{-3}, 1000)$ is the number 2002.

Exercises 2.2

1. Find the subsequential limits of the following sequences:

   (a) $\{\frac{1}{2}, -\frac{1}{2}, \frac{3}{4}, -\frac{3}{4}, \frac{7}{8}, \dots\}$
   (b) $\{\sin\left(\frac{n\pi}{2} + \frac{1}{n}\right)\}$.
   (c) $\{[(-1)^n + 1]n\}$.

2. Suppose $0 < a < 1$.

   (a) Show that $\{a^n\}$ is a decreasing sequence.
   (b) Show that $\{a^n\}$ converges.
   (c) Show that $\{a^n\}$ converges to 0. [*Hint:* Suppose that $\{a^n\} \to L$. Then $\{a^{n+1}\}$ is a subsequence of $\{a^n\}$ and $\lim a^{n+1} = a \lim a^n$].

3. Suppose $a > 1$.

   (a) Show that $\{a^n\}$ is an increasing sequence.
   (b) Show that $\{a^n\}$ diverges to ∞.

4. Suppose $a > 1$.

   (a) Show that $a^{1/n} \geq 1$ for every n.
   (b) Show that $\{a^{1/n}\}$ is an decreasing sequence.
   (c) Show that $\lim a^{1/n} = 1$.

5. Suppose $0 < a < 1$.

   (a) Show that $\{a^{1/n}\}$ is an increasing sequence.

(b) Show that $\lim a^{1/n} = 1$.

6. Define a sequence $\{a_n\}$ by $a_1 = \sqrt{2}$ and $a_{n+1} = \sqrt{2 + a_n}$.

   (a) Show that $a_n \le 2$ for every n.

   (b) Show that $\{a_n\}$ is an increasing sequence, and thus $\{a_n\}$ converges.

   (c) Show that $\lim a_n = 2$.

7. Define a sequence $\{a_n\}$ by $a_1 = \sqrt{2}$ and $a_{n+1} = \sqrt{2a_n}$.

   (a) Show that $\{a_n\}$ converges.

   (b) Find $\lim a_n$.

8. (a) Construct a sequence with exactly two subsequential limit points.

   (b) Construct a sequence in which every positive integer is a subsequential limit point.

   (c) Describe a sequence for which every real number is a subsequential limit point.

9. (a) Let $a_n = (1 + 1/2n)^{3n}$. Find $\lim a_n$ using the fact that $\lim(1 + 1/n)^n = e$.

   (b) Let $a_n = (1 + 1/\sqrt{n})^n$. Find $\lim a_n$.

10. Show that any sequence $\{a_n\}$ that has the property $|a_{n+1} - a_n| < b^n$ for $b < 1$ is a Cauchy sequence.

11. Let a_0 and a_1 be any two real numbers, and define
$$a_n = \frac{a_{n-1} - a_{n-2}}{2}.$$
Determine the convergence of the sequence $\{a_n\}$.

12. (a) Show that if $\{a_n\}$ converges to L and if $a_n \le L$ for infinitely many values of n, then there is a subsequence of $\{a_n\}$ that is increasing and converges to L.

   (b) Show that if $\{a_n\}$ converges to L, then there is either an increasing subsequence of $\{a_n\}$ that converges to L or a decreasing subsequence of $\{a_n\}$ that converges to L.

13. Let $k > 1$ and define a sequence $\{a_n\}$ by $a_1 = 1$ and $a_{n+1} = k(1 + a_n)/(k + a_n)$.

   (a) Show that $\{a_n\}$ converges.

   (b) Find $\lim a_n$.

14. Let $\{a_n\}$ be a sequence such that $\{a_{2n}\}$ converges to L and $\{a_{2n+1}\}$ converges to L. Show that $\{a_n\}$ converges to L.

15. The Fibonacci sequence $\{a_n\}$ is defined by $a_0 = 1$, $a_1 = 1$, and $a_n = a_{n-1} + a_{n-2}$ for $n \geq 2$. Let the sequence $\{f_n\}$ be defined by $f_n = a_n/a_{n-1}$ for $n \geq 1$. Assuming that $\{f_n\}$ converges (which it does), find $\lim f_n$.

16. Use the results of Exercises 9 and 10 of Section 1.2 and Exercises 4 and 5 of this section to show that for $a > 0$ and a given $\epsilon > 0$, there is a $\delta > 0$ such that if $|x| < \delta$, then $|a^x - 1| < \epsilon$. (This is sufficient to show l.u.b. A = g.l.b. B in Exercise 17, Section 1.3.)

2.3   The Bolzano-Weierstrass Theorem

In this section we prove a result known as the Bolzano-Weierstrass Theorem and use it to derive some properties of sequences. In later chapters we shall use the theorem in other ways. We shall give the proof in the one-dimensional case, but the theorem is true in any finite dimension, and the proof in higher dimensions is nearly identical to the one-dimensional case.

Definition: A number x is called a *limit point* (or *cluster point* or *accumulation point*) of a set of real numbers A if for any $\epsilon > 0$, the interval $(x - \epsilon, x + \epsilon)$ contains infinitely many points of A.

Notice that there is no requirement that a limit point of a set be an element of the set. For example, 0 is a limit point of the set $\{\frac{1}{2}, \frac{1}{3}, \frac{1}{4}, \dots\}$.

Recall that $[a, b]$ is the set $\{x | a \leq x \leq b\}$. In the proof of Theorem 2.12, when we say we divide the interval $[a, b]$ into two intervals of equal length, these intervals will be understood to be $[a, (a + b)/2]$ and $[(a + b)/2, b]$.

Theorem 2.12 (Bolzano-Weierstrass Theorem): Every bounded infinite set of real numbers has at least one limit point.

Proof: Let A be a bounded set of real numbers. Since A is bounded, there is a positive number M such that $A \subset [-M, M]$. Divide $[-M, M]$ into two closed intervals of equal length, $[-M, 0]$ and $[0, M]$. At least one of these intervals contains an infinite number of points of A. Choose one of the intervals that contains an infinite number of points of A–call it A_1. Notice that the length of A_1 is $M = 2M/2$.

Divide A_1 into two closed intervals of equal length. Choose one of the subintervals that contains infinitely many points of A, and call it A_2. The length of A_2 is $2M/2^2$. See Figure 2.5.

Continue the process inductively so that for each positive integer k, A_k is a closed interval of length $2M/2^k = M/2^{k-1}$, and A_k contains infinitely many

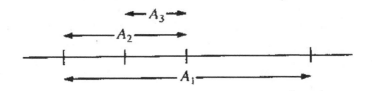

Figure 2.5
A sequence of intervals

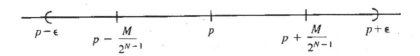

Figure 2.6
Intervals containing point p

points of A. Notice that $A_n \supset A_{n+1}$, and if $A_n = [a_n, b_n]$, then

$$\lim(b_n - a_n) = \lim \frac{M}{2^{n-1}} = 0.$$

Thus, by Theorem 2.7, $\cap_{n=1}^{\infty} A_n$ has exactly one point – call it p. We claim that p is a limit point of A.

Let $\epsilon > 0$ be given. We shall show that $(p - \epsilon, p + \epsilon)$ contains infinitely many points of A by showing that $(p - \epsilon, p + \epsilon) \supset A_N$ for some positive integer N.

Choose N so that $2M/2^N = M/2^{N-1} < \epsilon$. See Figure 2.6. Now $p \in A_N$ (since $p \in A_n$ for every n), and the length of A_N is $M/2^{N-1}$. Thus

$$(p - \epsilon, p + \epsilon) \supset \left[p - \frac{M}{2^{N-1}}, p + \frac{M}{2^{N-1}} \right] \supset A_N$$

which shows $(p - \epsilon, p + \epsilon)$ contains infinitely many points of A. ∎

We want to relate the subsequential limit points of a sequence and the limit points of a set. The *set* consisting of the terms of the sequence $\{a_n\}$ is the set of *different* numbers that appear in the sequence. For example, the set consisting of the terms of the sequence $\{0, 1, 0, 1, 0, 1, \dots\}$ is the set $\{0, 1\}$.

Theorem 2.13: Let $\{a_n\}$ be a sequence. Then the number L is a finite subsequential limit of $\{a_n\}$ if and only if L satisfies either of the following conditions:

(i) There are infinitely many terms of $\{a_n\}$ that equal L.

(ii) L is a limit point of the set consisting of the terms of $\{a_n\}$.

Proof: We give the idea of the proof and leave the formalization of the argument as Exercise 1, Section 2.3. According to Theorem 2.11, L is a subsequential limit of the sequence $\{a_n\}$ if and only if $(L - \epsilon, L + \epsilon)$ contains infinitely many terms of $\{a_n\}$ for any $\epsilon > 0$. But this occurs if and only if (i) or (ii) occurs. ∎

Example 2.10:

Let $\{a_n\} = \{0, 1, 0, 1, \dots\}$. Then the set $\{0, 1\}$ contains all of the elements of the terms of the sequence $\{a_n\}$. Since 0 and 1 each appear infinitely often, they are each subsequential limit points for the sequence $\{a_n\}$. They are not limit points of the set made up of the terms of $\{a_n\}$, since this set is $\{0, 1\}$ and as a finite set it has no limit points.

Example 2.11:

Let $\{a_n\} = \{1, \frac{1}{2}, 1, \frac{1}{4}, 1, \frac{1}{8}, \dots\}$. Here $\{1, \frac{1}{2}, 1, \frac{1}{4}, 1, \frac{1}{8}, \dots\}$ is the set of elements that makes up the terms of the sequence $\{a_n\}$. The number 1 appears infinitely often in the sequence, and 0 is the only limit point of the set $\{1, \frac{1}{2}, 1, \frac{1}{4}, 1, \frac{1}{8}, \dots\}$. Thus 0 and 1 are exactly the subsequential limit points of the sequence.

The next theorem uses the Bolzano-Weierstrass Theorem to prove an important characteristic of sequences.

Theorem 2.14: Every bounded sequence has a convergent subsequence.

Proof: Let $\{a_n\}$ be a bounded sequence. There are two possibilities:

Case 1: There is a number L for which an infinite number of terms of $\{a_n\}$ assume the value L. Suppose these are the $n_1, n_2, \dots$ terms of the sequence with $n_1 < n_2 < \cdots$. Then the subsequence $\{a_{n_1}, a_{n_2}, \dots\}$ converges to L.

Case 2: The terms of the sequence take on an infinite number of distinct values. (Case 1 does not exclude this possibility, but if Case 1 does not occur, then Case 2 must occur.) By the Bolzano-Weierstrass Theorem, there is a number p that has the property that $(p - \epsilon, p + \epsilon)$ contains an infinite number of terms of $\{a_n\}$ for any $\epsilon > 0$. By Theorem 2.11 there is a subsequence of $\{a_n\}$ that converges to p. ∎

Corollary 2.14: A bounded sequence that does not converge has more than one subsequential limit point.

The proof is left for Exercise 17, Section 2.3. ∎

Notice that if $\{a_n\}$ is a sequence for which $a \leq a_n \leq b$ for every n and if L is a subsequential limit point of $\{a_n\}$, then by Theorem 2.5, $a \leq L \leq b$.

We next examine a characteristic of unbounded sequences.

Theorem 2.15:

(a) A sequence that is unbounded above has a subsequence that diverges to ∞.

(b) A sequence that is unbounded below has a subsequence that diverges to $-\infty$.

Proof of (a): Let $\{a_n\}$ be a sequence that is unbounded above. We shall inductively construct a subsequence of $\{a_n\}$ that diverges to ∞.

Let $P(k)$ be the statement that there is a positive integer $n_k > n_{k-1}$ with $a_{n_k} > k$. We shall verify that $P(k)$ is true for each positive integer k.

Step 1: Since $\{a_n\}$ is unbounded above, there is a term, say a_{n_1} with $a_{n_1} > 1$. Thus $P(1)$ is true.

Step 2: There are infinitely many terms of $\{a_n\}$ that exceed 2. Why are there infinitely many? Because if there were only finitely many, then we could choose the largest number in this finite set to be an upper bound of $\{a_n\}$. The reason we need an infinite number of terms is that now we can choose a term a_{n_2} with $n_2 > n_1$.

Induction Step: Now assume $P(k)$ holds, that is, that we have found $a_{n_k} > k$ with $n_k > n_{k-1}$. Then, as in Step 2, there are infinitely many terms of $\{a_n\}$ larger than $k + 1$, so we can choose a term $a_{n_{k+1}} > k + 1$ with $n_{k+1} > n_k$. Thus $P(k + 1)$ is true.

By induction $P(k)$ is true for all k. Thus there is a sequence of positive integers

$$n_1 < n_2 < n_3 < \cdots$$

with $a_{n_k} > k$. This subsequence $\{a_{n_k}\}$ diverges to ∞. ∎

Example 2.12:

Let $\{a_n\}$ be the sequence $\{1, \frac{1}{2}, 2, \frac{1}{3}, 3, \frac{1}{4}, 4, \dots\}$. Then the subsequence $\{1, 2, 3, 4, \dots\}$ diverges to ∞.

Theorem 2.16 just puts some of our earlier results together to provide yet another way of characterizing convergent sequences. The proof amounts to citing these earlier results and is a useful exercise.

Theorem 2.16: A sequence $\{a_n\}$ converges if and only if it is bounded and has exactly one subsequential limit point. ∎

Since we have adopted the convention that ∞ and $-\infty$ are subsequential limit points for the sequence $\{a_n\}$, if there are subsequences that diverge to those values, then every sequence has at least one subsequential limit point. Using the fact that $-\infty < L < \infty$ for any real number L, we can make the following definition.

Definition: Let $\{a_n\}$ be a sequence of real numbers. Then $\limsup a_n = \overline{\lim} \, a_n$ is the least upper bound of the set of subsequential limit points of $\{a_n\}$, and $\liminf a_n = \underline{\lim}\, a_n$ is the greatest lower bound of the set of subsequential limit points of $\{a_n\}$.

Remark: We shall show in Exercise 16, Section 2.3 that the supremum of a set of limit points of a sequence is a limit point of the sequence, as is the infimum. Assuming this, we shall now refer to $\overline{\lim}\, a_n$ and $\underline{\lim}\, a_n$ as the largest and smallest respectively of the limit points of the sequence $\{a_n\}$.

Notice that $\overline{\lim}\, a_n$ is not necessarily the largest value of $\{a_n\}$, but the largest of the *limit points* of $\{a_n\}$. In fact if

$$\{a_n\} = \{1, \frac{1}{2}, \frac{1}{3}, \frac{1}{4}, \dots\},$$

then $\overline{\lim}\, a_n = 0$ even though every term of the sequence is larger than 0.

We next characterize $\overline{\lim}\, a_n$ and $\underline{\lim}\, a_n$ for a bounded sequence $\{a_n\}$.

Theorem 2.17: Let $\{a_n\}$ be a *bounded* sequence of real numbers. Then:

(a) $\overline{\lim}\, a_n = L$ if and only if, for any $\epsilon > 0$, there are infinitely many terms of $\{a_n\}$ in $(L-\epsilon, L+\epsilon)$ but only finitely many terms of $\{a_n\}$ with $a_n > L+\epsilon$.

(b) $\underline{\lim}\, a_n = K$ if and only if, for any $\epsilon > 0$, there are infinitely many terms of $\{a_n\}$ in $(K-\epsilon, K+\epsilon)$ but only finitely many terms of $\{a_n\}$ with $a_n < K-\epsilon$.

Proof of (a): Suppose $\overline{\lim}\, a_n = L$. Since L is a limit point of $\{a_n\}$, we know by Theorem 2.11 that there are infinitely many terms of $\{a_n\}$ in $(L - \epsilon, L + \epsilon)$. Thus we need to show that there are only finitely many terms of $\{a_n\}$ that exceed $L + \epsilon$ for any positive number ϵ. We do the proof by contradiction; that is, we shall suppose that for some $\epsilon > 0$ there are an infinite number of terms of $\{a_n\}$ that exceed $L + \epsilon$. We shall then show that there is a subsequential limit of $\{a_n\}$ larger than L. This will contradict the hypothesis that $L = \overline{\lim}\, a_n$ is the largest of the limit points of $\{a_n\}$.

Suppose there is an $\epsilon > 0$ for which there are an infinite number of terms of $\{a_n\}$ that exceed $L + \epsilon$. Since $\{a_n\}$ is bounded above, say by M, there are infinitely many terms of $\{a_n\}$ between $L + \epsilon$ and M. From those terms of $\{a_n\}$ that exceed $L + \epsilon$, we construct a subsequence $\{a_{n_k}\}$. (We leave the argument of how this may be done until Exercise 5, Section 2.3.) Now $\{a_{n_k}\}$ has a convergent subsequence, and since each term of $\{a_{n_k}\}$ is at least as large as $L + \epsilon$, this limit is at least as large as $L + \epsilon$. Thus we have shown the existence of a limit point of $\{a_n\}$ larger than L, contradicting the hypothesis.

Conversely, suppose $\{a_n\}$ is a bounded sequence such that for any $\epsilon > 0$ the interval $(L - \epsilon, L + \epsilon)$ contains infinitely many terms of $\{a_n\}$, but only finitely many terms of $\{a_n\}$ exceed $L+\epsilon$. Since $(L-\epsilon, L+\epsilon)$ contains infinitely many terms of $\{a_n\}$ for any $\epsilon > 0$, L is a limit point of $\{a_n\}$. Suppose $M > L$. We shall show that M is not a limit point of $\{a_n\}$ This will mean $L = \overline{\lim}\, a_n$.

Let $\epsilon = (M - L)/2$. (See Figure 2.7.) Then $(M - \epsilon, M + \epsilon)$ contains only finitely many terms of $\{a_n\}$ (since every number in $(M - \epsilon, M + \epsilon)$ exceeds

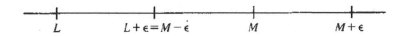

Figure 2.7
Intervals for the proof of Theorem 2.17

$L + \epsilon$); so M is not a limit point of $\{a_n\}$. ■

The proof of part (b) is similar and is left for Exercise 7, Section 2.3.

Corollary 2.17: A bounded sequence $\{a_n\}$ of numbers converges if and only if $\overline{\lim}\, a_n = \underline{\lim}\, a_n$.

Theorem 2.18: Let $\{a_n\}$ and $\{b_n\}$ be bounded sequences. Then

(a) $\overline{\lim}(a_n + b_n) \leq \overline{\lim}\, a_n + \overline{\lim}\, b_n$.

(b) $\underline{\lim}\, a_n + \underline{\lim}\, b_n \leq \underline{\lim}(a_n + b_n)$.

Remark: We first observe that equality does not always hold. If

$$\{a_n\} = \{(-1)^n\} \text{ and } \{b_n\} = \{(-1)^{n+1}\},$$

then

$$\overline{\lim}\, a_n = \overline{\lim}\, b_n = 1,$$

but

$$a_n + b_n = 0 \text{ for every } n.$$

Thus, $\overline{\lim}(a_n + b_n) = 0$. (Contrast this with Exercise 8 of Section 1.3.)

Proof: We give only the proof of the first inequality and leave the proof of the second inequality until Exercise 8, Section 2.3.

Let $K = \overline{\lim}\, a_n$ and $L = \overline{\lim}\, b_n$, and let $\epsilon > 0$ be given. To show that

$$\overline{\lim}(a_n + b_n) \leq \overline{\lim}\, a_n + \overline{\lim}\, b_n = K + L,$$

it is enough to show that there are at most finitely many terms of $\{a_n + b_n\}$ that exceed $K + L + \epsilon$, by Theorem 2.17.

By Theorem 2.17, there are at most finitely many terms of $\{a_n\}$ that exceed $K + \epsilon/2$. Call those that do exceed $K + \epsilon/2$ the $n_1, \ldots, n_r$ terms. At most finitely many terms of $\{b_n\}$ exceed $L + \epsilon/2$. Call those that do exceed $L + \epsilon/2$ the $m_1, \ldots, m_s$ terms. Now if $n \neq n1, \ldots, n_r, m_1, \ldots, m_s$, then

$$a_n + b_n < \left(K + \frac{\epsilon}{2}\right) + \left(L + \frac{\epsilon}{2}\right) = K + L + \epsilon. \qquad ■$$

Many of our applications of sequences will deal with functions. We close this chapter with such an application that will be used quite often when we discuss integration theory.

Definition: We say that a function f is *bounded* if the range of f is a bounded set.

If f is bounded, we denote l.u.b. $\mathcal{R}(f)$ by $\sup f$ and g.l.b. $\mathcal{R}(f)$ by $\inf f$.

Theorem 2.19: Let f and g be bounded functions with the same domain $\mathcal{D}$. Then

(a) $\sup(f + g) \leq \sup f + \sup g$.

(b) $\inf f + \inf g \leq \inf(f + g)$.

(Compare this with Theorem 2.18.)

Proof: We prove part (a). The proof of part (b) is similar. We then give an example to show that the equality need not hold. To prove part (a), we know by Theorem 2.8 that we can choose a sequence $\{y_n\}$ such that $y_n \in \mathcal{R}(f + g)$, $n = 1, 2, \ldots$ and

$$\lim y_n = \sup(f + g).$$

For each $y_n \in \mathcal{R}(f + g)$, there is an $x_n \in \mathcal{D}(f + g)$ such that

$$y_n = (f + g)(x_n) = f(x_n) + g(x_n).$$

But

$$f(x_n) \leq \sup f \text{ and } g(x_n) \leq \sup g$$

for all n. Thus

$$\sup(f + g) = \lim y_n \leq \sup f + \sup g. \qquad \blacksquare$$

Example 2.13:

Let

$$f(x) = \begin{cases} 0, & 0 \leq x \leq 1 \\ 1, & 1 < x \leq 2, \end{cases} \qquad g(x) = \begin{cases} 1, & 0 \leq x \leq 1 \\ 0, & 1 < x \leq 2. \end{cases}$$

Then $(f + g)(x) = 1$ for $0 \leq x \leq 2$. But $\sup f = 1$ and $\sup g = 1$, so $\sup(f + g) = 1 < 2 = \sup f + \sup g$.

Exercises 2.3

1. Write a formal proof of Theorem 2.13.

2. Show that a finite set can have no limit points.

3. Construct an infinite set that has no limit points.

4. Prove part (b) of Theorem 2.15.

5. (a) Show that if $\{a_n\}$ is a sequence that has infinitely many terms larger than $L + \epsilon$ for some $\epsilon > 0$, then there is a subsequence of $\{a_n\}$ all of whose terms are larger than $L + \epsilon$.

   (b) Show that if the subsequence of part (a) is bounded, then it has a subsequence that converges to a number at least as large as $L + \epsilon$.

6. Construct a sequence $\{a_n\}$ with $\underline{\lim}\, a_n = \infty$.

7. Prove part (b) of Theorem 2.17.

8. Prove part (b) of Theorem 2.18.

9. Prove part (b) of Theorem 2.19.

10. Suppose $\{a_n\}$ and $\{b_n\}$ are sequences such that for every n, $a_n \le b_n$. Prove that
$$\overline{\lim}\, a_n \le \overline{\lim}\, b_n \text{ and } \underline{\lim}\, a_n \le \underline{\lim}\, b_n.$$

11. (a) Suppose that $\{a_n\}$ and $\{b_n\}$ are bounded sequences of nonnegative terms with $\lim a_n = L$ and $\lim b_n = K$. Show that $\overline{\lim}(a_n b_n) = LK$.

    (b) Give an example where the result of part (a) fails if the terms of one of the sequences have negative values.

12. Show that if every open interval containing L contains a point of the set A distinct from L, then, for any $\epsilon > 0$, $(L - \epsilon, L + \epsilon)$ contains infinitely many points of A.

13. The purpose of this exercise is to prove that every Cauchy sequence is a convergent sequence. Let $\{a_n\}$ be a Cauchy sequence.

    (a) Show that $\{a_n\}$ is bounded.

    (b) Show that there is at least one subsequential limit point of $\{a_n\}$.

    (c) Prove there is no more than one subsequential limit point of $\{a_n\}$.

    (d) Show that $\{a_n\}$ converges.

14. Show that a convergent sequence is a Cauchy sequence.

15. Let A be an uncountable set of numbers.

    (a) Show that A has a (finite) limit point.

(b) Is it possible for A to have a most a finite number of limit points?

16. Show that the supremum (l.u.b.) of a set of limit points of a sequence $\{a_n\}$ is a limit point of $\{a_n\}$. Likewise, show that the infimum of a set of limit points of a sequence is a limit point of the sequence.

17. Show that a bounded sequence that does not converge has more than one subsequential limit point.

3

Topology of the Real Numbers

One of our primary objectives is to derive some of the properties of continuous real-valued functions on the real numbers. Even though we have not yet defined what a continuous function is, it is likely that you have some idea of the importance of continuous functions from elementary calculus. The study of sets and continuous functions is an area of mathematics called topology. In this chapter we define some of the topological properties that sets can have, and we determine which sets of real numbers have these these properties. Later we shall prove some results that describe how continuous functions behave on sets that have these properties. For example, one of the topological properties a set may have is compactness. As a consequence of this property, a continuous function on a closed bounded interval must attain its maximum and minimum values on the interval.

3.1 Topology of the Real Numbers

Open and Closed Sets

The topological properties with which we shall be concerned are defined in terms of open sets. The property of being open is not an inherent quality of a set, but in the setting in which we shall work, an open set is defined as follows:

Definition: A set U of real numbers is said to be *open* if for each $x \in U$ there is a number $\delta(x) > 0$) such that $(x - \delta(x), x + \delta(x)) \subset U$.

In the definition the number $\delta(x)$ depends on what x is. The idea is similar to some others we have encountered in that, to show a set U is open, we first choose $x \in U$ and then show there is a number $\delta(x) > 0$ (which may be different for different values of x) such that $(x - \delta(x), x + \delta(x)) \subset U$. Now that we have made the point that $\delta(x)$ is a number that depends on x, we shall frequently use the less cumbersome notation of δ instead of $\delta(x)$. Another way to state the condition $(x - \delta, x + \delta) \subset U$ is to say that if $|x - y| < \delta$, then $y \in U$.

Figure 3.1
Minimum distance between x and an endpoint of the interval (a, b).

Our first project will be to identify which sets are open. First, $\mathbb{R}$ is an open set, since for any $x \in \mathbb{R}, (x - 1, x + 1) \subset \mathbb{R}$. (This is the only nonempty set for which the same δ, in this case 1, works for every x). Also $\emptyset$ is an open set because the condition defining open sets is satisfied vacuously. That is, since there are no elements in $\emptyset$ then for any δ and $x \in \emptyset$, $(x - \delta, x + \delta) \subset \emptyset$. Some other open sets are identified in Theorem 3.1.

Theorem 3.1: The intervals $(a, b), (a, \infty)$ and $(-\infty, a)$ are open sets.

Remark: These are not the only open sets, but they are the building blocks from which other open sets are constructed.

Proof: We show that (a, b) is an open set. Pick $x \in (a, b)$. We must find a $\delta > 0$ such that $(x - \delta, x + \delta) \subset (a, b)$. Figure 3.1 may clarify how δ is selected. Choose $\delta = (1/2) \min \{x - a, b - x\}$. Then

$$x - \delta > x - (x - a) = a \text{ and } x + \delta < x + b - x = b$$

So $(x - \delta, x + \delta) \subset (a, b)$. Thus (a, b) is an open set.   ∎

Notice that as x changes on the interval, δ changes. The closer x is to an endpoint of an interval, the smaller δ must be.

Definition: A set A is closed if A^c is open.

Corollary 3.1: The intervals $(-\infty, a], [a, b]$ and $[b, \infty)$ are closed sets.

Again these are not the only closed sets.

Proof: Since $(-\infty, a] = (a, \infty)^c, (-\infty, a]$ is closed. The other sets are shown to be closed in a similar manner.   ∎

Theorem 3.1 and its corollary show why (a, b) is called an open interval and $[a, b]$ is called a closed interval.

We now give two ways to construct an open set from other open sets.

Theorem 3-2: The open sets of real numbers satisfy the following conditions:

(a) If $\{U_1, \ldots, U_n\}$ is a finite collection of open sets then $\cap_{k=1}^{n} U_k$ is an open set.

(b) If $\{U_\alpha\}$ is any collection of open sets, then $\cup_\alpha U_\alpha$ is an open set.

Proof: To see the first condition is satisfied, let $\{U_1, \ldots, U_n\}$ be a finite collection of open sets. Pick $x \in \cap\, U_k$. Now $x \in U_i$ for every $i = 1, \ldots, n$, and each U_i is open. Thus for every $i = 1, \ldots, n$, there is a number $\delta_i > 0$ such that $(x - \delta_i, x + \delta_i) \subset U_i$. Note that the $\delta_i s$ may be different. Now let

$$\delta = \min\{\delta_1, \ldots, \delta_n\}.$$

Since this is a finite set of positive numbers, $\delta > 0$. Then

$$(x - \delta, x + \delta) \subset U_i, i = 1, \ldots, n$$

so

$$(x - \delta, x + \delta) \subset \cap_{i=1}^{n} U_i.$$

Thus $\cap_{i=1}^{n} U_i$ is an open set.

To verify the second condition, let $\{U_\alpha\}$ be a collection of open sets. Let $x \in \cup_\alpha U_\alpha$. Then $x \in U_\alpha$ for some set in the collection, say $U_{\overline{\alpha}}$. Then there is a $\delta > 0$ such that

$$(x - \delta, x + \delta) \subset U_{\overline{\alpha}}$$

so that

$$(x - \delta, x + \delta) \subset \cup_\alpha U_\alpha.$$

Thus $\cup_\alpha U_\alpha$ is an open set. ∎

Note: Theorem 3.2 shows that the intersection of a finite number of sets is open, but the intersection of an infinite number of sets might not be. For example, if $E_n = (-1/n, 1 + 1/n)$ for n a positive integer, then $\cap E_n = [0, 1]$, which is not open. (Why?)

Theorem 3.3: The closed sets satisfy the following properties:

(a) $\emptyset$ and $\mathbb{R}$ are closed.

(b) If $\{A_\alpha\}$ is a collection of closed sets, then $\cap_\alpha A_\alpha$ is closed.

(c) If $\{A_1, \ldots, A_n\}$ is a finite collection of closed sets, then $\cup_{i=1}^{n} A_i$ is closed.

The proof is left until Exercise 16, Section 3.1. ∎

As we stated at the beginning of the chapter, our primary purpose for discussing topology is to enable us to study properties of continuous functions. It is often the case that the domain of a function is not all of the real numbers, but a subset. When discussing a particular function, we shall restrict our attention to its domain, and it is with this in mind that we make the next definition.

Definition: Let A be a set of real numbers. The set V is *open relative to* A if $V = A \cap U$ for some open set U.

The set V is *closed relative to* A if $V = A \cap U$ for some closed set U.

Example 3.1:

Let $f(x) = \sqrt{x}$. Then $\mathcal{D}(f) = [0, \infty)$. Any set of the form $[0, a)$ is an open set relative to $\mathcal{D}(f)$ because

$$[0, a) = (-\infty, a) \cap \mathcal{D}(f)$$

and $(-\infty, a)$ is an open set.

Theorem 3.4: Let A be a set of real numbers. The subsets of A that are open relative to A satisfy the following conditions:

(a) $\emptyset$ and A are open.

(b) If $\{V_1, ..., V_n\}$ is a finite collection of sets that are open relative to A, then $\cap_{i=1}^{n} V_i$ is open relative to A.

(c) If $\{V_\alpha\}$ is any collection of sets that are open relative to A then $\cup_\alpha V_\alpha$ is open relative to A.

The proof is left to Exercise 16, section 3.1. ■

Earlier we showed that certain intervals were open sets. Now we show exactly what an open set must be.

Theorem 3.5: A set is open if and only if it can be expressed as a countable union of disjoint open intervals.

Proof: Let A be an open set. Then if $x \in A$ there is a number $\delta > 0$ such that $(x - \delta, x + \delta) \subset A$. Let

$$\alpha = \sup\{\delta \mid [x, x + \delta) \subset A\} \text{ and } \beta = \sup\{\delta \mid (x - \delta, x] \subset A\}$$

It is possible that either α or β may be infinite. If, for example, β is infinite then $(-\infty, x] \subset A$. If α and β are both finite, then $(x - \beta, x + \alpha) \subset A$, but $x - \beta \notin A$ and $x + \alpha \notin A$ as we verify in Exercise 4, Section 3.1. Thus for each $x \in A$ we have this sort of *maximal interval* about x, $(x - \beta, x + \alpha)$ that is contained in A. Denote this *maximal interval* I_x. If $x, y \in A$, then either $I_x = I_y$ or $I_x \cap I_y = \emptyset$ as we verify in Exercise 5, Section 3.1. Now each interval I_x must contain a rational number, and since there are only countably many rational numbers, there can only be countably many maximal intervals. Also, each $x \in A$ is in some maximal interval I_x by our construction, and $I_x \subset A$ so we have

$$A = \cup_{x \in A} I_x.$$

Conversely, any set that is the countable union of disjoint open intervals is open by condition (b) of Theorem 3.2. ∎

Definition: Let A be a set of real numbers.

(a) If there is a $\delta > 0$ such that $(x - \delta, x + \delta) \subset A$ then x is said to be an *interior point* of A. The set of all interior points of A is called the *interior* of A and is denoted int(A).

(b) If, for every $\delta > 0$, the interval $(x - \delta, x + \delta)$ contains a point in A and a point not in A, then x is said to be a *boundary point* of A. The set of all boundary points of A is called the *boundary* of A and is denoted $b(A)$.

(c) If, for every $\delta > 0$, the interval $(x - \delta, x + \delta)$ contains a point of A distinct from x, then x is said to be a *limit point* (or *cluster point*, or *accumulation point*) of A.

(d) A point $x \in A$ is said to be an *isolated point* of A if there is a $\delta > 0$ such that $(x - \delta, x + \delta) \cap A = \{x\}$.

The definition of limit point is equivalent to the one given in Chapter 2 (as was proved in Exercise 12, Section 2.3).

Example 3.2:
Let $A = (0, 1]$. Then one can verify that the interior points are $(0, 1)$, the boundary points are $\{0, 1\}$, and the limit points are $[0, 1]$.

Example 3.3:
Let $A = $ rational numbers. There are no interior points. Every real number is a boundary point and a limit point.

Example 3.4:
Let $A = $ the integers. There are no interior points. The boundary points are A and there are no limit points. Each point of A is an isolated point.

Theorem 3.5 gave a fairly tangible characterization of open sets. Unfortunately no such characterization exists for closed sets, but we shall be able to determine some of their properties. It is often true that the simplest way to prove that a set is closed is to prove that its complement is open.

Theorem 3.6: A set is closed if and only if it contains all of its boundary points.
Proof: Let A be a closed set and let x be a boundary point of A. We shall show that x must be in A by supposing $x \notin A$ and showing this is impossible.

Suppose $x \notin A$. Then $x \in A^c$ and since A is closed, A^c is open. Thus there is a $\delta > 0$ such that $(x - \delta, x + \delta) \subset A^c$ This is an interval about x containing no points of A, so x cannot be a boundary point of A. Thus a closed set must contain all of its boundary points.

Conversely, suppose that A contains all of its boundary points. We shall show that A is closed by showing A^c is open. Pick $x \in A^c$. Then x is not a boundary point of A and $x \notin A$, so there must be some $\delta > 0$ such that $(x - \delta, x + \delta)$ contains no point of A. (If $x \in A^c$ and if, for every $\delta > 0$ there was a point of A in $(x - \delta, x + \delta)$, then x would be a boundary point of A.) Thus A^c is open. ∎

We make two observations that will be verified in Exercise 15, Section 3.1. First, if x is a limit point of A and $x \notin A$ then x is a boundary point of A. Likewise, if x is a boundary point of A and $x \notin A$, then x is a limit point of A. This does not imply that boundary points and limit points are the same, and some of the earlier examples show they are not. The second observation is that if x is a boundary point of A, then x is a boundary point of A^c. These observations make the proofs of the following corollaries almost immediate.

Corollary 3.6:

(a) A set is closed if and only if it contains all of its limit points.

(b) A set is closed if and only if it contains all of its boundary points.

Definition: Let A be a set of real numbers. The closure of A, denoted $\overline{A}$, is the set consisting of A and its limit points.

By our earlier observations, $\overline{A}$ could also be defined to be the set consisting of A and its boundary points. Thus in our previous examples, the closure of $(0, 1]$ is $[0, 1]$, the closure of the rational numbers is the real numbers, and the closure of the integers is the integers.

Theorem 3.7: Let A be a set of real numbers. Then $\overline{A}$ is a closed set.

Proof: We again show that a set is closed by showing its complement is open. Let $x \in (\overline{A})^c$. We must show that some interval about x contains no points of $\overline{A}$. Now $x \notin A$ and x is not a boundary point of A, so there is a $\delta > 0$ such that $(x - \delta, x + \delta)$ contains no points of A. We aren't quite done, because we need to show also that some interval about x contains no boundary points of A. It is possible to show that $(x - \delta, x + \delta)$ contains no boundary points of A, but it is a little easier to argue that $(x - \delta/2, x + \delta/2)$ contains no boundary points of A.

If $y \in (x - \delta/2, x + \delta/2)$ then

$$(y - \delta/2, y + \delta/2) \subset (x - \delta, x + \delta).$$

But $(x - \delta, x + \delta)$ contains no points of A, so y is not a boundary point of A. ∎

Compactness

Definition: Let A be a set. The collection of sets $\{I_\alpha\}_{\alpha \in \mathcal{A}}$ is said to be a cover of A if $A \subset \cup_{\alpha \in \mathcal{A}} I_\alpha$. If each set I_α is open, the collection is said to be an *open cover* of A. If the number of sets in the collection $\mathcal{A}$ is finite, then the collection is said to be a *finite cover*.

> ***Example 3.5:***
> Let $A = (0,1)$ and let $I_n = (1/n, 1 - 1/n)$ for each integer n larger than 2. Then each set I_n is open and $A \subset \cup_{n>2} I_n$. (Why?)

> ***Example 3.6:***
> Let A be the set of real numbers, and let our collection of open sets be the collection of all intervals centered at the rational numbers of rational radius; i.e. $\{(x-\delta, x+\delta) \mid x$ is a rational number and δ is a positive rational number$\}$. This is a countable collection of open sets and forms a cover of A.

Definition: A set A is said to be *compact* if every open cover of A has a finite subcover.

Note: An open cover of A has a finite subcover if there is a finite number of sets in the collection that covers A.

Compact sets will be important in studying properties of continuous functions as we mentioned at the beginning of the section. We want to find conditions that characterize compact sets. We first observe that neither of the sets in Examples 3.5 and 3.6 is compact.

In Example 3.5, if

$$\left\{ \left(\frac{1}{n_1}, 1 - \frac{1}{n_1} \right), \ldots, \left(\frac{1}{n_k}, 1 - \frac{1}{n_k} \right) \right\}$$

is a finite collection of the open cover and

$$N = \max\{n_1, \ldots, n_k\},$$

then $1/(2N)$ is not in any of the sets of the finite subcover.

In Example 3.6 if

$$\{(x_1 - \delta_1, x_1 + \delta_1), \ldots, (x_n - \delta_n, x_n + \delta_n)\}$$

is a finite subcollection, and if

$$x_0 = \max\{x_1 \ldots, x_n\} \text{ and } \delta_0 = \max\{\delta_1, \ldots, \delta_n\}$$

then $x_0 + 2\delta_0$ is a real number that is not in the finite subcollection.

Our task of characterizing compact sets is somewhat involved. We begin by proving a more general form of a result we derived in Theorem 2.7 concerning nested closed intervals.

Theorem 3.8 Let $\{A_1, A_2, ...\}$ be a countable collection of nonempty, closed, bounded sets of real numbers such that $A_i \supset A_j$ if $i \leq j$. Then $\cap A_i \neq \emptyset$.

Proof: Since each set A_i is nonempty, we can select a number $a_i \in A_i$. Since $A_1 \supset A_i$ for every i and A_1 is bounded, then $\{a_1, a_2, ...\}$ is a bounded sequence of real numbers. As we showed in Chapter 2, this means there is a subsequence $\{a_{i_k}\}$ that converges to some number p. We shall show that $p \in A_i$ for every i. Let $\epsilon > 0$ be given. Then

$$a_{i_k} \in (p - \epsilon, p + \epsilon) \text{ for } k \text{ sufficiently large.}$$

But by the nested property of the sets A_i, this mean $(p - \epsilon, p + \epsilon)$ contains a point of every A_i. Thus either $p \in A_i$ or p is a limit point of A_i for every A_i. But each A_i is closed and must therefore contain all of its limit points. Thus $p \in \cap A_i$ and thus $\cap A_i \neq \emptyset$. ■

Corollary 3.8: Let $\{A_1, A_2, ...\}$ be a countable collection of closed bounded sets of real numbers such that $A_i \supset A_j$ if $i < j$. If $\cap_{i=1}^{\infty} A_i = \emptyset$, then there is a positive integer N such that $\cap_{i=1}^{N} A_i = \emptyset$.

Proof: By Theorem 3.8, $\cap_{i=1}^{\infty} A_i = \emptyset$ only if *not* all the sets A_i are nonempty, that is, if at least one of the sets, say A_N, is empty. But then $\cap_{i=1}^{N} A_i = \emptyset$. ■

We are almost ready to prove the theorem that characterizes compact sets, but we need one more preliminary result.

Theorem 3.9: Let A be a set of real numbers. Then if $\{I_\alpha\}$ is an open cover of A, then some countable subcollection of $\{I_\alpha\}$ covers A. ■

This property of real numbers is called the *Lindelöf* property. We shall not prove the theorem, but leave it until Exercise 17, Section 3.1, with an outline of the steps required. The theorem is important because it enables us to deal with countable open covers instead of arbitrary open covers. This simplifies the proof of the next theorem.

Theorem 3.10 (Heine-Borel Theorem): If A is a closed bounded set of real numbers, then A is compact.

Proof: Suppose A is a closed bounded set of real numbers and $\{I_\alpha\}$ is an open cover of A. By Theorem 3.9 we may extract a countable subcover $\{I_1, I_2, ...\}$. From this countable subcover we construct an increasing sequence of open sets $\{J_1, J_2, ...\}$ where

$$J_1 = I_1$$

$$J_2 = I_1 \cup I_2$$

$$\vdots$$

$$J_n = I_1 \cup I_2 \cdots \cup I_n$$

$$\vdots$$

We now apply Corollary 3.8. We construct a *decreasing* sequence of *closed* subsets $\{K_1, K_2, \dots\}$ where

$$K_1 = A \setminus J_1$$

$$K_2 = A \setminus J_2$$

$$\vdots$$

$$K_n = A \setminus J_N$$

$$\vdots$$

(Why are the sets K_i closed?)

Since $\{I_1, I_2, \dots\}$ forms an open cover of A, so does $\{J_1, J_2, \dots\}$. Thus $A \setminus \cup_{i=1}^{\infty} J_i = \emptyset$. By DeMorgan's laws

$$\emptyset = A \setminus (\cup_{i=1}^{\infty} J_i) = \cap_{i=1}^{\infty}(A \setminus J_i) = \cap_{i=1}^{\infty} K_i.$$

By Corollary 3.8, there is an N such that $\cap_{i=1}^{N} K_i = \emptyset$. Thus

$$\emptyset = \cap_{i=1}^{N} K_i = \cap_{i=1}^{N}(A \setminus (J_i) = A \setminus (\cup_{i=1}^{N} J_i) = A \setminus (\cup_{i=1}^{N} I_i)$$

so $\{I_1, \dots, I_n\}$ is a finite subcover of A. ∎

Next we prove the converse of the Heine-Borel Theorem.

Theorem 3.11: A set that is compact is closed and bounded.

Proof: We shall prove the contrapositve of the theorem. That is, we shall show that if a set fails to be either closed or bounded, then it is not compact. Suppose A is not bounded. Let $I_n = (-n, n), n = 1, 2, \dots$. Then $\{I_n\}$ is an open cover of A, but there is no finite subcover. On the other hand, if A is not closed, then there at least one limit point of A that is not in A. Call that point x. Let

$$I_n = (-\infty, x - 1/n) \cup (x + 1/n, \infty), n = 1, 2, \dots$$

Then $\{I_n\}$ is an open cover of A that has no finite subcover. ∎

Example 3.7:
The Heine-Borel Theorem says that a closed interval $[a, b]$ is compact,

as is any finite union of intervals $[a_i, b_i]$. On the other hand, the intervals $(a, b), [a, b), (a, b]$, or unbounded intervals are not compact.

The next theorem gives another characterization of compact sets and will be very useful in examining the behavior of continuous functions on compact sets.

Theorem 3.12: A set A of real numbers is compact if and only if every infinite set of points in A has a limit point in A.

Proof: Suppose A is a compact set, and suppose that B is a subset of A containing infinitely many points. Since A is compact, it is bounded and therefore B is bounded. By the Bolzano-Weierstrass Theorem, B has a limit point, say p. Because p is a limit point of B, it is also a limit point of A. (Why?) Since A is compact, it is closed, so $p \in A$.

If A is not compact, either A is not closed or A is not bounded. We shall show that in either case there is an infinite set of points in A that does not have a limit point in A. If A is not closed, there is a limit point of A that is not in A, say x. Thus each interval $(x - 1/n, x + 1/n), n = 1, 2, ...$ contains some point of A distinct from x. Choose one such point from each interval and call it x_n. Then $\{x_1, x_2, ...\}$ contains infinitely many distinct points of A (although it is not necessarily true that $x_i \neq x_j$), and $\lim_{n \to \infty} x_n = x \notin A$.

If A is not bounded, then for each $n = 1, 2, ...$ there is a point $x_n \in A$ with $x_n \notin [-n, n]$. Then $\{x_n\}$ is an infinite set with no limit point.   ■

We now restate these results on compact sets.

Theorem 3.13: Let A be a set of real numbers. The following are equivalent.

(a) A is compact.

(b) A is closed and bounded.

(c) Every infinite set of points in A has a limit point in A.

(d) If $\{x_n\}$ is a sequence of numbers in A, there is a subsequence $\{x_{n_k}\}$ that converges to some point in A.

We show that parts (c) and (d) are equivalent in Exercise 19, Section 3.1.
■

Connectedness

Besides compactness, another property of sets that is useful when discussing properties of continuous functions is connectedness.

Definition: A set of real numbers A is said to be *connected* if there do not exist two open sets U and V such that

(i) $U \cap V = \emptyset$.

(ii) $U \cap A \neq \emptyset$ and $V \cap A \neq \emptyset$.

(iii) $(U \cap A) \cup (V \cap A) = A$.

The definition of connectedness is somewhat unusual in that it is stated as the negation of a condition. Occasionally, this makes it awkward to apply, but no one seems to have found a better way to define the concept.

Example 3.8:
Let A be the set of integers. Then A is not connected, for if $U = (-\infty, 1/2)$ and $V = (1/2, \infty)$ then

$$U \cap V = \emptyset, \quad U \cap A = \{0, -1, -2, \dots\},$$
$$V \cap A = \{1, 2, 3, \dots\} \text{ and } (U \cap A) \cup (V \cap A) = A.$$

Example 3.9:
Let A be the set of rational numbers. Then A is not connected as can be seen by letting $U = (-\infty, \sqrt{2}), V = (\sqrt{2}, \infty)$.

We now characterize connected sets.

Theorem 3.14: A set of real numbers with more than one element is connected if and only if it is an interval.
Proof: Let A be a set of real numbers with more than one element that is not an interval. We shall show that A is not connected. By the definition of an interval, if A is not an interval, there are numbers $r, s \in A$ with $r < s$ such that there is a number t with $r < t < s$ and $t \notin A$. Then

$$(-\infty, t) \cap A \neq \emptyset, \quad (t, \infty) \cap A \neq \emptyset$$

and

$$[(-\infty, t) \cap A] \cup [A \cap (t, \infty)] = A.$$

So A is not connected.

Conversely, suppose that A is an interval and that U and V are disjoint open sets such that (i) $A \subset U \cup V$ and (ii) $U \cap A \neq \emptyset$ and $V \cap A \neq \emptyset$ (i.e., we suppose that A is not connected.) We show this is impossible. Since U and V are open sets, there are countable collections of disjoint open intervals

$$\{U_i \mid i = 1, 2, \dots\} \text{ and } \{V_i \mid i = 1, 2, \dots\}$$

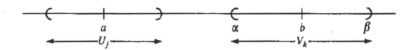

Figure 3.2
Intervals in the proof of Theorem 2.14

with

$$U = \bigcup_{i=1}^{\infty} U_i \text{ and } V = \bigcup_{i=1}^{\infty} V_i.$$

Suppose $a \in U \cap A, b \in V \cap A$. Then there is an interval U_j from the collection of disjoint open intervals whose union is U such that $a \in U_j$, and there is an interval V_k from the collection of disjoint open intervals whose union is V such that $b \in V_k$. Suppose $a < b$ and $V_k = (\alpha, \beta)$. See Figure 3.2.

We claim that α is in neither U nor V. Note that $a < \alpha < b$. First, $\alpha \notin U$ because if $\alpha \in U$ then there is a $\delta > 0$ with $(\alpha - \delta, \alpha + \delta) \subset U$ (since U is open). But then

$$(\alpha - \delta, \alpha + \delta) \cap V_k \neq \emptyset$$

which contradicts $U \cap V = \emptyset$. Second, $\alpha \notin V$ because $\alpha \notin V_k$, and if $\alpha \in V$ there must be an interval V_j in the collection of disjoint intervals whose union is V such that $\alpha \in V_j$. But then $V_k \cap V_j \neq \emptyset$ (why?), which contradicts the assumption that V_k and V_j are disjoint. Thus $\alpha \notin U$ and $\alpha \notin V$.

Now $a < \alpha < b$ and if $a, b \in A$ and A is an interval, then $\alpha \in A$. This contradicts our assumption that $A \subset U \cup V$. Thus A must be connected. ∎

One can use the property of connectedness to prove the Intermediate Value Theorem for continuous functions. This theorem states that if f is a continuous function on $[a, b]$ and x_1 and x_2 are in $[a, b]$ with $f(x_1) < f(x_2)$, then for any c with $f(x_1) < c < f(x_2)$, there is a point x_3 between x_1 and x_2 with $f(x_3) = c$.

Exercises 3.1

1. Give an example of an infinite number of closed sets whose union is not closed.

2. Show that $[a, b]$ is a closed set.

3. (a) If $A = [0, \infty)$, find the open intervals relative to A.
   (b) Repeat part (a) if $A = [1, 2]$.
   (c) Repeat part (a) if $A = (1, 2)$.

4. Let A be an open set containing x. Let

$$\alpha = \sup\{\delta \mid [x, x+\delta) \subset A\},$$

$$\beta = \sup\{\delta \mid (x-\delta, x] \subset A\}.$$

Assume that α and β are finite. Show that $(x-\beta, x+\alpha) \subset A$ but $x+\alpha \notin A$ and $x - \beta \notin A$.

5. Let I_x be the maximal interval containing x as defined in Theorem 3.5. Show that either $I_x = I_y$ or $I_x \cap I_y = \emptyset$. [*Hint:* Suppose $z \in I_x \cap I_y$. Show that $I_x = I_y$.]

6. (a) Show that if A is closed and B is open, then $A \setminus B$ is closed.

   (b) Show that if A is open and B is closed, then $A \setminus B$ is open.

7. (a) Let A be a bounded set of real numbers. Show that l.u.b. A and g.l.b. A are points of $\overline{A}$.

   (b) Is l.u.b. A necessarily a limit point of A?

8. Show that if $A \subset B$ then $\overline{A} \subset \overline{B}$.

9. (a) Show that $\overline{A \cup B} = \overline{A} \cup \overline{B}$.

   (b) Give an example where $\overline{(A \cap B)} \neq \overline{A} \cap \overline{B}$.

10. (a) Show that $\text{int}(A)$ is an open set.

    (b) Show that if B is an open set and $B \subset A$, then $B \subset \text{int}(A)$. This shows that $\text{int}(A)$ is the largest open set contained in A.

    (c) Show that $\text{int}(A \cap B) = \text{int}(A) \cap \text{int}(B)$.

    (d) Give an example where $\text{int}(A \cup B) \neq \text{int}(A) \cup \text{int}(B)$.

11. Tell whether the following sets A are (i) open or closed, (ii) connected, (iii) compact. Find (iv) the limit points of A, (v) $\text{int}(A)$, (vi) $\overline{A}$, (vii) the boundary of A.

    (a) $A =$ rational numbers
    (b) $A = (0, 1]$
    (c) $A = \{1, 1/2, 1/4, 1/8, \dots\}$.
    (d) $A = \{0, 1, 1/2, 1/4, 1/8, \dots\}$.
    (e) $A =$ integers
    (f) $A = (0, 1) \cup (1, 2]$

12. Show that the set of limit points of a set is closed.

13. Show that $\overline{A} \cap \overline{A^c} =$ boundary of A.

14. Show that the collection of intervals $\{(x - \delta, x + \delta)\}$ where x is a rational number and δ is a positive rational number, is countable.

15. (a) Show that if $x \notin A$ and x is a limit point of A, then x is a boundary point of A.

  (b) Show that if $x \notin A$ and x is a boundary point of A, then x is a limit point of A.

  (c) Show that if x is a boundary point of A, then x is a boundary point of A^c.

  (d) Prove Corollary 3.6(a).

  (e) Prove Corollary 3.6(b).

16. (a) Prove Theorem 3.3.

  (b) Prove Theorem 3.4.

17. This exercise shows that if $\{A_\alpha\}_{\alpha \in A}$ is a collection of open sets, then there is a countable subcollection A_n of $\{A_\alpha\}$ such that

$$\bigcup_{\alpha \in A} A_\alpha = \bigcup_{n=1}^{\infty} A_n.$$

  (a) Let $B = \cup_{\alpha \in A} A_\alpha$. Choose $x \in B$.

  (b) Let A_x be a set in the collection $\{A_\alpha\}$ containing x. Show that there is an interval with rational endpoints containing x, call it I_x, with $I_x \subset A_x$.

  (c) Show that the set of intervals with rational endpoints is countable.

  (d) Show that $B = \cup_{x \in B} I_x$. Notice that there are only countably many different I_x's.

  (e) For each of the countably many different I_x's in part (d), call them $\{I_1, I_2, ...\}$, there is an A_i with $I_i \subset A_i$. Show that $B = \cup_{i=1}^{\infty} A_i$.

18. (a) Give an example of a set A such that A is not connected, but $A \cup \{1\}$ is connected.

  (b) Show that in part (a), 1 must be a boundary point of A.

19. Prove that the following two conditions are equivalent:

  (i) Every infinite set of points in A has a limit point in A.

  (ii) If $\{x_n\}$ is a sequence of points in A, then there is a subsequence $\{x_{n_k}\}$ that converges to some point in A.

20. Show that a closed subset of a compact set is compact.

21. (a) Show that if A is compact and B is closed, then $A \cap B$ is compact.

(b) If A is compact and B is bounded, must $A \cap B$ be compact? Justify your answer.

22. Show that a compact set contains its supremum and infimum.

23. Show that a set A is open relative to V if, given any $x_0 \in A$, there is a $\delta > 0$ such that

$$(x_0 - \delta, x_0 + \delta) \cap V \subset A.$$

4

Continuous Functions

4.1 Limits and Continuity

In this section we define continuous functions and determine some of their properties. First we define what it means for a function to be continuous at a point and then what it means for a function to be continuous on a set. We shall find that there are two ways that a function can be continuous on a set. Throughout this chapter (and the rest of the text), unless otherwise stated, f will be a function from a subset of the real numbers into the real numbers and $\mathcal{D}(f)$ will denote the domain of f.

Limit of a Function

The property of continuity is intimately connected with the idea of the limit of a function, and this will be our starting point.

Definition: Let f be a function with domain $\mathcal{D}(f)$, and suppose that x_0 is a limit point of $\mathcal{D}(f)$. The *limit of f as x approaches x_0 is L* if, given any $\epsilon > 0$ there is a $\delta(\epsilon) > 0$ such that if $0 < |x - x_0| < \delta(\epsilon)$ and $x \in \mathcal{D}(f)$ then $|f(x) - L| < \epsilon$. In this case we write $\lim_{x \to x_0} f(x) = L$.

The idea is that $f(x)$ may be made arbitrarily close to L by making x sufficiently close to x_0. The numbers ϵ and $\delta(\epsilon)$ should be thought of as very small numbers. Again, the order of selection is important. First ϵ is chosen and then $\delta(\epsilon)$ is found. Usually $\delta(\epsilon)$ depends on ϵ.

Two other points are important: First, x_0 need not be in $\mathcal{D}(f)$ but only a limit point of $\mathcal{D}(f)$. Thus $f(x_0)$ need not be defined. Second, even if $x_0 \in \mathcal{D}(f)$, it is not required that $f(x_0) = \lim_{x \to x_0} f(x)$. In fact, the value of f at x_0 is completely irrelevant in view of the condition $0 < |x - x_0|$. Figure 4.1 represents these ideas from a graphical viewpoint.

In the proof of Theorem 4.1, we shall want to suppose that $\lim_{x \to x_0} f(x) \neq L$. This is equivalent to saying there is some positive number ϵ_0 such that for

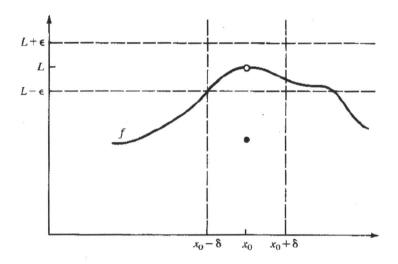

Figure 4.1
L is a limit point of the function f.

any $\delta > 0$ there is a number x for which

$$x \in (x_0 - \delta, x_0 + \delta) \cap \mathcal{D}(f), \ x \neq x_0$$

with $|f(x) - L| \geq \epsilon_0$.

Example 4.1:
    Let $f(x) = (2x^2 - 8)/(x - 2)$. We shall show that $\lim_{x \to 2} f(x) = 8$. Notice that $2 \notin \mathcal{D}(f)$. Let $\epsilon > 0$ be given. We must find a $\delta(\epsilon)$ such that

$$\left| \frac{2x^2 - 8}{x - 2} - 8 \right| < \epsilon \text{ if } 0 < |x - 2| < \delta(\epsilon).$$

We do this by working backward. That is, we begin with the inequality

$$\left| \frac{2x^2 - 8}{x - 2} - 8 \right| < \epsilon$$

and try to manipulate this inequality to get $|x - 2| < \delta(\epsilon)$ where this process gives an explicit form for $\delta(\epsilon)$. We then observe that our procedure is reversible; that is, we could have begun at the last step and ended up at the first. In this particular example, the procedure would go as follows:

$$\left| \frac{2x^2 - 8}{x - 2} - 8 \right| = |2(x + 2) - 8| \text{ if } x \neq 2.$$

and
$$|2(x+2) - 8| = 2|x - 2|.$$

Now $2|x - 2| < \epsilon$ if $|x - 2| < \epsilon/2$. Thus, in this example, we could take $\delta(\epsilon) = \epsilon/2$ (or any smaller positive value). We could now begin with $0 < |x - 2| < \epsilon/2$ and retrace our steps back to

$$\left|\frac{2x^2 - 8}{x - 2} - 8\right| < \epsilon \text{ if } 0 < |x - 2| < \delta(\epsilon)$$

but normally one merely observes that this can be done.

Example 4.2:

Let $f(x) = \sqrt{x + 1}$. We shall show that $\lim_{x \to -1} f(x) = 0$. Notice that $\mathcal{D}(f) = [-1, \infty)$. Thus, in choosing x close to -1, we must choose values of x larger than -1.

Let $\epsilon > 0$ be given. We must find a $\delta(\epsilon) > 0$ such that if $0 < |x - (-1)| < \delta(\epsilon)$ and $x > -1$ then $|\sqrt{x + 1} - 0| < \epsilon$. Now

$$|\sqrt{x + 1} - 0| = |\sqrt{x + 1}| = \sqrt{x + 1}.$$

and $\sqrt{x + 1} < \epsilon$ if $x + 1 < \epsilon^2$. Thus if

$$0 < |x + 1| = |x - (-1)| < \epsilon^2 \text{ and } x > -1,$$

then
$$|\sqrt{x + 1} - 0| < \epsilon.$$

Thus we could take $\delta(\epsilon) = \epsilon^2$ in this example.

The process of taking limits is often easier to approach from the point of view of sequences. We shall soon prove a theorem that legitimizes this viewpoint, but first we discuss some examples.

Suppose that $\{x_n\}$ is a sequence of numbers contained in the domain of the function f. Then $f(x_n)$ is a real number for each positive integer n, so $\{f(x_n)\}$ is a sequence of numbers. Now suppose that the sequence $\{x_n\}$ converges. Does that imply that the sequence $\{f(x_n)\}$ must converge? The answer depends on what $f(x)$ is, as the next two examples illustrate.

Example 4.3:

Let $f(x) = x^2$, and let $\{x_n\}$ be the sequence $\{2 - 1/n\}$. Then

$$f(x_n) = \left(2 - \frac{1}{n}\right)^2 = 4 - \frac{4}{n} + \frac{1}{n^2}.$$

So $\{x_n\} = \{2 - 1/n\}$ converges to 2, and $\{f(x_n)\} = \{4 - 4/n + 1/n^2\}$ converges to 4. In this case, not only does $\{f(x_n)\}$ converge, but

$$\lim f(x_n) = f(\lim x_n)$$

since

$$\lim f(x_n) = 4 = f(2) = f(\lim x_n).$$

Notice the difference between $\lim f(x_n)$ and $f(\lim x_n)$. In evaluating $\lim f(x_n)$, we first compute $f(x_n)$ for each n, and then find the limit of the sequence $\{f(x_n)\}$. To evaluate $f(\lim x_n)$, we find the limit of the sequence $\{x_n\}$ and then compute the value of f at that limit.

Example 4.4:
Let

$$f(x) = \begin{cases} \sin(1/x) & \text{if } x \neq 0 \\ 0 & \text{if } x = 0 \end{cases}$$

and let $\{x_n\}$ be defined by

$$x_n = \frac{2}{(2n+1)\pi}$$

so that

$$f(x_n) = \sin\left[(2n+1)\frac{\pi}{2}\right].$$

Then

$$f(x_1) = \sin\frac{3\pi}{2} = -1, \ \ f(x_2) = \sin\frac{5\pi}{2} = 1, \ \ f(x_3) = \sin\frac{7\pi}{2} = -1$$

and, in general, $f(x_n) = (-1)^n$. In this example, $\lim x_n = 0$, but $\{f(x_n)\} = \{(-1)^n\}$ diverges.

Theorem 4.1: Let f be a function with domain $\mathcal{D}(f)$, and suppose that x_0 is a limit point of $\mathcal{D}(f)$. Then $\lim_{x \to x_0} f(x) = L$ if and only if for every sequence $\{x_n\} \subset \mathcal{D}(f)$ with $x_n \neq x_0$ for every n and $\lim x_n = x_0$, the sequence $\{f(x_n)\}$ converges to L.

Remark: In Example 4.3 , we *cannot* say $\lim x^2 = 4$ from what we have done because we checked only one sequence of numbers that converges to 2. On the other hand, in Example 4.4 we can say $\lim_{x \to 0} \sin(1/x)$ does not exist.

Proof: We first suppose $\lim_{x \to x_0} f(x) = L$ and show that if $\{x_n\}$ is any sequence satisfying the hypotheses of the theorem, then $\lim f(x_n) = L$.

Let $\epsilon > 0$ be given. Then there is a $\delta > 0$ such that if $|x - x_0| < \delta$ and $x \in \mathcal{D}(f)$, then $|f(x) - L| < \epsilon$.

Since $\lim x_n = x_0$, there is a positive integer N such that if $n > N$ then $|x_n - x_0| < \delta$. But $x_n \neq x_0$, so

$$0 < |x_n - x_0| < \delta \text{ for } n > N$$

and

$$|f(x_n) - L| < \epsilon \text{ for } n > N.$$

Thus, $\lim f(x_n) = L$.

Conversely, suppose $\lim_{x \to x_0} f(x) \neq L$. We shall construct a sequence $\{x_n\}$ such that $\{x_n\} \subset \mathcal{D}(f)$, $x_n \neq x_0$ for any n, and $\lim x_n = x_0$, but $\lim f(x_n) \neq L$.

Since x_0 is a limit point of $\mathcal{D}(f)$ and $\lim_{x \to x_0} f(x) = L$, there is an $\epsilon_0 > 0$ such that for any $\delta > 0$ there is a number

$$\bar{x} \in \left(x_0 - \frac{1}{n}, x_0 + \frac{1}{n} \right) \cap \mathcal{D}(f), \text{ with } \bar{x} \neq x_0$$

for which $|f(\bar{x}) - L| \geq \epsilon_0$.

Thus for each positive integer n, there is a number x_n with

$$x_n \in \left(x_0 - \frac{1}{n}, x_0 + \frac{1}{n} \right) \cap \mathcal{D}(f), \, x_n \neq x_0$$

with $|f(x_n) - L| \geq \epsilon_0$. Therefore $\{x_n\} \subset \mathcal{D}(f)$, $x_n \neq x_0$ for any n, $\lim x_n = x_0$, and $\lim f(x_n) \neq L$. ∎

The following theorem, whose proof is left for Exercise 5, Section 4.1, is often useful in proving the existence of limits. It is sometimes called the "Pinching Theorem" or the "Squeeze Theorem." One instance of our use of this theorem will be in proving $\lim_{x \to 0} (\sin x)/x = 1$, which is a crucial fact in deriving the formulas for the derivatives of the trigonometric functions.

Theorem 4.2: Let f and g be functions with $\lim_{x \to a} f(x) = \lim_{x \to a} g(x) = L$. Suppose h is a function for which $f(x) \leq h(x) \leq g(x)$ if $x \in (a - \delta, a + \delta) \setminus \{a\}$ for some $\delta > 0$. Then $\lim_{x \to a} h(x)$ exists and $\lim_{x \to a} h(x) = L$. ∎

Continuous Functions

Definition: Let f be a function and $x_0 \in \mathcal{D}(f)$. We say that f is *continuous* at x_0 if, given any $\epsilon > 0$, there is a $\delta(\epsilon) > 0$ such that

$$|f(x) - f(x_0)| < \epsilon \text{ whenever } |x - x_0| < \delta(\epsilon) \text{ and } x \in \mathcal{D}(f).$$

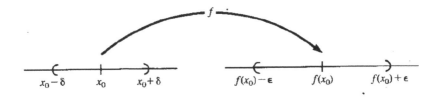

Figure 4.2
A continuous function f maps an interval containing x_0 into an interval containing $f(x_0)$.

If f is not continuous at x_0, then f is said to be *discontinuous* at x_0.
Note: Later in this section we will define "uniformly continous." The custom is to say that a pointwise continuous function is continuous and if it is uniformly continuous, to specify as such.

The definition of continuity asserts that if we pick any interval about $f(x_0)$, say $(f(x_0) - \epsilon, f(x_0) + \epsilon)$, then there is an interval about x_0, namely $(x_0 - \delta, x_0 + \delta)$ such that any point in $(x_0 - \delta, x_0 + \delta) \cap \mathcal{D}(f)$ is carried by f into the interval $(f(x_0) - \epsilon, f(x_0) + \epsilon)$. That is,

$$f((x_0 - \delta, x_0 + \delta) \cap \mathcal{D}(f)) \subset (f(x_0) - \epsilon, f(x_0) + \epsilon)$$

as illustrated in Figure 4.2.

The definition of continuity has several similarities to the limit of a function, but it also has some important differences. Here x_0 must be in $\mathcal{D}(f)$, so $f(x_0)$ must be defined. However, x_0 need not be a limit point in $\mathcal{D}(f)$. In fact, if there is an interval $(x_0 - \delta, x_0 + \delta)$ Such that

$$(x_0 - \delta, x_0 + \delta) \cap \mathcal{D}(f) = \{x_0\},$$

then f is continuous at x_0. If x_0 is a limit point of $\mathcal{D}(f)$, then the definition is equivalent to saying $\lim_{x \to x_0} f(x) = f(x_0)$.

Example 4.5:
Let $f(x) = (2x^2 - 8)/(x - 2)$. Then f is not continuous at $x = 2$ because $2 \notin \mathcal{D}(f)$. However, we showed earlier that $\lim_{x \to 2} f(x) = 8$; so if we redefine f to be

$$f(x) = \begin{cases} (2x^2 - 8)/(x - 2), & \text{if } x \neq 2 \\ 8, & \text{if } x = 2, \end{cases}$$

then f is continuous at $x = 2$.

Corollary 4.2: Let f be a function with domain $\mathcal{D}(f)$. Suppose $x_0 \in \mathcal{D}(f)$. Then f is continuous at x_0 if and only if $\lim f(x_n) = f(x_0)$ for very sequence $\{x_n\} \subset \mathcal{D}(f)$ with $\lim x_n = x_0$. ∎

Another way to state this corollary is that f is continuous at x_0 if and only if $\lim f(x_n) = f(\lim x_n)$ for any sequence $\{x_n\}$ satisfying the conditions of the corollary.

Example 4.6:
Let f be the function defined by

$$f(x) = \begin{cases} -1 & \text{if } x \text{ is rational} \\ 1 & \text{if } x \text{ is irrational.} \end{cases}$$

Then f is not continuous at any point x_0, because we can choose a sequence $\{x_n\}$ converging to x_0 such that the odd terms of $\{x_n\}$ are rational and the even terms are irrational. Then $\{f(x_n)\} = \{(-1)^n\}$, which diverges.

Example 4.7:
We shall show that the function

$$f(x) = \begin{cases} 0 & \text{if } x \text{ is irrational} \\ 1/n & \text{if } x = m/n \text{ in lowest terms, with } n > 0 \end{cases}$$

is continuous at the irrationals and discontinuous at the nonzero rationals. One reason this example is important is because it highlights the process of selecting ϵ and $\delta(\epsilon)$. Another reason is that this example runs counter to the intuition one often develops in elementary calculus about continuous functions.

We first show that $f(x)$ is not continuous at a rational number. Let $x_0 = m/n$ in lowest terms. Let $\epsilon = 1/(2n)$. See Figure 4.3.

Then no matter how small $\delta(\epsilon)$ is, there will be an irrational number x with $|x - x_0| < \delta(\epsilon)$. But $f(x) = 0$ if x is irrational, so

$$|f(x) - f(x_0)| = \left|0 - \frac{1}{n}\right| > \frac{1}{2n}.$$

It is a little easier to show that f is not continuous at the rational numbers using sequences, and we leave this project for Exercise 3, Section 4.1.

Next we show that f is continuous at the irrationals. Let x_0 be an irrational number and let $\epsilon > 0$ be given. There is a positive integer N such that $1/N < \epsilon$. Since x_0 is an irrational number, there is a number $\delta_1 > 0$ such that the interval $(x - \delta_1, x + \delta_1)$ contains no integers.

Likewise, there is a number $\delta_2 > 0$ such that the interval $(x - \delta_2, x + \delta_2)$ contains no rational numbers of the form $m/2$ in lowest terms. We continue this process; for each positive integer k there is a number $\delta_k > 0$ such that the interval $(x - \delta_k, x + \delta_k)$ contains no rational numbers of the form m/k in lowest terms.

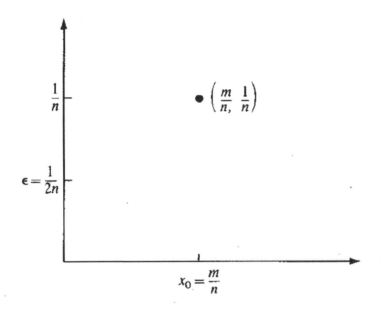

Figure 4.3
Graph of the point $(m/n, 1/n)$.

Now choose $\delta(\epsilon) = \min\{\delta_1, \ldots, \delta_N\}$. Notice $\delta(\epsilon) > 0$ since it is the smallest of a finite set of positive numbers. The interval $(x_0 - \delta(\epsilon), x_0 + \delta(\epsilon))$ contains no rational numbers whose denominator is less than or equal to N. Thus if $|x - x_0| < \delta(\epsilon)$, then

$$|f(x) - f(x_0)| = |f(x) - 0| < \frac{1}{N} < \epsilon.$$

In view of Corollary 4.1 and Theorem 2.4, we can easily show that if we combine continuous functions in certain ways, we get a continuous function.

Theorem 4.3: Let f and g be functions with domains $\mathcal{D}(f)$ and $\mathcal{D}(g)$, respectively. Suppose $x_0 \in \mathcal{D}(f) \cap \mathcal{D}(g)$ and both f and g are continuous at x_0. Then

(a) $\alpha f + \beta g$ is continuous at x_0 for any real numbers α and β.

(b) fg is continuous at x_0.

(c) f/g is continuous at x_0 if $g(x_0) \neq 0$.

Proof: We exhibit a proof of part (b) and leave the others for Exercise 4, Section 4.1.

Suppose $\{x_n\} \subset \mathcal{D}(f) \cap \mathcal{D}(g)$ and $\lim x_n = x_0$. Since f and g are continuous functions at x_0, $\lim f(x_n) = f(x_0)$ and $\lim g(x_n) = g(x_0)$. By Theorem 2.4

$$(fg)(x_0) = f(x_0)g(x_0) = \lim f(x_n) \lim g(x_n)$$

$$= \lim f(x_n)g(x_n) = \lim(fg)(x_n).$$

Thus by Corollary 4.1, fg is continuous at x_0. ∎

Next we show the composition of continuous functions is continuous. Here we have to be a little careful about the domain and range of the functions involved.

Theorem 4.4: Let f and g be continuous functions with domains $\mathcal{D}(f)$ and $\mathcal{D}(g)$, respectively, such that the range of f is contained in the domain of g. If $x_0 \in \mathcal{D}(f)$ and if f is continuous at x_0 and g is continuous at $f(x_0)$, then $g \circ f$ is continuous at x_0.

We shall give two proofs of the theorem. The first will use sequences and is simpler to write; the second uses an "ϵ, δ" argument and may provide more intuition.

First proof: Let $\{x_n\}$ be a sequence in $\mathcal{D}(f)$ with $\lim x_n = x_0$. Since $\mathcal{R}(f) \subseteq \mathcal{D}(g)$ then $\{f(x_n)\}$ is a sequence in $\mathcal{D}(g)$. We must show

$$\lim(g \circ f)(x_n) = \lim g(f(x_n)) = g(f(x_0)) = (g \circ f)(x_0).$$

Since f is continuous at x_0, then $\lim f(x_n) = f(\lim x_n)$. Since g is continuous at $f(x_0)$ and since $\{f(x_n)\}$ is a sequence that converges to $f(x_0)$, then

$$\lim g(f(x_n)) = g(\lim f(x_n)).$$

Thus

$$\lim g(f(x_n)) = g(\lim f(x_n)) = g(f(\lim x_n)) = g(f(x_0). ∎$$

Second proof: Before writing a formal proof, we give the ideas involved. Consider Figure 4.4. We basically want to show that if we pick any small interval about $g(f(x_0))$, there is a small interval about x_0 such that all the numbers in the interval about x_0 are carried by $g \circ f$ into the interval about $g(f(x_0))$. This works because the fact that g is continuous at $f(x_0)$ ensures that there is an interval about $f(x_0)$ such that all the numbers in the interval about $f(x_0)$ are carried into the interval about $g(f(x_0))$. Also, since f is continuous at x_0, there is an interval about x_0, so that all the numbers in the

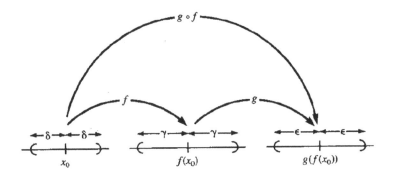

Figure 4.4
Composition of continuous functions

interval about x_0 are carried into the interval about $f(x_0)$. In summary, given any interval

$$(g(f(x_0)) - \epsilon, g(f(x_0)) + \epsilon)$$

there is an interval $(f(x_0) - \gamma, f(x_0) + \gamma)$ such that

$$g((f(x_0) - \gamma, f(x_0) + \gamma)) \subset (g(f(x_0)) - \epsilon, g(f(x_0)) + \epsilon)$$

and there is a $\delta > 0$ such that

$$f(x_0 - \delta, x_0 + \delta) \subset (f(x_0) - \gamma, f(x_0) + \gamma).$$

Thus

$$(g \circ f)(x_0 - \delta, x_0 + \delta) \subset g(f(x_0) - \gamma, f(x_0) + \gamma)$$
$$\subset (g(f(x_0)) - \epsilon, g(f(x_0)) + \epsilon).$$

We now formalize this argument.

Notice that $\mathcal{D}(g \circ f) = \mathcal{D}(f)$ since $\mathcal{R}(f) \subset \mathcal{D}(g)$. Let $\epsilon > 0$ be given. We need to show that there is a $\delta > 0$ such that if $|x - x_0| < \delta$ and $x \in \mathcal{D}(g \circ f) = \mathcal{D}(f)$, then

$$|g(f(x)) - g(f(x_0))| < \epsilon.$$

We are given that g is continuous at $f(x_0)$. Thus there is a $\gamma > 0$ such that if $y \in \mathcal{D}(g)$ and $|y - f(x_0)| < \gamma$, then

$$|g(y) - g(f(x_0))| < \epsilon.$$

Since f is continuous at x_0, there is a $\delta > 0$ such that if $|x - x_0| < \delta$ and $x \in \mathcal{D}(f)$, then

$$|f(x) - f(x_0)| < \delta.$$

Now put the pieces together. Suppose $x \in \mathcal{D}(f)$ and $|x - x_0| < \delta$. Then

$$|f(x) - f(x_0)| < \gamma$$

and $f(x) \in \mathcal{D}(g)$ since $\mathcal{R}(f) \subset \mathcal{D}(g)$. Thus

$$|g(f(x)) - g(f(x_0))| < \epsilon. \quad \blacksquare$$

Continuity on a Set

A function f is said to be continuous on a set A if f is continuous at each point in A.

There is some subtlety in this idea, which the next example highlights.

Example 4.8:

Let $f(x) = 1/x$ on $(0, 1)$. We show that f is continuous on $(0,1)$. Let $\epsilon > 0$ be given and choose $x_0 \in (0, 1)$. We must find a $\delta > 0$ such that if $|x - x_0| < \delta$ and $x \in (0, 1)$, then

$$\left| \frac{1}{x} - \frac{1}{x_o} \right| = \frac{|x_0 - x|}{|x||x_0|} < \epsilon.$$

Our first problem is to control the size of the denominator. We do this by requiring $x > x_0/2$, which will be true if $|x - x_0| < |x_0|/2$. Then

$$\frac{|x_0 - x|}{|x||x_0|} < \frac{|x_0 - x|}{|x_0/2||x_0|} = \frac{2}{|x_0^2|}|x - x_0|.$$

Now if $|x - x_0| < \epsilon |x_0^2|/2$, then

$$\frac{2}{|x_0^2|}|x - x_0| < \frac{2}{|x_0^2|} \frac{\epsilon |x_0^2|}{2} = \epsilon.$$

Thus if $\delta(\epsilon) = \min\{\epsilon |x_0^2|/2, |x_0|/2\}$ and if $|x - x_0| < \delta(\epsilon)$, then

$$\left| \frac{1}{x} - \frac{1}{x_0} \right| < \epsilon$$

which shows that $f(x) = 1/x$ is continuous at any $x \in (0, 1)$.

Notice, however, that $\delta(\epsilon)$ depends on x_0, and it *suggests* that no value of $\delta(\epsilon)$ works for every $x_0 \in (0, 1)$, even for a fixed value of ϵ. We show that, in fact, this is the case.

Let $\epsilon = 1$, and suppose there is a $\delta > 0$ such that if $|x - x_0| < \delta$ and $x, x_0 \in (0, 1)$, then

$$\left| \frac{1}{x} - \frac{1}{x_0} \right| < \epsilon.$$

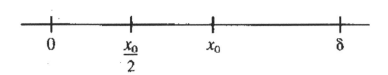

Figure 4.5
See Example 4.8.

That is, the same δ works for every x_0. Consider Figure 4.5. Choose $x_0 < \delta$ and suppose $x = x_0/2$. Then

$$|x - x_0| = \frac{x_0}{2} < \delta$$

but

$$\left|\frac{1}{x} - \frac{1}{x_0}\right| = \frac{2}{x_0} - \frac{1}{x_0} = \frac{1}{x_0} > 1 = \epsilon.$$

This shows that no δ can work for every x_0, even if ϵ is fixed.

This example prompts us to be a little more careful with our definition of continuity of a function on a set. We now formally define this idea.

Definition: Let f be a function with domain $\mathcal{D}(f)$ and $A \subset \mathcal{D}(f)$. We say that f is *continuous on* A if, given any $\epsilon > 0$ and $x_0 \in A$, there is a number $\delta(x_0, \epsilon)$ such that if $x \in \mathcal{D}(f)$ and $|x - x_0| < \delta(x_0, \epsilon)$, then $|f(x) - f(x_0)| < \epsilon$.

It is again important to realize the order in which things are done. First x_0 and ϵ are chosen, and then $\delta(x_0, \epsilon)$ is found. The number $\delta(x_0, \epsilon)$ will depend on both x_0 and ϵ.

Next we want to derive a condition that is equivalent to continuity of a function on a set. Before we do this, it may be worthwhile to review a definition from Chapter 1. Recall that if $f : X \to Y$ and if $B \subset Y$, then

$$f^{-1}(B) = \{x \in X | f(x) \in B\}.$$

Also $f^{-1}(\cup B_\alpha) = \cup f^{-1}(B_\alpha)$ where $\{B_\alpha\}$ is any collection of subsets of Y.

Theorem 4.5: Let $f : X \to Y$. Then f is continuous on $\mathcal{D}(f)$ if and only if for each open set $A \subset Y$, $f^{-1}(A)$ is an open set relative to $\mathcal{D}(f)$. That is, $f^{-1}(A) = B \cap \mathcal{D}(f)$ where B is an open set.

Remark: This is an abstract result. We prove it because it will give us a vehicle to derive some of the properties of continuous functions. Also, in more general settings, this is taken to be the definition of continuity.

Proof: The proof will be given for the setting in which we have been working; that is, X and Y are subsets of the real numbers. Assume f is continuous, and let A be an open set contained in Y. We must show that $f^{-1}(A) = \{x | f(x) \in A\}$ is open relative to $\mathcal{D}(f)$. Let $x_0 \in f^{-1}(A)$. (So $x_0 \in \mathcal{D}(f)$.) Then $f(x_0) \in A$. Since A is an open set, there is an interval

$$(f(x_0) - \epsilon, f(x_0) + \epsilon) \subset A \text{ for some } \epsilon > 0.$$

Since f is continuous at x_0, there is a $\delta > 0$ such that if $|x - x_0| < \delta$ and $x \in \mathcal{D}(f)$, then $|f(x) - f(x_0)| < \epsilon$. This is equivalent to saying that if $x \in (x_0 - \delta, x_0 + \delta) \cap \mathcal{D}(f)$, then

$$f(x) \in (f(x_0) - \epsilon, f(x_0) + \epsilon) \subset A.$$

Thus $(x_0 - \delta, x_0 + \delta) \cap \mathcal{D}(f) \subset f^{-1}(A)$, so $f^{-1}(A)$ is an open set relative to $\mathcal{D}(f)$.

Conversely, suppose that for each open set $A \subset Y$, $f^{-1}(A)$ is an open set relative to $\mathcal{D}(f)$. We shall show that f is continuous on $\mathcal{D}(f)$. Choose $x_0 \in \mathcal{D}(f)$, and let $\epsilon > 0$ be given. Then $(f(x_0) - \epsilon, f(x_0) + \epsilon)$ is an open set. The hypotheses state that

$$f^{-1}(f(x_0) - \epsilon, f(x_0) + \epsilon) = \{x \in \mathcal{D}(f) | f(x) \in (f(x_0) - \epsilon, f(x_0) + \epsilon)\}$$

is an open set relative to $\mathcal{D}(f)$. Now

$$x_0 \in f^{-1}(f(x_0) - \epsilon, f(x_0) + \epsilon)$$

so there is an interval $(x_0 - \delta, x_0 + \delta)$ such that

$$(x_0 - \delta, x_0 + \delta) \cap \mathcal{D}(f) \subset f^{-1}(f(x_0) - \epsilon, f(x_0) + \epsilon) \text{ for some } \delta > 0.$$

But this says that if $|x - x_0| < \delta$ and $x \in \mathcal{D}(f)$, then $|f(x) - f(x_0)| < \epsilon$. ■

The condition of Theorem 4.5 is often a difficult one to visualize and for the most part, we shall use it to determine properties of continuous functions rather than to examine specific examples. However, to get some idea of what is being said, consider the following examples.

Example 4.9:

Let f be a real-valued function on the real numbers defined by $f(x) = x^2$. We shall show that f is continuous using Theorem 4.5. Now

$$f^{-1}((a,b)) = \{x | f(x) \in (a,b)\} = \{x | a < x^2 < b\}.$$

(a) If $b \leq 0$, then $f^{-1}((a,b)) = \emptyset$.

(b) If $a < 0$ and $b > 0$, then $f^{-1}((a,b)) = \{x | x^2 < b\} = (-\sqrt{b}, \sqrt{b})$.

(c) If $a \geq 0$, then $f^{-1}((a,b)) = \{x | a < x^2 < b\} = (-\sqrt{b}, -\sqrt{a}) \cup (\sqrt{a}, \sqrt{b})$.

Thus $f^{-1}((a,b))$ is open. By Theorem 3.5, an open set A can be written $A = \cup(a_i, b_i)$. So

$$f^{-1}(A) = f^{-1}(\cup(a_i, b_i)) = \cup f^{-1}((a_i, b_i))$$

which is open because it is the union of open sets.

Example 4.10:

Let $f(x) = \sqrt{x}$. The domain of $f = [0, \infty)$. Consider $f^{-1}((a,b))$.

(a) If $b \leq 0$, then $f^{-1}((a,b)) = \emptyset$.

(b) If $a < 0$ and $b > 0$, then

$$f^{-1}((a,b)) = [0, b^2)$$

which is open relative to $\mathcal{D}(f)$, since $[0, b^2) = (-b^2, b^2) \cap [0, \infty)$.

(c) For $a \geq 0$, $f^{-1}((a,b)) = (a^2, b^2)$. Thus if $A = \cup(a_i, b_i)$, then

$$f^{-1}(A) = \cup f^{-1}((a_i, b_i))$$

is an open set.

We now have three ways to characterize a continuous function. We summmarize these in a theorem.

Theorem 4.6: Let f be a function with domain $\mathcal{D}(f)$. The following are equivalent to the condition that f is continuous on $\mathcal{D}(f)$.

(a) Given any $\epsilon > 0$ and $x_0 \in \mathcal{D}(f)$, there is a number $\delta(\epsilon, x_0) > 0$ such that if $|x - x_0| < \delta(\epsilon, x_0)$ and $x \in \mathcal{D}(f)$, then $|f(x) - f(x_0)| < \epsilon$.

(b) If $x_0 \in \mathcal{D}(f)$ and $\{x_n\}$ is any sequence in $\mathcal{D}(f)$ with $\{x_n\} \to x_0$, then $\{f(x_n)\} \to f(x_0)$; i.e., $f(\lim x_n) = \lim f(x_n)$.

(c) If A is an open set of real numbers, then $f^{-1}(A)$ is an open set relative to $\mathcal{D}(f)$. ∎

Properties of continuous functions

We are now ready to prove some of the familiar properties of continuous functions.

Theorem 4.7: Let $f : X \to Y$ be a continuous function where X and Y are subsets of $\mathbb{R}$. If A is a compact subset of X, then $f(A)$ is a compact set.

Proof: Suppose A is a compact subset of X. Then $f(A) \subset Y$. Suppose that $\{U_\alpha\}$ is an open cover of $f(A)$ (so that $\{U_\alpha\}$ is an open set in Y). Since f is continuous, $f^{-1}(U_\alpha)$ is an open set in X. We claim $\{f^{-1}(U_\alpha)\}$ is an open cover of A. Choose $x_0 \in A$. Then $f(x_0) \in f(A)$, so $f(x_0) \in U_\alpha$ for some α. Then $x_0 \in f^{-1}(U_\alpha)$ for that α, by the definition of $f^{-1}(U_\alpha)$. Thus $\{f^{-1}(U_\alpha)\}$ is an open cover of A.

Since A is compact, some finite subcollection $\{f^{-1}(U_1), \ldots, f^{-1}(U_n)\}$ covers A. We claim this means that $\{U_1, \ldots, U_n\}$ covers $f(A)$. Suppose $y_0 \in f(A)$. Then there is an $x_0 \in A$ with $f(x_0) = y_0$. But $x_0 \in f^{-1}(U_i)$ for some $i = 1, \ldots, n$, so $y_0 = f(x_0) \in U_i$. Thus $\{U_1, \ldots, U_n\}$ is a finite cover of $f(A)$, so $f(A)$ is compact. ∎

Corollary 4.7(a):

A continuous function on a compact set of real numbers is bounded.

Proof: If A is compact, then $f(A)$ is compact. But compact sets of real numbers are closed and bounded. ∎

Corollary 4.7(b):

A continuous function on a compact set A attains its supremum and infimum on A.

Proof: We show that f attains its supremum on A. Let $y_0 = $ l.u.b. $f(A)$. There is a sequence $\{y_n\} \subset f(A)$ with $\lim y_n = y_0$. Since $y_n \in f(A)$, there is an $x_n \in A$ with $f(x_n) = y_n$. Because A is compact, there is a subsequence of $\{x_n\}$, say $\{x_{n_k}\}$, that converges to some $x_0 \in A$. Since f is continuous at $x_0 \in A$

$$f(x_0) = f(\lim x_{n_k}) = \lim f(x_{n_k}) = \lim y_{n_k} = y_0$$

so the supremum of f on the set A is attained at $x_0 \in A$

Likewise, f attains its infimum on A. ■

In Exercise 24, Section 4.1, we show that it is *not* always the case that a continuous function on a bounded set is bounded.

Note: If a function attains its supremum (infimum) on a set, we shall say that it attains its maximum (minimum) on the set.

Theorem 4.8: If f is a continuous function at $x = c$ and if $f(c) > 0$, then there is a $\delta > 0$ such that $f(x) > 0$ if $x \in (c - \delta, c + \delta) \cap \mathcal{D}(f)$. ■

Corollary 4.8: If f is a continuous function at $x = c$ and if $f(c) < 0$ then there is a $\delta > 0$ such that $f(x) < 0$ if $x \in (c - \delta, c + \delta) \cap \mathcal{D}(f)$.

We prove a stronger version of Theorem 4.8 and Corollary 4.8 in Exercise 12, section 4.1.

Theorem 4.9: If f is a continuous function on $[a, b]$ and if $f(a)$ and $f(b)$ are of opposite signs, then there is a number $c \in (a, b)$ for which $f(c) = 0$.

Proof: We do the proof in the case $f(a) > 0$ and $f(b) < 0$. The proof of the other case is nearly identical.

Let $A = \{t | f(x) > 0 \text{ if } x \in [a, t]\}$. Now $A \neq \emptyset$ since $a \in A$, and b is an upper bound of A. Let $c = $ l.u.b. A. Note that $a < c < b$ because f is continuous and $f(a) > 0$ and $f(b) < 0$ so there is a $\bar{\delta} > 0$ such that $f(x) > 0$ on $[a, a + \bar{\delta})$ and $f(x) < 0$ on $(b - \bar{\delta}, b]$ by Theorem 4.8 and its corollary.

We claim that $f(c) = 0$. Suppose this is not the case. If $f(c) < 0$, then there is a $\delta > 0$ such that $f(x) < 0$ if $x \in (c - \delta, c + \delta)$. But then $c - (\delta/2)$ would be an upper bound of A, contradicting $c = $ l.u.b. A. If $f(c) > 0$, then there is a $\delta > 0$ such that $f(x) > 0$ if $x \in (c - \delta, c + \delta)$. If this is the case, then $f(x) > 0$ on $[a, c - (\delta/2]$ and $f(x) > 0$ on $(c - \delta, c + \delta)$ so that $f(x) > 0$ on $[a, c + (\delta/2)]$. This contradicts that c is an upper bound of A. ■

Corollary 4.9 (The Intermediate Value Theorem):

If f is a continuous function on $[a, b]$ and α is a number between $f(a)$ and $f(b)$, then there is a number $c \in (a, b)$ where $f(c) = \alpha$.

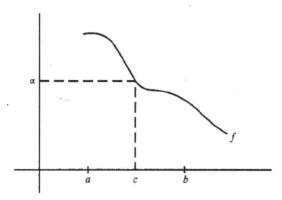

Figure 4.6
Illustration of the Intermediate Value Theorem.

Proof: For definiteness, assume $f(a) > f(b)$. Then $f(b) < \alpha < f(a)$. (See Figure 4.6). Let $g(x) = f(x) - \alpha$. Then $g(a) = f(a) - \alpha > 0$ and $g(b) = f(b) - \alpha < 0$. Thus, by Theorem 4.9, there is a number $c \in (a, b)$ for which $g(c) = f(c) - \alpha = 0$. That is, $f(c) = \alpha$. ∎

Definition: Let f be a function defined on an interval I. We say that f has the *intermediate value property* if whenever $x_1, x_2 \in I$ with $f(x_1) \neq f(x_2)$, then for any number α between $f(x_1)$ and $f(x_2)$ there is a number x_3 between x_1 and x_2 with $f(x_3) = \alpha$.

It may seem reasonable to conjecture that a function that satisfies the intermediate value property must be continuous. However, this is not the case as the function

$$f(x) = \begin{cases} \sin(1/x) & \text{if } x \neq 0 \\ 0 & \text{if } x = 0 \end{cases}$$

shows. (See Exercise 6, Section 4.1.)

Uniform Continuity

As a final point in this section, we return to an earlier example in which we considered the function $f(x) = 1/x$. We showed that that this function was continuous on $(0, 1)$, but as the point x_0 became closer to 0, the number

$$\delta(\epsilon) = \min\left\{\frac{x_0}{2}, \frac{\epsilon x_0^2}{2}\right\}$$

became smaller, even with ϵ fixed. In this example, suppose we had restricted our set to be $[1/2, 1)$ instead of $(0, 1)$. Then

$$\delta(\epsilon) = \min\left\{\frac{x_0}{2}, \frac{\epsilon x_0^2}{2}\right\} \leq \min\left\{\frac{1}{4}, \frac{\epsilon}{8}\right\}$$

(taking $x_0 = 1/2$, the smallest it can be). Now we *do* have a situation where it is possible to get a $\delta(\epsilon)$ that works for every x in the set, once ϵ is chosen. The difference in behavior turns out to be very important. One use of it occurs in our study of integration.

Definition: Let f be a function from a set A to the real numbers. We say that f is *uniformly continuous* on A if given $\epsilon > 0$ there is a $\delta(\epsilon) > 0$ such that if $x, y \in A$ and $|x - y| < \delta(\epsilon)$ then $|f(x) - f(y)| < \epsilon$.

The difference between continuity and uniform continuity is as follows. In uniform continuity, once ϵ is given, it is possible to find a $\delta(\epsilon)$ that works for every $x \in A$. That is, the $\delta(\epsilon)$ does not depend on the $x \in A$ that was chosen. The difference may seem subtle at first, but it is an important one. If we refer back to the example preceding the previous definition, it may help to highlight the difference. The function $f(x) = 1/x$ is continuous on $(0, 1)$ but not uniformly continuous there. If, however, we restrict our attention to $[1/2, 1)$, the function becomes uniformly continuous on that interval. If we consider the graph of $f(x) = 1/x$ on $(0, 1)$ in Figure 4.7, we can see what is happening.

Notice how small the interval about x_0 needs to be to ensure that the value of f in the interval is between $f(x_0) - \epsilon$ and $f(x_0) + \epsilon$. On the other hand, the interval about x_1 which is needed to ensure that the value of f in the interval is between $f(x_1) - \epsilon$ and $f(x_1) + \epsilon$ is comparatively large. In *uniform* continuity there is one size of interval that works for every $\bar{x}$ to ensure that $f(x)$ is between $f(\bar{x}) - \epsilon$ and $f(\bar{x}) + \epsilon$ if x is in the interval centered at $\bar{x}$.

Theorem 4.10: If f is a continuous function on a compact set A, then f is uniformly continuous on A.

Proof: Let $\epsilon > 0$ be given. Since f is continuous on A, given $x \in A$, there is a number $\delta(x)$ such that if $y \in A$ and $|x-y| < \delta(x)$, then $|f(x)-f(y)| < \epsilon/2$. Now if it were possible to choose

$$\bar{\delta} = \inf\{\delta(x)|x \in A\}$$

so that $\bar{\delta} > 0$, we would be done. However, we are taking the infimum of an *infinite* set of positive numbers, which may be 0. If we could somehow reduce

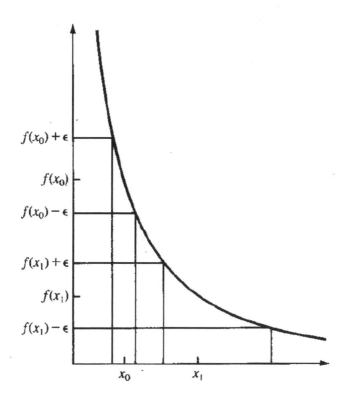

Figure 4.7
The function f is not uniformly continuous.

this to a *finite* set of positive numbers, we could then select the minimum of this set, which would be positive. This is what compactness allows us to do.

For each $x \in A$.

$$\left(x - \frac{\delta(x)}{2}, x + \frac{\delta(x)}{2} \right)$$

is an open set containing x so,

$$\left\{ \left(x - \frac{\delta(x)}{2}, x + \frac{\delta(x)}{2} \right) \mid x \in A \right\}$$

is an open cover of A. Since A is compact, some finite subcollection, say

$$\left\{ \left(x_1 - \frac{\delta_1}{2}, x_1 + \frac{\delta_1}{2} \right), \ldots, \left(x_n - \frac{\delta_n}{2}, x_n + \frac{\delta_n}{2} \right) \right\}$$

covers A. Let

$$\delta = \min \left\{ \frac{\delta_1}{2}, \ldots, \frac{\delta_n}{2} \right\}.$$

Now $\delta > 0$, since this set is finite. We claim that if x' and $x'' \in A$ with $|x' - x''| < \delta$, then $|f(x') - f(x'')| < \epsilon$. To show this, note that

$$x' \in \left(x_j - \frac{\delta_j}{2}, x_j + \frac{\delta_j}{2} \right) \text{ for some } j = 1, \ldots, n$$

from our finite subcover. Then if

$$|x' - x''| < \delta \leq \frac{\delta_j}{2},$$

it follows that $x'' \in (x_j - \delta_j, x_j + \delta_j)$. Thus $|x'' - x_j| < \delta_j$, so

$$|f(x'') - f(x_j)| < \frac{\epsilon}{2} \text{ and likewise } |f(x') - f(x_j)| < \frac{\epsilon}{2}.$$

Hence

$$|f(x') - f(x'')| \leq |f(x') - f(x_j)| + |f(x_j) - f(x'')|$$
$$< \frac{\epsilon}{2} + \frac{\epsilon}{2} = \epsilon. \quad \blacksquare$$

Corollary 4.10: A continuous function on an interval $[a, b]$ is uniformly continuous.

Exercises 4.1

1. Use the definition of limit to prove the following:

   (a) $\lim_{x \to 4} 3x + 7 = 19$

   (b) $\lim_{x \to 2} 4x^2 + 3x - 1 = 21$

   (c) $\lim_{x \to 1} (x^3 - 1)/(x - 1) = 3$

2. Use the ϵ, δ definition of continuity to show that the following functions are continuous on their domain:

   (a) $f(x) = x$

   (b) $f(x) = \sqrt{x}$

   (c) $f(x) = |x|$

   (d) $f(x) = x^2$

   (e) $f(x) = \begin{cases} x^2 & \text{if } x \leq 1 \\ x & \text{if } x > 1 \end{cases}$

3. Let $f(x) = \begin{cases} 0 & \text{if } x \text{ is irrational or } x = 0 \\ 1/n & \text{if } x = m/n \text{ in lowest terms.} \end{cases}$

   Use sequences to show that f is not continuous at the nonzero rational numbers.

4. Prove parts (a) and (c) of Theorem 4.3.

5. Suppose that f, g, and h are functions such that $f(x) \le g(x) \le h(x)$ for every x except possibly $x = a$. Show that if

$$\lim_{x \to a} f(x) = \lim_{x \to a} h(x) = L,$$

then $\lim_{x \to a} g(x) = L$.

6. Show that the function

$$f(x) = \begin{cases} \sin(1/x) & \text{if } x \ne 0 \\ 0 & \text{if } x = 0 \end{cases}$$

satisfies the intermediate value property on $[0, 1]$.

7. (a) Show that x^n is continuous for any positive integer n.

   (b) Prove that any polynomial $f(x) = a_n x^n + a_{n-1} x^{n-1} + \cdots + a_0$ is continuous.

8. Give examples of the following.

   (a) A function on the interval $[0,1]$ that does not assume its supremum.

   (b) A continuous function of $[1, \infty)$ that does not assume its infimum.

   (c) A continuous function on $(0,1)$ that does not assume its supremum or infimum.

9. Show that any function whose domain is the integers is uniformly continuous.

10. Show that

$$f(x) = \begin{cases} x \sin(1/x) & \text{if } x \ne 0 \\ 0 & \text{if } x = 0 \end{cases}$$

is continuous at $x = 0$.

11. (a) Show that if the function f is continuous, then $|f|$ is continuous.

    (b) Give an example where $|f|$ is continuous, but f is not continuous.

12. Suppose that f is a continuous function whose domain is the real numbers.

    (a) If $f(a) > 0$, show there is a $\delta > 0$ such that $f(x) \ge \frac{1}{2} f(a)$ if $x \in (a - \delta, a + \delta)$.

    (b) If $f(a) < 0$, show that there is a $\delta > 0$ such that $f(x) \le \frac{1}{2} f(a)$ if $x \in (a - \delta, a + \delta)$.

13. (a) Show that $f(x) = \sin x$ is a continuous function. You may use the fact that $|\sin x| \le |x|$.

    (b) Show that $f(x) - \cos x$ is a continuous function.

    (c) Show that $f(x) = \tan x$ is continuous for $x \in (-\pi/2, \pi/2)$.

14. Show that $f(x) = 3x^3 + \sin x - 1$ has a root (i.e., a value of x where $f(x) = 0$) between -1 and 1.

15. (a) Let $f : [0,1] \to [0,1]$ be continuous. Show there is an $x \in [0,1]$ where $f(x) = x$.

    (b) Show there is a value of x for which $\cos x = x$.

    (c) Suppose f is a continuous function on $[0,2]$ with $f(0) = f(2)$. Show there is an $x \in [0,1]$ where $f(x) = f(x+1)$.

16. Show that if $a > 0$ and n is a positive integer, there is a number b for which $b^n = a$.

17. (a) Show that a function f is continuous if and only if for each closed set A, $f^{-1}(A)$ is closed relative to $\mathcal{D}(f)$.

    (b) Sow that if f is a continuous function, then $\{x | f(x) = a\}$ is a closed set relative to $\mathcal{D}(f)$ for any real number a.

18. (a) Suppose f and g are continuous functions on $\mathbb{R}$ with $f(x) = g(x)$ except possibly at $x_1, \ldots, x_n$. Show that $f(x_i) = g(x_i)$ for $i = 1, \ldots, n$.

    (b) Show that if f and g are continuous functions on $\mathbb{R}$ with $f(x) = g(x)$ for any rational number x, then $f(x) = g(x)$ for all $x \in \mathbb{R}$.

19. Suppose f is continuous on $[a,b]$ with $f(a) = f(b)$. Let

$$M = \text{l.u.b.} \ \{f(x) | x \in [a,b]\}$$

and

$$m = \text{g.l.b.} \ \{f(x) | x \in [a,b]\}.$$

Show that if $m < c < M$, then there exist $x_1, x_2 \in [a,b]$, $x_1 \neq x_2$, where $f(x_1) = f(x_2) = c$.

20. Suppose f is a continuous function on $\mathbb{R}$.

    (a) Show that $g(x) = f(x+c)$ is continuous on $\mathbb{R}$.

    (b) Show that $h(x) = f(ax+b)$ is continuous on $\mathbb{R}$.

21. Show that $f(x) = a^x$ is a continuous function for any $a > 0$.

22. Suppose f is uniformly continuous on (a,b) and is continuous at $x = a$ and $x = b$. Show that f is uniformly continuous on $[a,b]$.

23. Suppose f is continuous on (a,b) and that $\lim_{x \to a} f(x)$ and $\lim_{x \to b} f(x)$ exist. Show that the function F defined by

$$F(x) = \begin{cases} f(x) & \text{if } x \in (a,b) \\ \lim_{x \to a} f(x) & \text{if } x = a \\ \lim_{x \to b} f(x) & \text{if } x = b \end{cases}$$

is uniformly continuous on $[a,b]$.

24. (a) Show that if f is uniformly continuous on a bounded interval I, then f is bounded on I.

    (b) Give an example of a continuous function on $(0, 1)$ that is not bounded.

    (c) Give an example of a bounded continuous function on $(0, 1)$ that is not uniformly continuous.

25. (a) Show that $f(x) = x$ is uniformly continuous on $(-\infty, \infty)$.

    (b) Show that $f(x) = x^2$ is *not* uniformly continuous on $(-\infty, \infty)$. Thus the product of uniformly continuous functions is not always uniformly continuous.

    (c) Show that the sum of two uniformly continuous functions is uniformly continuous.

    (d) Show that the product of two uniformly continuous functions on a bounded interval is uniformly continuous.

26. Show that if f is a continuous function on $[a, b]$, then $f([a, b])$ is either a point or a closed, bounded interval.

27. Show that if f is uniformly continuous on (a, b) and $\{x_n\} \subset (a, b)$ is a Cauchy sequence, then $\{f(x_n)\}$ is a Cauchy sequence.

28. (a) Give an example of a function f that is continuous on $(0, 1)$ and a Cauchy sequence $\{x_n\} \subset (0, 1)$ for which $\{f(x_n)\}$ is not a Cauchy sequence.

    (b) Show that if f is a function that is continuous but not uniformly continuous on (a, b), then there is a Cauchy sequence $\{x_n\} \subset (a, b)$ for which $\{f(x_n)\}$ is not a Cauchy sequence.

    (c) Is the result in part (b) necessarily true if the interval is (a, ∞)?

29. (a) Show that if f is a continuous function and A is a connected set, then $f(A)$ is a connected set.

    (b) Use the result of part (a) to prove the Intermediate Value Theorem.

30. Suppose f and g are continuous functions on $[a, b]$. Define

$$(f \vee g)(x) = \max\{f(x), g(x)\}$$

$$(f \wedge g)(x) = \min\{f(x), g(x)\}.$$

    Show that $f \vee g$ and $f \wedge g$ are continuous on $[a, b]$.

31. A function f is said to be *convex* on (a, b) if for $x_1, x_2 \in (a, b)$ and $0 \le \lambda \le 1$,

$$f(\lambda x_1 + (1 - \lambda)x_2) \le \lambda f(x_1) + (1 - \lambda)f(x_2).$$

    (a) Give an example of a convex function.

    (b) Show that a convex function on an open interval is continuous.

4.2   Monotone and Inverse Functions

In this section we continue our examination of limits and continuous functions. We begin by defining the limit of a function when either the independent variable or the function becomes infinite. We then define the left and right limit of a function, which leads to the idea of left and right continuity. The idea of one-sided continuity is important for several reasons, and we shall use it to help classify the ways in which a function may be discontinuous. We then investigate increasing and decreasing functions and find there is only one type of discontinuity such functions can have. Finally, we use this information to examine the properties of the inverse of a continuous 1-1 function.

Limits Involving Infinity

In Section 4.1 we defined what it means to say, "The limit of the function f as x approaches x_0 is L." In that definition x_0 and L were assumed to be real numbers. We now extend the idea of the limit of a function to the cases where x_0 or L may be infinite.

Definition: Let f be a function with domain $\mathcal{D}(f)$, and suppose that x_0 is a limit point of $\mathcal{D}(f)$. We say *the limit of $f(x)$ as x approaches x_0 is ∞* if, given any number M, there is a number $\delta(M) > 0$ such that if $0 < |x - x_0| < \delta(M)$ and $x \in \mathcal{D}(f)$, then $f(x) > M$.

In this case we write $\lim_{x \to x_0} f(x) = \infty$. In a similar manner we can define what it means for the limit of f as x approaches x_0 to be $-\infty$. It is a useful exercise to formulate the definition.

The number M in the definition should be thought of as a very large positive number and $\delta(M)$ as a very small positive number. Figure 4.8 illustrates this idea. The order of selection of the two numbers is that first M is chosen and then $\delta(M)$ is found; $\delta(M)$ will depend on M. The relationship between M and $\delta(M)$ is that as M becomes larger, $\delta(M)$ becomes smaller.

Example 4.11:

Let $f(x) = 1/(1 - x)^2$. Then $\mathcal{D}(f) = \{x | x \neq 1\}$. We shall show $\lim_{x \to 1} = \infty$.

Choose any number $M > 0$. (If $M < 0$, any number works for $\delta(M)$). If $0 < |x - 1| < 1/\sqrt{M}$ then

$$f(x) = \frac{1}{(x-1)^2} > M$$

which is what we needed to show. Thus we could take $\delta(M) = 1/\sqrt{M}$.

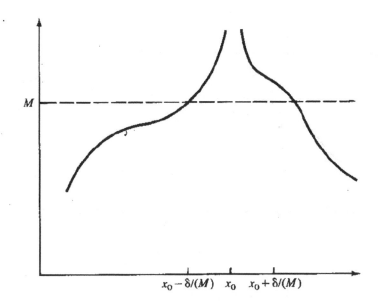

Figure 4.8
A function which diverges to infinity at x_0.

Definition: Let f be a function with domain $\mathcal{D}(f)$ such that

$$\mathcal{D}(f) \cap (M, \infty) \neq \emptyset \text{ for any number } M.$$

Then we say *the limit of f as x goes to ∞ is L* provided that, given any $\epsilon > 0$, there is a number $N(\epsilon)$ such that if $x > N(\epsilon)$ and $x \in \mathcal{D}(f)$, then $|f(x) - L| < \epsilon$.

In this case we write $\lim_{x \to \infty} f(x) = L$. Likewise, we could define the limit of f as x approaches $-\infty$.

In the definition, ϵ should be thought of as a very small positive number and $N(\epsilon)$ as a very large number. See Figure 4.9. The number ϵ is chosen first and then $N(\epsilon)$ is found; $N(\epsilon)$ will often depend on ϵ. The purpose of requiring $\mathcal{D}(f) \cap (M, \infty) \neq \emptyset$ is to guarantee that we can always find an $x \in \mathcal{D}(f)$ with $x > N(\epsilon)$ no matter how large $N(\epsilon)$ is.

Example 4.12:
We show that
$$\lim_{x \to \infty} \frac{x^2 - 1}{x^2 + 1} = 1.$$
Let $\epsilon > 0$ be given. We need to find a number $N(\epsilon)$ such that if $x > N(\epsilon)$,

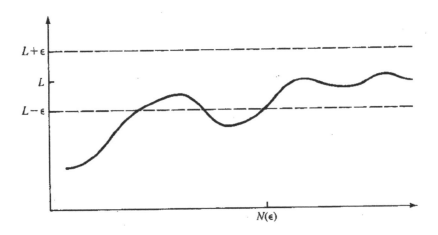

Figure 4.9

then

$$\left| \frac{x^2 - 1}{x^2 + 1} - 1 \right| < \epsilon.$$

Now

$$\left| \frac{x^2 - 1}{x^2 + 1} - 1 \right| = \left| \frac{(x^2 - 1) - (x^2 + 1)}{x^2 + 1} \right| = \left| \frac{-2}{x^2 + 1} \right| = \frac{2}{x^2 + 1}.$$

Also, $2/(x^2 + 1) < \epsilon$ if $(2/\epsilon) < x^2 + 1$, that is, if $x > \sqrt{(1/\epsilon) - 1}$. (We assume $\epsilon < 2$ so that $(2/\epsilon) - 1 > 0$.) Thus we may take $N(\epsilon) = \sqrt{(2/\epsilon) - 1}$.

Definition: Let f be a function with domain $\mathcal{D}(f)$ such that $\mathcal{D}(f) \cap (M, \infty) \neq \emptyset$ for any number M. We say that *the limit of f as x approaches ∞ is ∞* provided that given any number K, there is a number $N(K)$ such that if $x > N(K)$ and $x \in \mathcal{D}(f)$, then $f(x) > K$

In this case we write $\lim_{x \to \infty} f(x) = \infty$. Likewise, we could define

$$\lim_{x \to \infty} f(x) = -\infty, \quad \lim_{x \to -\infty} f(x) = \infty, \quad \text{and} \quad \lim_{x \to -\infty} f(x) = -\infty.$$

In the definition, the numbers K and $N(K)$ should both be thought of as very large positive numbers. First K is chosen, then $N(K)$ is found and will depend on K.

Example 4.13:
We show that $\lim_{x \to \infty} \sqrt{x} = \infty$. Let $K > 0$ be given. We must find a number $N(K)$ such that if $x > N(K)$ then $\sqrt{x} > K$. But if $x > K^2$, then $\sqrt{x} > K$ so we could take $N(K) = K^2$.

Right- and Left-Hand Limits

Next we discuss the notions of right- and left-hand limits, which may be familiar topics from elementary calculus. Here, we look at the behavior of a function as x approaches a given point specifically from the right or the left. As was mentioned in the introduction, we shall use these properties to help classify the types of discontinuities a function may have.

Definition: Let f be a function with domain $\mathcal{D}(f)$. Let x_0 be a limit point of $\mathcal{D}(f) \cap [x_0, \infty)$. We say *the limit of f as x approaches x_0 from the right is L* if, given any $\epsilon > 0$, there is a $\delta(\epsilon) > 0$ such that if $0 < x - x_0 < \delta(\epsilon)$ and $x \in \mathcal{D}(f)$, then $|f(x) - L| < \epsilon$.

In this case we write

$$\lim_{x \downarrow x_o} f(x) = L \text{ or } \lim_{x \to x_0^+} f(x) = L.$$

The difference between the right-hand limit and the limit of a function at x_0 is that in the former idea we are interested in how f behaves only for those values of x larger than x_0. We could also define

$$\lim_{x \downarrow x_o} f(x) = \infty \text{ or } \lim_{x \downarrow x_o} f(x) = -\infty$$

patterned after our earlier definitions.

Definition: Let f be a function with domain $\mathcal{D}(f)$. Let x_0 be a limit point of $(-\infty, x_0] \cap \mathcal{D}(f)$. We say *the limit of f as x approaches x_0 from the left is L* if, given any $\epsilon > 0$, there is a $\delta(\epsilon) > 0$ such that if $0 < x_0 - x < \delta(\epsilon)$ and $x \in \mathcal{D}(f)$, then $|f(x) - L| < \epsilon$.

In this case we write

$$\lim_{x \uparrow x_0} f(x) = L \text{ or } \lim_{x \to x_0^-} f(x) = L.$$

Example 4.14:
Let $f(x) = [x]$, the *greatest integer function*. That is, $[x] = n$ if n is the integer such that $n \le x < n + 1$. The graph is shown in Figure 4.10.
Here $\lim_{x \uparrow 3}[x] = 2$ since for $2 \le x < 4$, $[x] = 2$. On the other hand, $\lim_{x \downarrow 3}[x] = 3$ since for $3 \le x < 4$, $[x] = 3$. Thus both the left- and right-hand limits of $[x]$ exist at $x = 3$, but they are not the same.

We leave the proof of Theorem 4.11 until Exercise 5, Section 4.2.

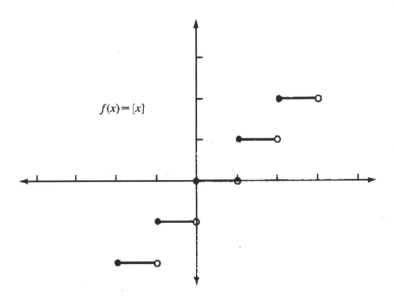

$f(x) = [x]$

Figure 4.10
The greatest integer function

Theorem 4.11: Let f be a function with domain $\mathcal{D}(f)$, and suppose x_0 is a limit point of $\mathcal{D}(f) \cap [x_0, \infty)$ and $\mathcal{D}(f) \cap (-\infty, x_0]$. Then $\lim_{x \to x_0} f(x) = L$ if and only if

$$\lim_{x \downarrow x_0} f(x) = L \text{ and } \lim_{x \uparrow x_0} f(x) = L. \quad \blacksquare$$

Definition: Let f be a function with domain $\mathcal{D}(f)$, and suppose $x_0 \in \mathcal{D}(f)$. We say f *is continuous from the right (left) at* x_0 if, given $\epsilon > 0$, there is a $\delta(\epsilon) > 0$ such that if $0 \le x - x_0 < \delta(\epsilon)$ $(0 \le x_0 - x < \delta(\epsilon))$ and $x \in \mathcal{D}(f)$, then $|f(x) - f(x_0)| < \epsilon$.

Thus in our example with the greatest integer function, $[x]$ is continuous from the right at $x = 3$ (since $[3] = 3$) but not from the left.

Corollary 4.11: Let f be a function with domain $\mathcal{D}(f)$, and suppose $x_0 \in \mathcal{D}(f)$. Then f is continuous at x_0 if and only if it is continuous from the right and left at x_0.

Using our work in Section 4.1 as a model, we could prove the following result.
Theorem 4.12: Let f be a function with domain $\mathcal{D}(f)$ with $x_0 \in \mathcal{D}(f)$. Then f is continuous from the right (left) at x_0 if and only if for every sequence $\{x_n\} \subset \mathcal{D}(f)$ with $x_n \ge x_0$ $(x_n \le x_0)$ and $\{x_n\} \to x_0$, $\{f(x_n)\} \to f(x_0)$. $\blacksquare$

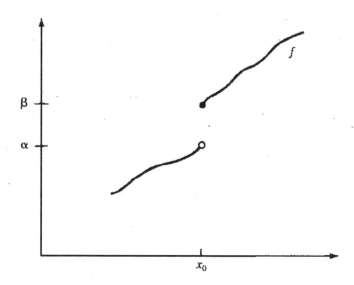

Figure 4.11
An increasing function

Corollary 4.12: Let f be a function with domain $\mathcal{D}(f)$ with $x_0 \in \mathcal{D}(f)$. Then f is continuous from the right (left) at x_0 if and only if for every decreasing (increasing) sequence $\{x_n\} \subset \mathcal{D}(f)$ with $\{x_n\} \to x_0$, $\{f(x_n)\} \to f(x_0)$. ∎

Monotone Functions

Next we define what it means for a function to be monotone. In our discussion of sequences, we defined the concept of monotonicity, and the idea with functions is very much the same. Monotone functions have important applications; one possibly familiar example is the distribution function in probability theory.

Definition: A function f is said to be *monotone increasing (decreasing)* if for any $x_1, x_2 \in \mathcal{D}(f)$ with $x_1 < x_2$, $f(x_1) \le f(x_2)$ $(f(x_1) \ge f(x_2))$. If, for any $x_1, x_2 \in \mathcal{D}(f)$ with $x_1 < x_2$, $f(x_1) < f(x_2)$ $(f(x_1) > f(x_2))$, f is *strictly monotone increasing (decreasing)*. A function that is either monotone increasing or decreasing is said to be monotone.

Monotone functions have some nice properties that arbitrary functions do not necessarily possess. Your understanding of some of these properties may become more intuitive if we examine the graph of such a function. (Later we shall prove, in theorems or exercises, that our ideas were correct.)

Suppose f is an increasing function as shown in Figure 4.11. It seems reasonable that either the graph has no breaks (is continuous) or, if a break does occur, then it must be similar to what happens at x_0 in Figure 4.11. In this case,

$$\lim_{x \uparrow x_0} f(x) \text{ and } \lim_{x \downarrow x_0} f(x)$$

both exist and are α and β, respectively. Also $\alpha < \beta$ and $f(x)$ does not assume any value between α and β except $f(x_0)$. Thus, if f is not continuous at some value, then it will not have the intermediate value property. It also appears that a strictly monotone function must be 1-1.

Theorem 4.13: Let f be a monotone function with domain an open interval (a, b). Then $\lim_{x \uparrow x_0} f(x)$ and $\lim_{x \downarrow x_0} f(x)$ exist for each $x_0 \in (a, b)$.

Proof: We do the proof in the case that f is an increasing function. Choose $x_o \in (a, b)$. We shall show $\lim_{x \uparrow x_0} f(x)$ exists. Since f is increasing, if $a < x < x_0$, then $f(x) \le f(x_0)$. Thus $f(x_0)$ is an upper bound for the set

$$A = \{f(x) | a < x < x_o\}.$$

Since A is bounded above, it has a least upper bound α. We shall show $\lim_{x \uparrow x_0} f(x) = \alpha$. (Notice $\alpha \le f(x_0)$.)

Let $\epsilon > 0$ be given. By definition of least upper bound, there is an x_1 with $a < x_1 < x_0$ such that $\alpha - \epsilon < f(x_1) \le \alpha$. Since f is increasing, if $x_1 < x < x_0$, then $f(x_1) \le f(x) \le \alpha$, so that $\alpha - \epsilon < f(x) \le \alpha$. Thus if $x_1 < x < x_0$, $|f(x) - \alpha| < \epsilon$. Take $\delta(\epsilon) = x_0 - x_1$. If $0 < x_0 - x < \delta(\epsilon)$, then $x > x_1$ and $|f(x) - \alpha| < \epsilon$, which proves $\lim_{x \uparrow x_0} f(x) = \alpha$.

Similarly, $\{f(x) | x \in (x_0, b)\}$ is bounded below by $f(x_0)$. It is a useful exercise to show that $\lim_{x \downarrow x_0} f(x)$ is the greatest lower bound of this set. ∎

if f is a decreasing function, then $-f$ is an increasing function. Thus, if we can prove the theorem for increasing functions, the result for decreasing functions follows immediately. This observation is relevant for the next theorem also.

Theorem 4.14: A monotone function f with $\mathcal{D}(f) \supset (a, b)$ can have a most countably many discontinuities on (a, b).

Proof: We again do the proof in the case that f is an increasing function. We show that with each point of discontinuity we can associate a distinct rational number. Notice that in our proof of Theorem 4.13 for increasing functions we showed

$$\lim_{x \uparrow x_0} f(x) = \alpha = \text{ l.u.b. } \{f(x) | x < x_0\} \le f(x_0) \le \beta$$

$$= \text{ g.l.b. } \{f(x) | x > x_0\} = \lim_{x \downarrow x_0} f(x).$$

Thus if f is an increasing function, there is a discontinuity at x_0 if and only if $\beta > \alpha$.

Suppose f has discontinuities at x_0 and x_1 with $x_0 < x_1$. Let $\alpha_0 = \lim_{x \uparrow x_0} f(x)$ and $\beta_0 = \lim_{x \uparrow x_0} f(x)$. So $\alpha_0 < \beta_0$. Similarly, let $\alpha_1 = \lim_{x \downarrow x_0} f(x)$ and $\beta_1 = \lim_{x \downarrow x_1} f(x)$. Then $\alpha_1 < \beta_1$. Further, since $x_0 < x_1$, $\alpha_0 < \beta_0 \le \alpha_1 < \beta_1$. Choose rational numbers q_0 and q_1 with $\alpha_0 < q_0 < \beta_0$ and $\alpha_1 < q_1 < \beta_1$. Now $q_0 < \beta_0 \le \alpha_1 < q_1$. Thus with every point of discontinuity we can associate a distinct rational number. Since the rational numbers are countable, it follows that there are only countably many discontinuities. ∎

Types of Discontinuities

In Example 4.4, we considered the function $f(x) = \sin(1/x)$ and showed that there was a sequence $\{x_n\} \subset \mathcal{D}(f)$ with $\lim x_n = 0$ such that $\{f(x_n)\} = \{(-1)^n\}$. Thus $\sin(1/x)$ is not continuous (i.e., is discontinuous) at $x = 0$ no matter how $f(0)$ is defined. On the other hand, the function $g(x) = (2x^2 - 8)/(x - 2)$ is not continuous at $x = 2$, but it can be made continuous at $x = 2$ by defining $g(2)$ to be 8. The functions f and g fail to be continuous for different reasons. We want to classify the different reasons why a function may fail to be continuous.

Definition: A function f is said to have a *removable* discontinuity at x_0 if $\lim_{x \to x_0} f(x)$ exists, but $\lim_{x \to x_0} f(x) \ne f(x_0)$ or $f(x_0)$ does not exist.

Notice this includes the possibility that $x_0 \notin \mathcal{D}(f)$.

Definition: A function f is said to have a *jump* discontinuity at x_0 if $\lim_{x \uparrow x_0} f(x)$ and $\lim_{x \downarrow x_0} f(x)$ both exist but $\lim_{x \uparrow x_0} f(x) \ne \lim_{x \downarrow x_0} f(x)$.

Definition: A function f is said to have a *discontinuity of the third type* at x_0 if x_0 is a limit point of $\mathcal{D}(f)$ and if f is not continuous at x_0, but the discontinuity is neither a removable nor a jump discontinuity.

A discontinuity of the third type occurs if either

(i) x_0 is a limit point of $(-\infty, x_0] \cap \mathcal{D}(f)$, but $\lim_{x \uparrow x_0} f(x)$ does not exist; or

(ii) x_0 is a limit point of $\mathcal{D}(f) \cap [x_0, \infty)$, but $\lim_{x \downarrow x_0} f(x)$ does not exist.

(In each of these definitions, by $\lim f(x)$ we mean the limit as a real number.)

Example 4.15:
The function

$$f(x) = \begin{cases} (x^2 - 4)/(x - 2) & \text{if } x \neq 2 \\ 6 & \text{if } x = 2 \end{cases}$$

has a removable discontinuity at $x = 2$. Removable discontinuities are so named because we can make the function continuous at the point by redefining the function at that point. For example, if we redefine $f(2)$ to be 4, then $f(x)$ would be continuous at $x = 2$.

Example 4.16:
The greatest integer function $[x]$ has a jump discontinuity at each integer. An examination of the graph of $[x]$ gives the best idea of why this type of discontinuity is so named.

Example 4.17:
The function

$$f(x) = \begin{cases} \sin(1/x) & \text{if } x \neq 0 \\ 0 & \text{if } x = 0 \end{cases}$$

has a discontinuity of the third type at x = 0.

Earlier we gave an example of a function $(f(x) = \sin(1/x))$ that satisfies the intermediate value property but is not continuous. The next theorem says that if a discontinuity occurs in such a function, then it must be of the third type.

Theorem 4.15: If f is a function defined on (a, b) that satisfies the intermediate value property, then f can have neither a removable nor a jump discontinuity.
The proof is left for Exercise 12, Section 4.2. ■

Continuity of the Inverse Function

The main result that we want to prove now is that the inverse of a continuous 1-1 function defined on an interval $[a, b]$ is continuous.

Consider Figure 4.12, which is the graph of a function and its inverse. Very roughly speaking, the idea that will guide us is that the graph of f^{-1} is the reflection of the graph of f about the line $y = x$. We think that if f is continuous, then it should have no breaks in its graph. Then f^{-1} will have no breaks in its graph and must be continuous.

Recall that in Chapter 1 we defined the inverse of a function f, denoted f^{-1}, to be the function that satisfied $f^{-1}(f(x)) = x$ for all $x \in \mathcal{D}(f)$ and $f(f^{-1}(y)) = y$ for all $y \in \mathcal{D}(f^{-1})$. We also argued that in order for f^{-1} to exist, f must be 1-1 and when f^{-1} exists, $\mathcal{D}(f) = \mathcal{R}(f^{-1})$ and $\mathcal{R}(f) = \mathcal{D}(f^{-1})$.

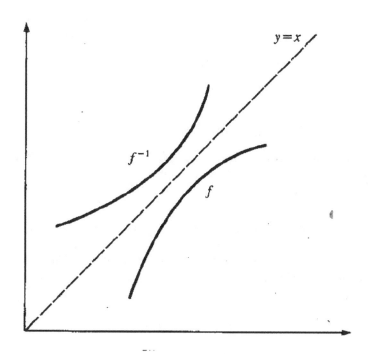

Figure 4.12
A function and its inverse

In the exercises we shall show that a function that is continuous and 1-1 on an interval must be monotone and that a 1-1 function that satisfies the intermediate value property must be continuous. Our main results are given in the next theorem.

Theorem 4.16: Suppose f is continuous and strictly monotone on $[a, b]$. Then

(a) f^{-1} is strictly monotone on its domain;

(b) f^{-1} is continuous on its domain.

Proof: We suppose that f is a strictly increasing function.

(a) Since f is strictly increasing on $[a.b]$, f is a 1-1 function (why?), so f^{-1} exists. The domain of f^{-1} is $[f(a), f(b)]$, since the domain of f^{-1} is the range of f, and the range of f is $[f(a), f(b)]$ by the Intermediate Value Theorem.

Let $y_1, y_2 \in [f(a), f(b)]$ with $y_1 < y_2$. We want to show

$$f^{-1}(y_1) < f^{-1}(y_2).$$

Let x_1 and x_2 be the unique numbers in $[a, b]$ with $f(x_1) = y_1$ and $f(x_2) = y_2$. Now $x_1 < x_2$, for if $x_1 \geq x_2$, then, since f is increasing

$$y_1 = f(x_1) \geq f(x_2) = y_2.$$

So

$$f^{-1}(y_1) = f^{-1}(f(x_1)) = x_1 < x_2 = f^{-1}(f(x_2)) = f^{-1}(y_2).$$

Thus f^{-1} is strictly increasing.

(b) We continue to assume that f is increasing. Fix $y_0 \in (f(a), f(b)]$, and let $\{y_n\}$ be an increasing sequence of numbers in $[f(a), f(b)]$ with $\lim y_n = y_0$. We shall show

$$\lim f^{-1}(y_n) = f^{-1}(y_0).$$

For each n, choose the unique $x_n \in [a, b]$ with $f(x_n) = y_n$, that is, $x_n = f^{-1}(y_n)$. Since $\{y_n\}$ is an increasing sequence, and f^{-1} is an increasing function, then the sequence $\{f^{-1}(y_n)\} = \{x_n\}$ is increasing. Now $\{x_n\}$ is bounded above by b, so $\lim x_n$ exists. Let $x_0 = \lim x_n$. Note that $x_0 \in [a, b]$.

By the continuity of f,

$$\lim f(x_n) = f(\lim x_n) = f(x_0)$$

but

$$f(x_n) = y_n \text{ and } \lim y_n = y_0.$$

Thus

$$y_0 = \lim y_n = \lim f(x_n) = f(x_0).$$

Hence

$$f^{-1}(y_0) = x_0 = \lim x_n = \lim f^{-1}(y_n).$$

By Corollary 4.12, it follows that $\lim_{y \uparrow y_0} f^{-1}(y_0) = f^{-1}(y_0)$.

Similarly, by using decreasing sequences, we can show

$$\lim_{y \downarrow y_0} f^{-1}(y) = f^{-1}(y_0) \text{ for } y_0 \in [f(a), f(b)).$$

Thus f^{-1} is continuous on $[f(a), f(b)]$. ∎

Example 4.18:

The function $f(x) = x^{1/n}$ is continuous on $[0, \infty)$ for n a positive integer, since $g(x) = x^n$ is continuous and 1-1 on $[0, \infty)$ and $g^{-1} = f$.

Notice how much easier this is than trying to show that $f(x) = x^{1/n}$ is continuous using an ϵ, δ argument.

Exercises 4.2

1. Describe the types of discontinuities of the following functions at the given points:

    (a)
    $$f(x) = \begin{cases} 1 & \text{if } x \text{ is rational} \\ 0 & \text{if } x \text{ is irrational} \end{cases} \quad \text{at any point x.}$$

    (b) $f(x) = (\sin x)/x$ at $x = 0$.

    (c) $f(x) = 1/x$ at $x = 0$.

    (d)
    $$f(x) = \begin{cases} |x|/x & \text{if } x \neq 0 \\ 1 & \text{if } x = 0 \end{cases} \quad \text{at } x = 0.$$

    (e)
    $$f(x) = \begin{cases} (x^2 - 9)/(x - 3) & \text{if } x \neq 3 \\ 7 & \text{if } x = 3 \end{cases} \quad \text{at } x = 3.$$

2. Use the definition of the appropriate limit to show

    (a) $\lim_{x \to \infty} (2x^2 + 4x - 1)/(3x^3 + 1) = 0$.

    (b) $\lim_{x \to \infty} \sqrt{3x^2 + 4}/x = \sqrt{3}$.

    (c) $\lim_{x \to -\infty} \sqrt{3x^2 + 4}/x = -\sqrt{3}$.

    (d) $\lim_{x \downarrow 0} (1/x - 1/x^2) = -\infty$.

3. (a) Suppose that f is continuous at $x = a$ and $f(a) > 0$. Show that for any positive integer n

    $$\lim_{x \to a} \sqrt[n]{f(x)} = \sqrt[n]{f(a)}.$$

    (b) Suppose that $\lim_{x \to a} f(x)$ exists and is positive. Show that for any positive integer n

    $$\lim_{x \to a} \sqrt[n]{f(x)} = \sqrt[n]{\lim_{x \to a} f(x)}.$$

4. (a) Find $\lim_{x \downarrow 0} [1/x]/(1/x)$.

    (b) Find $\lim_{x \downarrow 0} [x] \cdot [1/x]$.

5. Prove Theorem 4.11.

6. Prove Corollary 4.11.

7. (a) Prove Theorem 4.12.

    (b) Prove Corollary 4.12.

8. Show that if $\lim_{x \uparrow c} f(x)$ exists and $\lim_{x \uparrow c} f(x) < \theta$, then there is a $\delta > 0$ such that if $x \in [c - \delta, c)$, then $f(x) < \theta$.

9. (a) Show that if $a > 0$ and $a \neq 1$, the function $f(x) = a^x$ is a 1-1 function.

   (b) For $a > 0$ and $a \neq 1$, define $f(x) = \log_a x$ to be the inverse of a^x. Show that $\log_a x$ is a continuous function.

10. (a) If $f(x) > 0$ and $\lim_{x \uparrow a} f(x) = 0$, show $\lim_{x \uparrow a} (1/f(x)) = \infty$.

    (b) Give an example of a function f for which $\lim_{x \uparrow a} f(x) = 0$, but $\lim_{x \uparrow a} (1/f(x)))$ is neither ∞ or $-\infty$.

11. Suppose that f is uniformly continuous on $[0, M)$ for all $M > 0$ and $\lim_{x \to \infty} f(x) = L$. Show that f is uniformly continuous on $[0, \infty)$. Is this true if we do not require $\lim_{x \to \infty} f(x) = L$?

12. Here we show that a function defined on an interval (a, b) and satisfying the intermediate value property cannot have a removable or a jump discontinuity.

    (a) Suppose f has a jump discontinuity at $x_0 \in (a, b)$ and

    $$\lim_{x \uparrow x_0} f(x) < \lim_{x \downarrow x_0} f(x).$$

    Choose θ such that

    $$\lim_{x \uparrow x_0} f(x) < \theta < \lim_{x \downarrow x_0} f(x) \text{ and } \theta \neq f(x_0).$$

    In Exercise 8 we showed there is an interval $[x_0 - \delta, x_0)$ such that $f(x) < \theta$ if $x \in [x_0 - \delta, x_0)$. Likewise, there is an interval $(x_0, x_0 + \delta]$ such that $f(x) > 0$ if $x \in (x_0, x_0 + \delta]$. Conclude that f does not satisfy the intermediate value property on $[x_0 - \delta, x_0 + \delta]$.

    (b) Suppose f has a removable discontinuity at

    $$x_0 \in (a, b) \text{ and } \alpha = \lim_{x \to x_0} f(x) < f(x_0).$$

    Show that there is an interval $[x_0 - \delta, x_0)$ such that

    $$f(x) < \alpha + \frac{1}{2}[f(x_0) - \alpha] \text{ if } x \in [x_0 - \delta, x_0].$$

    Conclude that f does not satisfy the intermediate value property.

13. Show that if f is a 1-1 function defined on $[a, b]$ that satisfies the intermediate value property, then f is a strictly monotone function.

14. Show that if f is a monotone function defined on $[a, b]$ that satisfies the intermediate value property, then f is continuous.

15. Show that if f is a uniformly continuous function on (a, b), then $\lim_{x \downarrow a} f(x)$ and $\lim_{x \uparrow b} f(x)$ exist.

5

Differentiation

In this chapter we develop the theory of differentiation that one uses in elementary calculus. There are two ways in which the derivative is typically introduced: as the slope of the line tangent to a curve, or as the instantaneous rate of change of a moving particle. We shall not pursue either of these approaches, since we presume they are familiar interpretations. We shall use these notions in the discussion of the Mean Value Theorems and the derivative of the inverse of a function where the idea of the derivative being the slope of the tangent line gives some geometric intuition to our results.

In Section 5.1 we derive the formulas for differentiation and establish the properties of the derivative that one uses in elementary calculus for graphing and solving max-min problems. In Section 5.2 we prove several theorems, which are called Mean Value Theorems, and two important results based on these theorems, Taylor's Theorem and L'Hôpital's Rule. We shall assume some of the formulas for differentiation of functions from elementary calculus in some of our examples. These will be developed rigorously either in the exercises or in later sections.

5.1 The Derivative of a Function

For most of our work, we shall assume that f is a function whose domain contains an open interval (a, b) and define the derivative of f at a point in this interval. This assumption is not necessary. The derivative can be defined at a point that is a limit point of the domain of f, $\mathcal{D}(f)$, and is in $\mathcal{D}(f)$. Our approach somewhat simplifies the notation and is sufficiently general for most purposes.

Definition: Let f be a function defined on an interval (a, b), and suppose that $c \in (a, b)$. We say that f *is differentiable at* c if the limit

$$\lim_{x \to c} \frac{f(x) - f(c)}{x - c} \tag{5.1}$$

exists and is finite. If this is the case, this limit is called *the derivative of f at* c and is denoted $f'(c)$.

It is often more convenient (and equivalent) to formulate the definition of the derivative of f at c as

$$\lim_{h \to 0} \frac{f(c + h) - f(c)}{h}$$

which can be done by letting $h = x - c$. We shall use both formulations during the discussion and the exercises.

In terms of an ϵ, δ definition, we can alternatively define the derivative of f as follows:

Definition: Let f be a function defined on an interval (a, b), and suppose that $c \in (a, b)$. We say that the real number L is the *derivative of f at c* if, given $\epsilon > 0$ there is a number $\delta(\epsilon) > 0$ such that if $0 < |x - c| < \delta(\epsilon)$ then

$$\left| \frac{f(x) - f(c)}{x - c} - L \right| < \epsilon$$

In this case, we write $f'(c)$ for L.

The alternative definition is, of course, just a restatement of the definition of the limit in the first definition. In view of Theorem 4.1, which relates the behavior of sequences and limits, we have the following result:

Theorem 5.1: Let f be a function defined on an interval (a, b) with $c \in (a, b)$. Then f is differentiable at $x = c$ with derivative $f'(c)$ if and only if, for every sequence $\{x_n\} \subset \mathcal{D}(f)$ with $\{x_n\} \to c$ and $x_n \neq c$ for all n,

$$\lim_{n \to \infty} \frac{f(x_n) - f(c)}{x_n - c} = f'(c). \quad \blacksquare$$

Now we have a new function f' whose domain is a subset of $\mathcal{D}(f)$; i.e. $\mathcal{D}(f') = \{x \in \mathcal{D}(f) | f'(x) \text{ exists}\}$. Then we can define the derivative of f' at points that are interior to $\mathcal{D}(f')$ and obtain yet another function f'', called the *second derivative of f*. We can repeat this procedure as long as the resulting functions are sufficiently well behaved to allow it. The nth derivative of f will be denoted $f^{(n)}$.

The first theorem we prove in this section relates the properties of differentiability and continuity.

Theorem 5.2: Let f be a function defined on (a,b) and suppose f is differentiable at $c \in (a,b)$. Then f is continuous at c.

Proof: Let $\epsilon > 0$ be given. By our second definition, there is a $\delta(\epsilon) > 0$ such that if $0 < |x - c| < \delta(\epsilon)$, then

$$\left| \frac{f(x) - f(c)}{x - c} - f'(c) \right| < \epsilon \tag{5.2}$$

so that

$$|f(x) - f(c) - f'(c)(x - c)| < \epsilon |x - c|$$

or

$$|f(x) - f(c)| < \epsilon |x - c| + |f'(c)||x - c| = (\epsilon + |f'(c)|)|x - c|.$$

Now we can make $(\epsilon + |f'(c)|)|x - c|$ smaller than ϵ by taking $|x - c|$ smaller than both $\delta(\epsilon)$ (so that (5.2) holds) and

$$\frac{\epsilon}{\epsilon + |f'(c)|}.$$

Thus if

$$0 < |x - c| < \min\left\{ \delta(\epsilon), \frac{\epsilon}{\epsilon + |f'(c)|} \right\} = \bar{\delta}$$

then $|f(x) - f(c)| < \epsilon$. Of course, if $x = c$, then $|f(x) - f(c)| = 0$ so that if $|x - c| < \bar{\delta}$ then $|f(x) - f(c)| < \epsilon$, which proves f is continuous at c. ∎

One might ask if the converse is true. That is, are continuous functions necessarily differentiable? The next example shows that the answer is no.

Example 5.1:
Let $f(x) = |x|$. Then f is continuous at $x = 0$ (why?), but we shall see that f is not differentiable at $x = 0$.

Let $\{x_n\}$ be a sequence of *positive* numbers converging to 0. Then $f(x_n) = |x_n| = x_n$ so that

$$\lim_{n \to \infty} \frac{f(x_n) - f(0)}{x_n - 0} = \frac{x_n - 0}{x_n - 0} = 1.$$

On the other hand, if $\{x_n\}$ is a sequence of *negative* numbers converging to 0, then $f(x_n) = |x_n| = -x_n$ so that

$$\lim_{n \to \infty} \frac{f(x_n) - f(0)}{x_n - 0} = \frac{-x_n - 0}{x_n - 0} = -1.$$

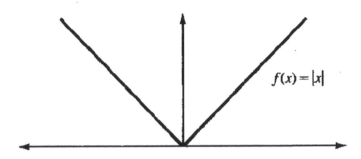

Figure 5.1
The absolute value function

Thus
$$\lim_{x \to 0} \frac{f(x) - f(0)}{x - 0}$$
does not exist.

From a geometric point of view, one way to decide whether a continuous function f is not differentiable at $x = c$ is if the graph of f has a "kink" or "corner" at the point $x = c$. (See Figure 5.1.)

This is the case with $f(x) = |x|$ at $x = 0$. This shows up analytically in the fact that

$$\lim_{h \downarrow 0} \frac{f(c + h) - f(c)}{h} \neq \lim_{h \uparrow 0} \frac{f(c + h) - f(c)}{h}$$

This is an oversimplification of the most general case. (Consider $f(x) = x \sin(1/x)$ at $x = 0$, for example.) However, it may be helpful in the beginning to develop some intuition about why continuous functions are not differentiable at certain values.

You may be surprised that there are functions that are continuous everywhere but differentiable nowhere on an interval. We shall not exhibit such a function here, but note that constructing a function with this property requires forming a series of functions. (See Chapter 8.)

We now have the string of implications: Differentiability of f at $x = c$ implies continuity of f at $x = c$, and continuity of f at $x = c$ implies the limit of f at $x = c$ exists, but neither of the implications can be reversed.

Example 5.2:
Let $f(x) = a$. We show that f is differentiable and $f'(x) = 0$. We have

$$\lim_{x \to c} \frac{f(x) - f(c)}{x - c} = \lim_{x \to c} \frac{a - a}{x - c} = 0.$$

Example 5.3:
We show that the function $f(x) = \sqrt{x}$ is differentiable on the interval $(0, \infty)$.
Choose $c \in (0, \infty)$. Then, if $x > 0$,

$$\frac{f(x) - f(c)}{x - c} = \frac{\sqrt{x} - \sqrt{c}}{x - c} = \frac{(\sqrt{x} - \sqrt{c})(\sqrt{x} + \sqrt{c})}{(x - c)(\sqrt{x} + \sqrt{c})} = \frac{(x - c)}{(x - c)} \frac{1}{(\sqrt{x} + \sqrt{c})}.$$

Now

$$\lim_{x \to c} \frac{f(x) - f(c)}{x - c} = \lim_{x \to c} \frac{1}{(\sqrt{x} + \sqrt{c})} = \frac{1}{2\sqrt{c}}$$

since $\lim_{x \to c} \sqrt{x} = \sqrt{c}$ by continuity. Thus $f'(c) = 1/(2\sqrt{c})$.

Next we prove the rules for differentiation of the sum, product and quotient of differentiable functions that are used in elementary calculus.

Theorem 5.3: Let f and g be functions defined on (a, b) that are differentiable at $c \in (a, b)$. Then

(a) $(\alpha f + \beta g)'(c) = \alpha f'(c) + \beta g'(c)$ for any real numbers α and β.

(b) $(fg)'(c) = f(c)g'(c) + g(c)f'(c)$

(c)

$$\left(\frac{f}{g}\right)'(c) = \frac{g(c)f'(c) - f(c)g'(c)}{[g'(c)]^2} \text{ if } g(c) \neq 0.$$

Proof:

(a) We need to show that

$$\lim_{h \to 0} \frac{(\alpha f + \beta g)(h + c) - (\alpha f + \beta g)(c)}{h}$$

$$= \alpha \lim_{h \to 0} \frac{f(h + c) - f(c)}{h} + \beta \lim_{h \to 0} \frac{g(h + c) - g(c)}{h}.$$

This is immediate from properties of limits. We have

$$\lim_{h \to 0} \frac{(\alpha f + \beta g)(h + c) - (\alpha f + \beta g)(c)}{h}$$

$$= \lim_{h \to 0} \frac{\alpha f(h+c) + \beta g(h+c) - \alpha f(c) - \beta g(c)}{h}$$

$$= \lim_{h \to c} \left[\frac{\alpha f(h+c) - \alpha f(c)}{h} + \frac{\beta g(h+c) - \beta g(c)}{h} \right]$$

$$= \alpha \lim_{h \to 0} \left[\frac{f(h+c) - f(c)}{h} \right] + \beta \lim_{h \to 0} \left[\frac{g(h+c) - g(c)}{h} \right].$$

(b) Consider

$$\lim_{h \to 0} \frac{(fg)(h+c) - (fg)(c)}{h} = \lim_{h \to 0} \frac{f(h+c)g(h+c) - f(c)g(c)}{h}$$

$$= \lim_{h \to 0} \frac{f(h+c)g(h+c) - g(c)f(h+c) + g(c)f(h+c) - f(c)g(c)}{h}$$

$$= \lim_{h \to 0} \left[\frac{f(h+c)g(h+c) - g(c)f(h+c)}{h} + \frac{g(c)f(h+c) - f(c)g(c)}{h} \right].$$

Using the properties of limits, we can write this as

$$\lim_{h \to 0} f(h+c) \cdot \lim_{h \to 0} \frac{g(h+c) - g(c)}{h} + g(c) \lim_{h \to 0} \frac{f(h+c) - f(c)}{h}. \qquad (5.3)$$

Since f is differentiable at c, f is continuous at c so that $\lim_{h \to 0} f(h+c) = f(c)$. Therefore in the limit, (5.3) becomes

$$f(c)g'(c) + g(c)f'(c).$$

(c) Consider

$$\lim_{h \to 0} \frac{(f/g)(h+c) - (f/g)(c)}{h} = \lim_{h \to 0} \frac{f(h+c)/g(h+c) - f(c)/g(c)}{h}$$

$$= \lim_{h \to 0} \frac{g(c)f(h+c) - f(c)g(h+c)}{g(c)g(h+c)h}.$$

We are assuming that g is differentiable and thus continuous at $x = c$, and that $g(c) \neq 0$. Thus, for h sufficiently small, $g(h+c) \neq 0$. (See Theorem 4.8); so the denominator is nonzero.

Now

$$\lim_{h \to 0} \frac{g(c)f(h+c) - f(c)g(h+c)}{g(c)g(h+c)h}$$

$$= \lim_{h \to 0} \frac{g(c)f(h+c) - g(c)f(c) + g(c)f(c) - f(c)g(h+c)}{g(c)g(h+c)h}$$

$$= \frac{g(c)}{g(c)} \lim_{h \to 0} \frac{1}{g(h+c)} \left[\lim_{h \to 0} \frac{f(h+c) - f(c)}{h} \right] - \frac{f(c)}{g(c)} \lim_{h \to 0} \frac{1}{g(h+c)} \left[\lim_{h \to 0} \frac{g(h+c) - g(c)}{h} \right.$$

$$= \frac{g(c)f'(c) - f(c)g'(c)}{[g(c)]^2}$$

where we are using the fact that since g is continuous at $x = c$ and $g(c) \neq 0$,

$$\lim_{h \to 0} \frac{1}{g(h + c)} = \frac{1}{g(c)}. \quad \blacksquare$$

We next want to state and prove the Chain Rule for differentiation of real-valued functions. Before doing so, we establish some notation and make some observations that will simplify the proof. Suppose g is a function defined on an open interval I, and g is differentiable at $d \in I$. If we define a function G on I by

$$G(y) = \begin{cases} \frac{g(y) - g(d)}{y - d}) & \text{if } y \neq d \\ g'(d) & \text{if } y = d, \end{cases}$$

then G is continuous at $y = d$ since

$$\lim_{y \to d} G(y) = \lim_{y \to d} \frac{g(y) - g(d)}{y - d} = g'(d) = G(d).$$

Also for *any* $y \in I$, even $y = d$,

$$g(y) - g(d) = G(y)(y - d).$$

Theorem 5.4 (Chain Rule): Let f be a function defined on (a, b), and suppose that $f'(c)$ exists for some $c \in (a, b)$. Suppose g is defined on an open interval containing the range of f, and suppose that g is differentiable at $f(c)$. Then $g \circ f$ is differentiable at c and

$$(g \circ f)'(c) = g'(f(c))f'(c).$$

Proof: Consider

$$\frac{(g \circ f)(x) - (g \circ f)(c)}{x - c} = \frac{g(f(x)) - g(f(c))}{x - c}.$$

Using the function G previously defined and letting $y = f(x)$ and $d = f(c)$ we have

$$G(f(x)) = \frac{g(f(x)) - g(f(c))}{f(x) - f(c)}$$

so that

$$G(f(x))(f(x) - f(c)) = g(f(x)) - g(f(c)).$$

Thus

$$\frac{g(f(x)) - g(f(c))}{x - c} = G(f(x))\frac{f(x) - f(c)}{x - c}.$$

Let $\{x_n\}$ be any sequence in $\mathcal{D}(f)$ with $x_n \neq c$ for every n and $\lim x_n = c$. Then

$$\lim_{n \to \infty} \frac{(g \circ f)(x_n) - (g \circ f)(c)}{x_n - c} = \lim_{n \to \infty} G(f(x_n)) \frac{f(x_n) - f(c)}{x_n - c}.$$

Now $\lim f(x_n) = f(c)$, since f is continuous at c (by the differentiability of f at c). Also by our earlier observation, G is continuous at $f(c)$. Thus $\lim_{n \to \infty} G(f(x_n)) = G(f(c))$. Furthermore, f is differentiable at $x = c$, so

$$\lim_{n \to \infty} \frac{f(x_n) - f(c)}{x_n - c} = f'(c).$$

Thus

$$\lim_{n \to \infty} \frac{(g \circ f)(x_n) - (g \circ f)(c)}{x_n - c} = \lim_{n \to \infty} G(f(x_n)) \lim_{n \to \infty} \frac{f(x_n) - f(c)}{x_n - c}$$

$$= G(f(c)) \cdot f'(c) = g'(f(c)) \cdot f'(c)$$

since $G(f(c)) = g'(f(c))$. ∎

Example 5.4:
Let $F(x) = (7x^2 + 2)^9$. Then $F = g \circ f$ where $g(x) = x^9$ and $f(x) = 7x^2 + 2$. Now $g'(x) = 9x^8$ and $f'(x) = 14x$ so

$$F'(x) = g'(f(x))f'(x) = 9(7x^2 + 2)^8 14x$$

Some of the applications of the derivative that one encounters in elementary calculus are in graphing and max-min problems. We want to prove the theorems that are the basis for the techniques used in these applications.

Definition: A function f has a *relative maximum (minimum)* at a point $c \in \mathcal{D}(f)$ if there is a number $\delta > 0$ such that $f(c) \geq f(x)$ $((f(c) \leq f(x))$ for every $x \in (c - \delta, c + \delta) \cap \mathcal{D}(f)$.

A relative extremum is either a relative maximum or relative minimum.

We now prove a group of theorems that relates the behavior of a function to the derivative of the function. These theorems describe how the value of a function near a point c compares with the value of the function at c if $f'(c)$ exists and $f'(c) \neq 0$. We also show that if a function f has a relative maximum or minimum at an interior point of $\mathcal{D}(f)$ and $f'(c)$ exists, then $f'(c) = 0$.

Theorem 5.5: Let f be a function defined on (a, b). Suppose that f is differentiable at $c \in (a, b)$ and $f'(c) > 0$. then there is a number $\delta > 0$ such that $f(c) < f(x)$ if $x \in (c, c + \delta)$ and $f(x) < f(c)$ if $x \in (c - \delta, c)$.

Proof: By definition of the derivative, given $\epsilon > 0$ there is a $\delta(\epsilon) > 0$ such that

$$\left| \frac{f(c+h) - f(c)}{h} - f'(c) \right| < \epsilon$$

if $0 < |h| < \delta(\epsilon)$. (Notice that it is possible to choose h small enough so that $c + h \in (a, b)$.) Take $\epsilon = f'(c)/2 > 0$. Then there is a $\delta > 0$ such that

$$\left| \frac{f(c+h) - f(c)}{h} - f'(c) \right| < \frac{f'(c)}{2}$$

or

$$-\frac{f'(c)}{2} < \frac{f(c+h) - f(c)}{h} - f'(c) < \frac{f'(c)}{2}$$

if $0 < |h| < \delta$. Then

$$\frac{f'(c)}{2} < \frac{f(c+h) - f(c)}{h} < 3\frac{f'(c)}{2}.$$

If $0 < h < \delta$, then

$$0 < f\frac{f'(c)}{2}h < f(c+h) - f(c)$$

so that $f(c) < f(c+h)$ for $c < c + h < c + \delta$. If $0 > h > -\delta$,

$$0 > \frac{f'(c)}{2}h > f(c+h) - f(c)$$

so that $f(c) > f(c+h)$ for $c - \delta < c + h < c$. ∎

Corollary 5.5:
Let f be a function defined on (a, b). Suppose that f is differentiable at $c \in (a, b)$ and $f'(c) < 0$. Then there is a number $\delta > 0$ such that $f(c) < f(x)$ if $x \in (c - \delta, c)$ and $f(c) > f(x)$ if $x \in (c, c + \delta)$. ∎

The proof is a useful exercise. It can be accomplished by making slight modifications in the proof of the theorem, or by considering the function $-f$.

One must be a little careful interpreting this theorem. It does *not* say that if $f'(c) > 0$ there is an interval about c on which f is increasing. For example, let

$$f(x) = \begin{cases} x + 2x^2 \sin(1/x) & \text{if } x \neq 0 \\ 0 & \text{if } x = 0. \end{cases}$$

(See Olmstead and Gelbaum, p.37.) Then

$$f'(x) = \begin{cases} 1 + 4x \sin(1/x) - 2\cos(1/x) & \text{if } x \neq 0 \\ 1 & \text{if } x = 0. \end{cases}$$

Thus $f'(0) > 0$, so there is an interval $(-\delta, \delta)$ about 0 such that $f(x) < f(0)$ if $x \in (-\delta, 0)$ and $f(0) < f(x)$ if $x \in (0, \delta)$. In particular, one may take $\delta = 1/2$. (See Exercise 9, Section 5.1.) However, $f'(x)$ takes on both positive and negative values arbitrarily close to 0, so there is no interval about 0 on which f is increasing.

Theorem 5.6: Suppose f is defined on (a, b) and has a relative extremum at $x_0 \in (a, b)$. If f is differentiable at x_0, then $f'(x_0) = 0$.

Proof: We do the proof in the case where f has a relative maximum at x_0. Since f is differentiable at $x_0 \in (a, b)$, then

$$\lim_{h \to 0} \frac{f(x_0 + h) - f(x_0)}{h}$$

exists and is $f'(x_0)$. Thus

$$\lim_{h \downarrow 0} \frac{f(x_0 + h) - f(x_0)}{h} = \lim_{h \uparrow 0} \frac{f(x_0 + h) - f(x_0)}{h} = f'(x_0)$$

since the limit at an interior point exists if and only if the left- and right-hand limits exist and are equal. We shall show the left-hand limit is greater than or equal to 0 and the right-hand limit is less than or equal to 0. We shall then be forced to conclude that the limit, which is $f'(x_0)$, is 0.

If f has a relative maximum at $x_0 \in (a, b)$, then there is a $\delta > 0$ such that $f(x_0) \geq f(x)$ for $x \in (x_0 - \delta, x_0 + \delta)$. Suppose $0 < h < \delta$. Then $f(x_0 + h) \leq f(x_0)$ so that

$$\frac{f(x_0 + h) - f(x_0)}{h} \leq 0$$

and thus

$$\lim_{h \downarrow 0} \frac{f(x_0 + h) - f(x_0)}{h} \leq 0$$

since the inequality holds for all sufficiently small positive values of h. Now suppose $-\delta < h < 0$. Again $f(x_0 + h) \leq f(x_0)$ so that

$$\frac{f(x_0 + h) - f(x_0)}{h} \geq 0 \text{ and thus } \lim_{h \uparrow 0} \frac{f(x_0 + h) - f(x_0)}{h} \geq 0. \quad \blacksquare$$

Exercises 5.1

1. Use the definition of the derivative to find the derivative of the following functions:

   (a) $f(x) = \sqrt{x+2}$.

   (b) $f(x) = \frac{1}{x}$.

   (c) $f(x) = 3x - \frac{1}{x}$.

2. (a) Use the definition of the derivative and the Binomial Theorem to show that the derivative of $f(x) = x^n$ is $f'(x) = nx^{n-1}$ for any positive integer n.

   (b) Use induction to prove the result in part (a).

3. In statistics it is important to know the value of x at which

$$f(x) = \sum_{i=1}^{n} (x - a_i)^2$$

is minimized, where $a_1, a_2, \ldots, a_n$ are constants. Find this value.

4. The *left-hand derivative* of f at c is defined as follows: Let c be a limit point of $(c - \delta, c] \cap \mathcal{D}(f)$ for some $\delta > 0$, and suppose $c \in \mathcal{D}(f)$. The left-hand derivative of f at c is

$$\lim_{h \uparrow 0} \frac{f(c+h) - f(c)}{h}$$

provided this limit exists and is finite. The right-hand derivative is defined similarly.

   (a) Show that the derivative of f at c exists if and only if the left- and right-hand derivatives exist and are equal.

   (b) Show that if

$$f(x) = \begin{cases} x^2 & \text{if } x \leq 1 \\ x & \text{if } x > 1, \end{cases}$$

then f has left- and right-hand derivatives at $x = 1$, but is not differentiable at $x = 1$.

5. Here we derive the formulas for the derivatives of the trigonometric functions.

   (a) The area of a circle of radius r subtended by an angle θ is given by $\frac{1}{2}r^2\theta$, and the length of the subtended arc is given by $r\theta$, where θ *is measured in radians*. Consider a circle of radius 1 with central angle θ. Use Figure 5.2 to argue that

$$\frac{1}{2}\sin\theta\cos\theta < \frac{1}{2} < \frac{1}{2}\tan\theta.$$

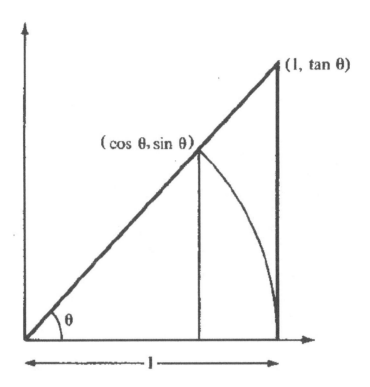

Figure 5.2
Comparison of areas for finding the derivative of the sine function.

(b) Show that
$$\lim_{\theta \to 0} \frac{1 - \cos \theta}{\theta} = 0.$$

(c) Use the fact that $\lim_{\theta \to 0} \cos \theta = 1$ to show that
$$\lim_{\theta \to 0} \frac{\sin \theta}{\theta} = 1.$$

(d) Use the formula $\sin(\alpha + \beta) = \sin \alpha \cos \beta + \sin \beta \cos \alpha$ to show that if $f(x) = \sin x$ then $f'(x) = \cos x$.

(e) Use the fact that $\cos x = \sin(\pi/2 - x)$ to show that if $f(x) = \cos x$ then $f'(x) = -\sin x$.

(f) Derive the formulas for the derivatives of the other trigonometric functions.

6. Show that $d/dx \sin x = \frac{\pi}{180} \cos x$ if x is measured in degrees.

7. Show that if f is differentiable on (a, b) and $f'(c) > 0$ where $c \in (a, b)$ and

f' is continuous at $x = c$, then there is an interval about c on which f is increasing.

8. (a) Suppose $|f(x + h) - f(x)| \le Kh^\alpha$ for some constant K and $\alpha > 0$. Show that f is continuous.

   (b) Suppose $|f(x + h) - f(x)| \le Kh^\alpha$ for some constant K and $\alpha > 1$. Show that f is differentiable and $f'(x) = 0$.

   (c) Suppose $|f(h) - f(0)| \le h$. Must f be differentiable at $x = 0$? *Hint:* Consider

$$f(x) = \begin{cases} \frac{1}{2}x & \text{if } x \text{ is rational} \\ x & \text{if } x \text{ is irrational.} \end{cases}$$

9. Verify that if

$$f(x) = \begin{cases} x + 2x^2 \sin(1/x) & \text{if } x \ne 0 \\ 0 & \text{if } x = 0, \end{cases}$$

then $f'(x) = 1$ if $x = 0$. Also show that $f'(x)$ takes on positive and negative values arbitrarily close to 0 and that $f(0) < f(x)$ if $x < \frac{1}{2}$.

10. Show that

$$f(x) = \begin{cases} x^2 \sin(1/x) & \text{if } x \ne 0 \\ 0 & \text{if } x = 0 \end{cases}$$

is differentiable, but the derivative is not continuous at $x = 0$.

11. For what values of α is the function

$$f(x) = \begin{cases} x^\alpha & \text{if } x \text{ is rational} \\ 0 & \text{if } x \text{ is irrational} \end{cases}$$

differentiable at $x = 0$? Show that 0 is the only point at which f can be differentiable.

12. Show that

$$f(x) = \begin{cases} x \sin(1/x) & \text{if } x \ne 0 \\ 0 & \text{if } x = 0 \end{cases}$$

is continuous but not differentiable at $x = 0$.

13. Suppose f and g have nth order derivatives on (a, b). Let $h = f \cdot g$. Show that, for $c \in (a, b)$,

$$h^{(n)}(c) = \sum_{k=0}^{n} \binom{n}{k} f^{(k)}(c) g^{(n-k)}(c).$$

This is called Leibniz's Rule.

14. Suppose f is differentiable and $f(x + y) = f(x)f(y)$. Show that $f'(x) = f'(0)f(x)$. What is $f(0)$?

15. Suppose f is differentiable on (a, b), and let $c \in (a, b)$. Let $\{\alpha_n\}$ and $\{\beta_n\}$ be sequences that converge to c with $\alpha < \alpha_n < c < \beta_n < b$. Show that

$$\lim_{n \to \infty} \frac{f(\beta_n) - f(\alpha_n)}{\beta_n - \alpha_n} = f'(c).$$

Note: This result does not necessarily hold if we do not require $\alpha_n < c < \beta_n$. For example, take $f(x) = x^2 \sin(\pi/(2x))$ with $\alpha_n = 1/n$ and $\beta_n = (1/n) + (1/n^2)$.

5.2   Some Mean Value Theorems

In this section we prove several results that are used in elementary calculus, probably the best known of which is L'Hôpital's Rule. The proof of the first group of theorems, called Mean Value Theorems, relies on the following simple facts: If f is a continuous function and there are numbers a and b where $f(a) = f(b)$, then somewhere between a and b, there is a point where $f(x)$ assumes its maximum or minimum value. If this point is c and if f is differentiable at c then $f'(c) = 0$. This is Rolle's Theorem. The proofs of some of the theorems that follow it consist mostly of constructing a function that satisfies the hypotheses of Rolle's Theorem. If this can be kept in mind, it may make the proofs of these theorems less mysterious. The Mean Value Theorems have several applications, including the proofs of L'Hôpital's Rule and the Fundamental Theorem of Calculus.

Mean Value Theorems

Theorem 5.7 (Rolle's Theorem): Suppose that f is continuous on $[a, b]$ and is differentiable on (a, b). If $f(a) = f(b)$, then there is a number $c \in (a, b)$ for which $f'(c) = 0$.
   Note: To say f is continuous on $[a, b]$ means that f is continuous at any $x \in (a, b)$, continuous from the right at $x = a$, and continuous from the left at $x = b$.

   Proof: We show that either f is constant on $[a, b]$, in which case $f'(c) = 0$ for all $c \in (a, b)$, or f has an extremum at $c \in (a, b)$, in which case $f'(c) = 0$ by Theorem 5.6.
   First note that since f is continuous on $[a, b]$, it is bounded and assumes its maximum value M and minimum value m somewhere on $[a, b]$. If $M = m$ then f is constant on $[a, b]$ and $f'(x) = 0$ for all $x \in (a, b)$.

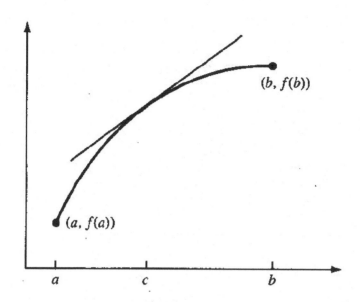

Figure 5.3
A tangent line guaranteed by the Mean Value Theorem.

Otherwise $M \neq m$ and since $f(a) = f(b)$, at least one of M or m must differ from $f(a)$ and $f(b)$.

Suppose $M > f(a)$. Then there is some $c \in (a, b)$ where $f(c) = M$ and thus f has a relative maximum at $x = c$ so $f'(c) = 0$. A similar argument works if $m < f(a)$. ∎

In the exercises we construct examples that show Rolle's Theorem fails if any of the hypotheses is omitted.

Theorem 5.8 (Mean Value Theorem): Suppose f is continuous on $[a, b]$ and differentiable on (a, b). Then there is a number $c \in (a, b)$ for which

$$f'(c) = \frac{f(b) - f(a)}{b - a}.$$

Proof: First notice that $\frac{f(b)-f(a)}{b-a}$ is the slope of the line joining the points $(a, f(a))$ and $(b, f(b))$. We are to prove that there is a point c at which the tangent line is parallel to this line. (See Figure 5.3.)

Let

$$g(x) = f(x) - \left[\frac{f(b) - f(a)}{b - a} \right] x.$$

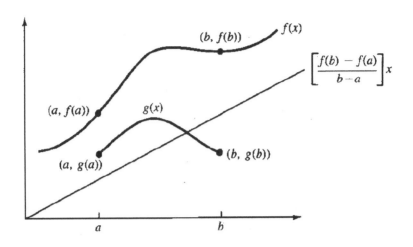

Figure 5.4
Using Rolle's Theorem to prove the Mean Value Theorem.

The graph of
$$\left[\frac{f(b) - f(a)}{b - a)}\right] x$$
is a straight line passing through the origin parallel to the line joining $(a, f(a))$
and $(b, f(b))$, so it appears that $g(a) = g(b)$. (See Figure 5.4.) We show that
this is the case.
    Now
$$g(a) = f(a) - \left[\frac{f(b) - f(a)}{b - a}\right] a = \frac{bf(a) - af(b)}{b - a}$$
and
$$g(b) = f(b) - \left[\frac{f(b) - f(a)}{b - a}\right] b = \frac{bf(a) - af(b)}{b - a}.$$
Also g is continuous on $[a, b]$ and differentiable on (a, b), since f satisfies these
hypotheses. Thus $g(x)$ satisfies the hypotheses of Rolle's Theorem, so there is
a point $c \in (a, b)$ where $g'(c) = 0$.
    But
$$g'(c) = f'(c) - \left[\frac{f(b) - f(a)}{b - a}\right] = 0$$
so
$$f'(c) = \frac{f(b) - f(a)}{b - a}. \qquad \blacksquare$$

Corollary 5.8(a):
Suppose f is continuous on $[a, b]$ and differentiable on (a, b). If $f'(x) = 0$ on (a, b), then f is constant on $[a, b]$.

Proof: Choose $c \in (a, b]$. We shall show that $f(c) = f(a)$. By the Mean Value Theorem, there is a point $\alpha \in (a, c)$ with

$$f(c) - f(a) = f'(\alpha)(c - a).$$

But $f'(\alpha) = 0$, so $f(c) = f(a)$. ∎

Corollary 5.8(b):
Suppose f and g are continuous on $[a, b]$ and differentiable on (a, b). If $f'(x) = g'(x)$ on (a, b), then f and g differ by a constant.

Proof: Let $h(x) = f(x) - g(x)$. Then h satisfies the hypotheses of Corollary 5.8(a), so $h(x)$ is constant. ∎

Example 5.5:
Suppose f and g are differentiable functions with $f(a) = g(a)$ and $f'(x) < g'(x)$ for $x > a$. We shall show that $f(b) < g(b)$ for $b > a$. To do this, define $h(x) = g(x) - f(x)$. Then $h(a) = 0$ and $h'(x) > 0$ for $x > a$. Thus

$$h(b) - h(a) = h'(c)(b - a) \text{ for some } c \in (a, b).$$

Now $h'(c) > 0$ and $b - a > 0$ so $h(b) > h(a) = 0$. But $h(b) = g(b) - f(b)$.

Example 5.6:
We show that $\ln(1 + x) < x$ for $x > 0$. Let $f(x) = \ln(1 + x)$ and $g(x) = x$. Then

$$f(0) = g(0) = 0, \ f'(x) = \frac{1}{1 + x}, \ g'(x) = 1$$

so $f'(x) < g'(x)$ for $x > 0$. Thus, by Example 5.5,

$$f(x) = \ln(1 + x) < g(x) = x \text{ for } x > 0.$$

Example 5.7:
If f is differentiable at $x = c$ and g is differentiable on an interval containing $f(c + h)$ for h sufficiently small, and if g' is continuous at $f(c)$, then we can use the Mean Value Theorem to give an easy proof of the Chain Rule. For h sufficiently small

$$\frac{g(f(c + h)) - g(f(c))}{h} = g'(\theta)\frac{f(c + h) - f(c)}{h}.$$

for some θ between $f(c + h)$ and $f(c)$ by the Mean Value Theorem. Now f is differentiable at $x = c$, so

$$\lim_{h \to 0} \frac{f(c + h) - f(c)}{h} = f'(c).$$

Also f is continuous at $x = c$, so $\lim_{h \to 0} f(c + h) = f(c)$, and thus

$$\lim_{h \to 0} \theta = f(c)$$

(since θ is between $f(c + h)$ and $f(c)$). Now g' is continuous at $f(c)$, so

$$\lim_{h \to 0} g'(\theta) = g'(f(c)).$$

Thus

$$g'(f(c))f'(c) = \lim_{h \to 0} g'(\theta) \lim_{h \to 0} \frac{f(c + h) - f(c)}{h}$$

$$= \lim_{h \to 0} g'(\theta) \frac{f(c + h) - f(c)}{h}$$

$$= \lim_{h \to 0} \frac{g(f(c + h)) - g(f(c))}{h} = (g \circ f)'(c).$$

Theorem 5.9 (Generalized Mean Value Theorem): Suppose f and g are continuous functions on $[a, b]$ and dfferentiable on (a, b). Then there is a point $c \in (a, b)$ where

$$[f(b) - f(a)]g'(c) = [g(b) - g(a)]f'(c).$$

Remarks: Notice this reduces to the Mean Value Theorem in the case $g(x) = x$. The geometric interpretation of the theorem is as follows: If we describe a curve parametrically in the $x - y$ plane by $x = f(t), y = g(t)$ for $t \in [a, b]$, then there is a point on the curve where the line tangent to the curve is parallel to the line joining the points $(f(a), g(a))$ and $(f(b), g(b))$. This curve is not necessarily the graph of a function.

Proof: We want to describe a function $h(x)$ that satisfies the hypotheses of Rolle's Theorem so that the conclusion of Rolle's Theorem will provide a proof of this theorem. To this end, define

$$h(x) = f(x) - \alpha g(x)$$

where α is a constant that will be chosen to force $h(a) = h(b)$. Now

$$h(a) = f(a) - \alpha g(a) \text{ and } h(b) = f(b) - \alpha g(b).$$

So, if $h(a) = h(b)$, then

$$f(a) - \alpha g(a) = f(b) - \alpha g(b) \text{ or } f(b) - f(a) = \alpha(g(b) - g(a)).$$

Now if $g(b) = g(a)$ and $f(b) \neq f(a)$, then no such α exists. However, if $g(b) \neq g(a)$ then we let

$$\alpha = \frac{f(b) - f(a)}{g(b) - g(a)} \text{ and so } h(x) = f(x) - \left[\frac{f(b) - f(a)}{g(b) - g(a)}\right] g(x).$$

Now $h(x)$ satisfies the hypotheses of Rolle's Theorem, so there is a number $c \in (a, b)$ where $g'(c) = 0$. Then

$$h'(c) = f'(c) - \left[\frac{f(b) - f(a)}{g(b) - g(a)}\right] g'(c) = 0$$

or

$$f'(c)[g(b) - g(a)] = g'(c)[f(b) - f(a)].$$

Now suppose $g(b) = g(a)$. Then g satisfies the hypotheses of Rolle's Theorem, so there is a point $c \in (a, b)$ where $g'(c) = 0$. Then

$$f'(c)[g(b) - g(a)] = g'(c)[f(b) - f(a)]$$

because both sides of the equation are 0. ∎

Example 5.8:
Let $f(x) = 4x^3 - 7x^2$ and $g(x) = x^4 - 5$ on $[0, 2]$. Then

$$f(0) = 0, \quad f(2) = 4, \quad g(0) = -5, \quad g(2) = 11,$$

$$f'(x) = 12x^2 - 14x \text{ and } g'(x) = 4x^3.$$

Theorem 5.9 guarantees a value of x between 0 and 2 where

$$(12x^2 - 14x)[11 - (-5)] = 4x^3(4 - 0).$$

This value is $6 - \sqrt{22}$.

Taylor's Theorem

We next discuss a result known as Taylor's Theorem that gives another application of the Mean Value Theorem and will preview a later topic, Taylor series.

Suppose that f is a function with derivatives $f', f'', \ldots, f^{(n)}$ that are defined and continuous on an open interval containing $[a, b]$, and suppose that $f^{(n+1)}$ exists for $a < x < b$. We want to approximate f by a polynomial $P_n(x)$ of degree n in $(x - a)$. That is,

$$P_n(x) = c_0 + c_1(x - a) + c_2(x - a)^2 + \cdots + c_n(x - a)^n.$$

How should the constants c_i be chosen so that $P_n(x)$ will provide the best approximation possible of $f(x)$ near $x = a$? To answer this question we would need to know what is meant by "the best approximation" but some things that might reasonably be required are

$$f(a) = P_n(a), \ f'(a) = P_n'(a), ..., f^{(n)}(a) = P_n^{(n)}(a).$$

Now

$$P_n(a) = c_0, \ P_n'(a) = c_1, \ P_n''(a) = 2c_2, ..., P_n^{(n)}(a) = n!c_n$$

so that

$$c_k = \frac{P_n^{(k)}(a)}{k!} = \frac{f^{(k)}(a)}{k!}, \ k = 0, 1, ..., n.$$

With these ideas in mind, we state and prove Taylor's Theorem.

Theorem 5.10 (Taylor's Theorem): Let f be a function such that $f, f'f'', ..., f^{(n)}$ are continuous on some open interval containing $[a, b]$, and suppose $f^{(n+1)}$ exists on (a, b). Then there is a number $c \in (a, b)$ for which

$$f(b) = f(a) + f'(a)(b - a) + \cdots + \frac{f^{(n)}(a)}{n!}(b - a)^n + \frac{f^{(n+1)}(c)}{(n + 1)!}(b - a)^{n+1}.$$

Proof: Because $b - a \neq 0$, there is a number J for which

$$\frac{(b - a)^{n+1}}{(n + 1)!}J = f(b) - \left\{ f(a) + \sum_{k=1}^{n} \frac{f^{(k)}(a)}{k!}(b - a)^k \right\}.$$

We shall show that $J = f^{(n+1)}(c)$ for some $c \in (a, b)$. Consider

$$F(x) = f(b) - \left\{ f(x) + \sum_{k=1}^{n} \frac{f^{(k)}(x)}{k!}(b - x)^k + \frac{(b - x)^{n+1}}{(n + 1)!}J \right\}.$$

Now F is continuous on $[a, b]$, differentiable on (a, b), and $F(a) = F(b) = 0$; that is, F satisfies the hypotheses of Rolle's Theorem. Thus there is a number $c \in (a, b)$ where

$$0 = F'(c) = -\left\{ f'(c) + \sum_{k=1}^{n} \left[\frac{f^{(k+1)}(c)}{k!}(b - c)^k - \frac{f^{(k)}(c)}{k!}k(b - c)^{k-1} \right] - \frac{(b - c)^n}{n!}J \right\}$$

Now

$$f'(c) + \sum_{k=1}^{n} \frac{f^{(k+1)}(c)}{k!}(b - c)^k = \sum_{k=0}^{n} \frac{f^{(k+1)}(c)}{k!}(b - c)^k$$

and

$$\sum_{k=1}^{n} \frac{f^{(k)}(c)}{k!} k(b-c)^{k-1} = \sum_{k=1}^{n} \frac{f^{(k)}(c)}{(k-1)!}(b-c)^{k-1}$$

$$= \sum_{k=0}^{n-1} \frac{f^{(k+1)}(c)}{(k)!}(b-c)^{k}.$$

Thus

$$0 = F'(c) = \frac{f^{n+1}(c)}{n!}(b-c)^{n} - \frac{(b-c)^{n}}{n!}J$$

so

$$J = f^{(n+1)}(c). \quad \blacksquare$$

Corollary 5.10: Let f be a function that satisfies the hypotheses of Taylor's Theorem. Then, if $x \in (a,b)$ there is a number $c \in (a,x)$ for which

$$f(x) = f(a) + f'(a)(x-a) + \cdots + \frac{f^{(n)}(a)}{n!}(x-a)^{n} + \frac{f^{(n+1)}(c)}{(n+1)!}(x-a)^{n+1}. \quad \blacksquare$$

Note: Now the number c will depend on x. We shall refer to either Theorem 5.10 or Corollary 5.10 as Taylor's Theorem.

Example 5.9:
Let $f(x) = \cos(x)$ and take $a = 0$ in Taylor's Theorem. Then $f'(x) = -\sin x$, $f''(x) = -\cos x$, $f'''(x) = \sin x$ and $f^{(4)}(x) = \cos x$. So $f(0) = 1, f'(0) = 0, f''(0) = -1, f'''(0) = 0$. Taylor's Theorem says that if $x > 0$,

$$f(x) = \cos x = 1 + 0(x-0) + \frac{-1(x-0)^2}{2!} + \frac{0(x-0)^3}{3!} + \frac{\cos(c)(x-0)^4}{4!}$$

$$= 1 - \frac{x^2}{2} + \frac{\cos(c)x^4}{4!}$$

for some c between 0 and x.

Taylor's Theorem gives a very useful way to find the approximate value of a function near the value a. In our example, if $x = .1$, then

$$\cos(.1) = 1 + \frac{(.1)^2}{2} + \frac{\cos(c)(.1)^4}{4!}.$$

But $|\cos c| \leq 1$, so that the error we incur by taking $1 - \frac{(.1)^2}{2}$ for the value of $\cos(.1)$ is no more than $\frac{(1)(.1)^4}{4!}$ or approximately .000006.

L'Hôpital's Rule

In evaluating limits in elementary calculus, one finds that it is usually simple to find the limit of the quotient of two functions f and g at $x = a$ unless

$$\lim_{x \to a} f(x) = \lim_{x \to a} g(x) = 0 \text{ or } \lim_{x \to a} |f(x)| = \lim_{x \to a} |g(x)| = \infty.$$

In the first encounter with limits, one usually deals with such problems via algebraic manipulations as in the case with $\lim_{x \to 2}[(x^2 - 4)/(x - 2)]$ or geometric estimates as in the case $\lim_{x \to 0}(\sin x)/x$. Later one learns that there is sometimes an easier approach to such problems using a result known as L'Hôpital's Rule. Our next project is to prove this theorem. The proof consists of applying the Generalized Mean Value Theorem.

Theorem 5.11 (L'Hôpital's Rule): Let f and g be differentiable functions on an interval (a, b) with $g'(x) \neq 0$ on (a, b).
(a) Suppose

$$\lim_{x \downarrow a} f(x) = 0, \quad \lim_{x \downarrow a} g(x) = 0, \quad \text{and} \quad \lim_{x \downarrow a} \frac{f'(x)}{g'(x)} = L.$$

Then

$$\lim_{x \downarrow a} \frac{f(x)}{g(x)} = L.$$

(b) Suppose

$$\lim_{x \downarrow a} f(x) = \pm\infty, \quad \lim_{x \downarrow a} g(x) = \pm\infty, \quad \text{and} \quad \lim_{x \downarrow a} \frac{f'(x)}{g'(x)} = L.$$

Then

$$\lim_{x \downarrow a} \frac{f(x)}{g(x)} = L.$$

Remarks:

(1) In the hypothesis a may be $-\infty$. The theorem is also true if $\lim_{x \downarrow a}$ is replaced by $\lim_{x \uparrow b}$, and in this case b may be ∞. Also, L need not be finite.

(2) This is not the most general form of the theorem. It is possible to show

$$\underline{\lim}_{x \downarrow a} \frac{f'(x)}{g'(x)} \leq \underline{\lim}_{x \downarrow a} \frac{f(x)}{g(x)} \leq \overline{\lim}_{x \downarrow a} \frac{f(x)}{g(x)} \leq \overline{\lim}_{x \downarrow a} \frac{f'(x)}{g'(x)}$$

(see Randolph, p. 375). The form stated here is sufficient for most applications.

(3) Because of the various forms that the theorem can assume, there are several cases to be considered. We shall give the proof of four of the possible forms. First, we prove part (a), when a and L are finite; second, we prove part (a) when a is finite and L is infinite; third, we prove part (a) when a is infinite and L is finite; and finally, we prove part (b) when a and L are finite. The proofs of other forms can be patterned after these.

Proof: We first do the proof for part (a) in the case that a and L are finite. We shall want to apply the Generalized Mean Value Theorem. To do so, f and g will need to be continuous at a, so define $f(a) = g(a) = 0$. Now if $x \in (a, b)$, f and g are differentiable on (a, x) and continuous on $[a, x]$. Since $g'(t) \neq 0$ for $t \in (a, b)$, then $g(x) \neq g(a)$ for $x \in (a, b)$ by Rolle's Theorem. Thus, by the Generalized Mean Value Theorem, if $x \in (a, b)$, there is a number $\overline{x} \in (a, x)$ such that

$$\frac{f(x) - f(a)}{g(x) - g(a)} = \frac{f'(\overline{x})}{g'(\overline{x})}$$

and since $f(a) = g(a) = 0$

$$\frac{f(x)}{g(x)} = \frac{f'(\overline{x})}{g'(\overline{x})}.$$

Now let $\epsilon > 0$ be given. Since

$$\lim_{x \downarrow a} \frac{f'(x)}{g'(x)} = L,$$

there is a $\delta > 0$ such that if $0 < x - a < \delta$, then

$$\left| \frac{f'(x)}{g'(x)} - L \right| , < \epsilon.$$

Suppose $0 < x - a < \delta$; then

$$\left| \frac{f(x)}{g(x)} - L \right| = \left| \frac{f'(\overline{x})}{g'(\overline{x})} - L \right|$$

where $\overline{x} \in (a, x)$. Then $0 < \overline{x} - a < \delta$ so

$$\left| \frac{f'(\overline{x})}{g'(\overline{x})} - L \right| < \epsilon$$

and thus

$$\left| \frac{f(x)}{g(x)} - L \right| < \epsilon$$

if $0 < x - a < \delta$, which proves

$$\lim_{x \downarrow a} \frac{f(x)}{g(x)} = L.$$

Now suppose

$$\lim_{x \downarrow a} \frac{f'(x)}{g'(x)} = \infty.$$

Choose any number N. Then there is a $\delta > 0$ such that if $0 < x - a < \delta$, then $f'(x)/g'(x) > N$. So if $0 < x - a < \delta$, there is an $\overline{x} \in (a, x)$ (which forces $0 < \overline{x} - a < \delta$) with

$$\frac{f(x)}{g(x)} = \frac{f'(\overline{x})}{g'(\overline{x})} > N$$

so that

$$\lim \frac{f(x)}{g(x)} = \infty.$$

The proof for $\lim[f'(x)/g'(x)] = -\infty$ is similar.

We next consider the case where $a = -\infty$. To do this, we define functions $\psi(t) = f(-1/t)$ and $\phi(t) = g(-1/t)$, so that the domain of ψ and ϕ is $(0, c)$ for some positive number c. Thus ψ and ϕ are the composition of differentiable functions so ψ and ϕ are differentiable functions on $(0, c)$. Also

$$\lim_{t \downarrow 0} \psi(t) = \lim_{t \downarrow 0} f(-1/t) = \lim_{x \to -\infty} f(x) = 0,$$

$$\lim_{t \downarrow 0} \phi(t) = \lim_{t \downarrow 0} g(-1/t) = \lim_{x \to -\infty} g(x) = 0,$$

and $\phi'(t) = g'(-1/t)(1/t^2) \neq 0$ on $(0, c)$, if $g'(-1/t) \neq 0$ on $(0, c)$. Thus

$$\lim_{t \downarrow 0} \frac{\psi'(t)}{\phi'(t)} = \lim_{t \downarrow 0} \frac{f'(-1/t)(1/t^2)}{g'(1/t)(1/t^2)} = \lim_{t \downarrow 0} \frac{f'(-1/t)}{g'(-1/t)} = \lim_{x \to -\infty} \frac{f'(x)}{g'(x)} = L.$$

Applying the first case with $a = 0$, we have

$$\lim_{t \downarrow 0} \frac{\psi(t)}{\phi}(t) = L \text{ so that } \lim_{x \to \infty} \frac{f(x)}{g(x)} = L.$$

(b) We give the proof in case (b) when L and a are finite. The proof of (b) is much more delicate, because we cannot define $f(a)$ and $g(a)$ as we did for part (a) to form

$$\frac{f(x) - f(a)}{g(x) - g(a)}.$$

Let $\epsilon > 0$ be given. There is a $\delta > 0$ such that if $0 < x - a < \delta$ (i.e. if $x \in (a, a + \delta)$), then

$$\left| \frac{f'(x)}{g'(x)} - L \right| < \frac{\epsilon}{3}.$$

Let x_1 and x_2 be such that $a < x_1 < x_2 < a + \delta$. Figure 5.5 may help you remember the relative size of the numbers that we shall be using.

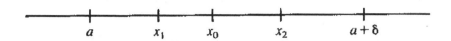

Figure 5.5
Relative locations of points in the proof of L'Hopital's Rule.

Since f and g satisfy the hypotheses of the Generalized Mean Value Theorem, there is a point $x_0 \in (x_1, x_2)$ at which

$$\frac{f(x_1) - f(x_2)}{g(x_1) - g(x_2)} = \frac{f'(x_0)}{g'(x_0)}$$

or

$$\frac{f(x_1)(1 - f(x_2)/f(x_1))}{g(x_1)(1 - g(x_2)/g(x_1))} = \frac{f'(x_0)}{g'(x_0)}.$$

(Since $\lim_{x \downarrow a} f(x) = \infty$ and $\lim_{x \downarrow a} g(x) = \infty$, δ may be chosen so that $f(x) > 0$ and $g(x) > 0$ if $x \in (a, a + \delta)$). Thus

$$\frac{f(x_1)}{g(x_1)} = \frac{f'(x_0)}{g'(x_0}\frac{[1 - g(x_2)/g(x_1)]}{[1 - f(x_2)/f(x_1)]}. \tag{5.4}$$

Now hold x_2 fixed and let x_1 vary. Since

$$\lim_{x_1 \downarrow a} g(x_1) = \lim_{x_1 \downarrow a} f(x_1) = \infty$$

then

$$\lim_{x_1 \downarrow a} \left[1 - \frac{g(x_2)}{g(x_1)} \right] = \lim_{x_1 \downarrow a} \left[1 - \frac{f(x_2)}{f(x_1)} \right] = 1.$$

So

$$\lim_{x_1 \downarrow a} \frac{1 - g(x_2)/g(x_1)}{1 - f(x_2)/f(x_1)} = 1. \tag{5.5}$$

Now let

$$h(x_1) = \frac{1 - g(x_2)/g(x_1)}{1 - f(x_2)/f(x_1)}.$$

(Remember, x_2 is fixed.) Now by (5.5) there is a $\delta_2 > 0$ such that if $0 < x_1 - a < \delta_2$ (i.e., $x_1 \in (a, a + \delta_2)$), then

$$|h_1(x) - 1| < \frac{\epsilon}{2(|L| + (\epsilon/3))}.$$

Now by (5.4),

$$\left| \frac{f(x_1)}{g(x_1)} - L \right| = \left| \frac{f'(x_0)}{g'(x_0)} \cdot h(x_1) - L \right|$$

$$\le \left| \frac{f'(x_0)}{g'(x_0)} \cdot h(x_1) - \frac{f'(x_0)}{g'(x_0)} \right| + \left| \frac{f'(x_0)}{g'(x_0)} - L \right|$$

$$= \left| \frac{f'(x_0)}{g'(x_0)} \right| |h(x_1) - 1| + \left| \frac{f'(x_0)}{g'(x_0)} - L \right|$$

$$< \left(|L| + \frac{\epsilon}{3} \right) |h(x_1) - 1| + \left| \frac{f'(x_0)}{g'(x_0)} - L \right|$$

$$\left(\text{since } \left| \frac{f'(x_0)}{g'(x_0)} - L \right| < \frac{\epsilon}{3} \text{ forces } \left| \frac{f'(x_0)}{g'(x_0)} \right| < |L| + \frac{\epsilon}{3} \right)$$

$$< \left(|L| + \frac{\epsilon}{3} \right) \frac{\epsilon}{2(|L| + \epsilon/3)} + \frac{\epsilon}{3} < \epsilon \text{ if } 0 < x_1 - a < \delta_2.$$

So

$$\lim_{x_1 \downarrow a} \frac{f(x_1)}{g(x_1)} = L. \quad \blacksquare$$

Example 5.10:

$$\lim_{x \to 0} \frac{x - \sin x}{x^3} = \lim_{x \to 0} \frac{1 - \cos x}{3x^2} = \lim_{x \to 0} \frac{\sin x}{6x} = \lim_{x \to 0} \frac{\cos x}{6} = \frac{1}{6}.$$

In this case we have to apply L'Hôpital's Rule three times before the function is not of the form $0/0$ in the limit. One implication of this calculation is that for x near 0, $x - \sin x$ is approximately equal to $x^3/6$.

Example 5.11:

Here we assume some of the properties of the natural logarithm function, including continuity. These will be developed in Chapter 6.

Find $\lim_{x \downarrow 0} x^x$. As this function is written, we cannot apply L'Hôpital's Rule. However, with some manipulations we can use L'Hôpital's Rule. Let

$$y = x^x.$$

Then

$$\ln y = \ln x^x = x \ln x.$$

We want to compute $\lim_{x \downarrow 0} \ln x^x$. Since $\ln x$ is a continuous function, $\lim_{x \downarrow 0} \ln x^x = \ln(\lim_{x \downarrow 0} x^x)$. As $x \downarrow 0$, $x \ln x$ has the indeterminant form $0 \cdot \infty$, but

$$x \ln x = \frac{\ln x}{1/x}$$

which is of the form $-\infty/\infty$ as $x \downarrow 0$. Now we can apply L'Hôpital's rule, so

$$\lim_{x \downarrow 0} \frac{\ln x}{1/x} = \lim_{x \downarrow 0} \frac{1/x}{-1/x^2} = \lim_{x \downarrow 0} (-x) = 0.$$

Thus $\lim_{x \downarrow 0} \ln y = 0$, so

$$\lim_{x \downarrow 0} x^x = \lim_{x \downarrow 0} y = \lim_{x \downarrow 0} e^{\ln y} = e^{\lim_{x \downarrow 0} \ln y} = e^0 = 1.$$

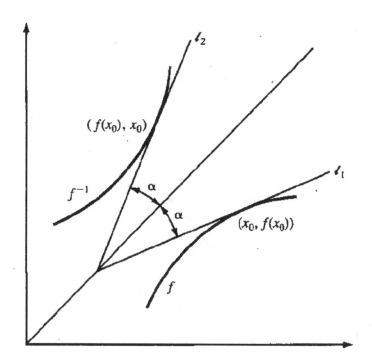

Figure 5.6
See Example 5.11.

Derivative of the Inverse Function

The final topic we address in this chapter is the derivative of the inverse of a function. Suppose f is a function defined on (a, b). If f^{-1} is to exist then f must be a $1 - 1$ function. If $(f^{-1})'$ is to exist, then f^{-1} must be continuous, and thus $(f^{-1})^{-1} = f$ must be continuous. Now, as we noted in Chapter 4, a continuous 1.1 function must be strictly monotone on an interval. Thus we know that for $(f^{-1})'$ to exist, f must be a strictly monotone continuous function.

Before presenting a proof, we sketch a geometric and an analytic argument for the result. Examine the graph in Figure 5.6. Suppose l_1 is the line tangent to the graph of f at x_0 and l_2 is the line tangent to the graph of f^{-1} at $f(x_0)$. Since the graph of f^{-1} is the reflection of the graph of f about the line $y = x$, each line makes an angle α with the line $y = x$ as shown. The slope of l_1 is

$$\tan\left(\frac{\pi}{4} - \alpha\right) = \frac{1}{\tan\left(\frac{\pi}{4} + \alpha\right)}$$

for $0 < \alpha < \pi/4$ so the slopes of l_1 and l_2 are reciprocals of one another.

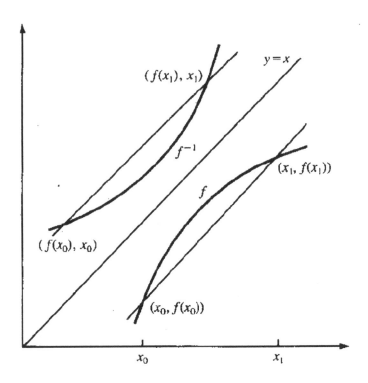

Figure 5.7
Secant lines in the graphs of f and f^{-1}.

The next argument is an analytic one and models the proof that we shall give. Consider Figure 5.7.

One way to interpret the derivative of a function f at a point x_0 is as the limit of the slopes of secant lines between the points $(x_0, f(x_0))$ and $(x_1, f(x_1))$ as x_1 approaches x_0. Now the slope of any such line is $[f(x_1)-f(x_0)]/(x_1-x_0)$. The reflection of the points $(x_0, f(x_0))$ and $(x_1, f(x_1))$ about the line $y = x$ is the points

$$(f(x_0), x_0) = (f(x_0), f^{-1}(f(x_0))) \text{ and } (f(x_1), x_1) = (f(x_1), f^{-1}(f(x_0)))$$

respectively. The slope of the line joining the reflected points is

$$\frac{x_1 - x_0}{f(x_1) - f(x_0)} = \frac{f^{-1}(f(x_1)) - f^{-1}(f(x_0))}{f(x_1) - f(x_0)}$$

which is the inverse of the slope of the secant line joining $(x_0, f(x_0))$ and $(x_1, f(x_1))$. Thus one might expect that if

$$\lim_{x_1 \to x_0} \frac{f(x_1) - f(x_0)}{x_1 - x_0}$$

exists as a finite nonzero number L, then

$$\lim_{x_1 \to x_0} \frac{x_1 - x_0}{f(x_1) - f(x_0)} = \lim_{x_1 \to x_0} \frac{f^{-1}(f(x_1)) - f^{-1}(f(x_0))}{f(x_1) - f(x_0)} = \frac{1}{L}.$$

Theorem 5.12: Suppose f is a strictly monotone continuous function on (a, b). If $f'(x_0)$ exists and is not zero for some $x_0 \in (a, b)$, then f^{-1} is differentiable at $f(x_0)$ and

$$(f^{-1})'(f(x_0)) = \frac{1}{f'(x_0)}.$$

Proof: Let $y_0 = f(x_0)$ and let $\{y_n\}$ be a sequence of points in the domain of f^{-1} (i.e., the range of f) such that $\{y_n\} \to y_0$ and $y_n \neq y_0$ for all n. For each y_n, there is an $x_n \in \mathcal{D}(f)$ with $f(x_n) = y_n$. The number x_n is unique, since f is 1-1. Now $\{x_n\} = \{f^{-1}(y_n)\} \to f^{-1}(y_0) = x_0$ (since f^{-1} is continuous, because f is continuous). Thus

$$\lim_{n \to \infty} \frac{f^{-1}(y_n) - f^{-1}(y_0)}{y_n - y_0} = \lim_{n \to \infty} \frac{x_n - x_0}{f(x_n) - f(x_0)}$$

$$= \lim_{n \to \infty} \frac{1}{\frac{f(x_n) - f(x_0)}{x_n - x_0}} = \frac{1}{f'(x_0)}. \tag{5.6}$$

So

$$\lim_{n \to \infty} \frac{f^{-1}(y) - f^{-1}(y_0)}{y - y_0}$$

exists and by definition is $(f^{-1})'(y_0)$. But from Equation (5.6) we then have

$$(f^{-1})'(y_0) = (f^{-1})'(f(x_0) = 1/f'(x_0). \quad \blacksquare$$

Example 5.12:
Let $f(x) = x^n$ for n a positive integer. Then f is monotone for n odd, and for n even, f is monotonic on the intervals $(-\infty, 0]$ and $[0, \infty)$. We shall consider f on the interval $(0, \infty)$. Hence $f^{-1}(y) = y^{1/n}$. Thus $f'(x) \neq 0$ so $(f^{-1})'$ exists and

$$(f^{-1})'(f(x)) = \frac{1}{f'(x)} = \frac{1}{nx^{n-1}}$$

$$= \frac{1}{n}x^{1-n} = \frac{1}{n}(x^n)^{1/n-1} = \frac{1}{n}(f(x))^{1/n-1}.$$

Thus, with $y = f(x)$,

$$(y^{\frac{1}{n}})' = \frac{1}{n}y^{1/n-1}.$$

Example 5.13:

For $f(x) = x^{m/n}$ with m and n positive integers and $x > 0$, we have

$$f'(x) = (x^{m/n})' = [(x^{1/n})^m]' = m(x^{1/n})^{m-1}\frac{1}{n}x^{1/n-1}$$

by the Chain Rule and Example 5.12. So

$$f'(x) = \frac{m}{n}x^{m/n-1/n+1/n-1} = \frac{m}{n}x^{m/n-1}.$$

Exercises 5.2

1. Give an example where Rolle's Theorrem fails if the following hypothesis is omitted:

   (a) f is continuous on $[a, b]$ and $f(a) = f(b)$, but f is not differentiable on (a, b).

   (b) f is differentiable (a, b) and $f(a) = f(b)$ but f is not continuous on $[a, b]$.

   (c) f is continuous on $[a, b]$ and differentiable on (a, b), but $f(a) \neq f(b)$.

2. (a) Show that if f is differentiable and $|f'(x)| \leq M$, then f is uniformly continuous.

   (b) Give an example of a function that is differentiable and uniformly continuous on $(0, 1)$ but whose derivative is unbounded on $(0, 1)$.

3. Show that if f is continuous on $[a, b]$ and if $f'(x) > 0$ on (a, b), then the minimum value of f on $[a, b]$ occurs at $x = a$ and the maximum value occurs at $x = b$.

4. (a) Suppose f is continuous on $[a, b]$ and differentiable on $(a, b) \setminus \{c\}$ where $c \in (a, b)$.
   (i) Show that if $f'(x) > 0$ for $x < c$ and $f'(x) < 0$ for $x > c$, then f has a relative maximum at $x = c$.
   (ii) Show that if $f'(x) < 0$ for $x < c$ and $f'(x) > 0$ for $x > c$, then f has a relative minimum at $x = c$.

   (b) Find the relative extrema of $f(x) = x^{2/3}(8 - x)^2$ on $[-10, 10]$ and classify them as maxima or minima.

5. (a) Show that $|\sin x - \sin y| \leq |x - y|$. Note that this implies $|\sin x| \leq |x|$.

   (b) Use part (a) to show $\cos x \geq 1 - x^2/2$ for $x > 0$.

   (c) Use part (b) to show $\sin x \geq x - x^3/6$ for $x > 0$.

   (d) Use part (c) to show $\cos x \leq 1 - x^2/2 + x^4/24$ for $x > 0$.

6. (a) Suppose $f'(x)$ exists and is continuous on (a, b) and $c \in (a, b)$. Show that

$$\lim_{h \to 0} \frac{f(c + h) - 2f(c) + f(c - h)}{h^2} = f''(c).$$

(b) Let $f(x) = x|x|$. Show that

$$\lim_{h \to 0} \frac{f(0+h) - 2f(0) + f(0-h)}{h^2} = 0$$

but $f''(0)$ does not exist.

7. (a) Suppose f is a function whose derivative exists for every value of x and suppose that f has n distinct roots. Show that f' has at least $n - 1$ distinct roots.

   (b) Is it possible that f' has more roots than f?

8. Find a bound on the maximum difference of $\sin x$ and $x - x^3/6 + x^5/120$ on $[0, 1]$. *Hint:* In Taylor's Theorem, let $a = 0$ and $x = 1$. What is the maximum value that $[f^{(n+1)}(c)/(n+1)!](x-a)^{n+1}$ can take on?

9. Use Taylor's Theorem to find $\sqrt{8}$ accurate to 0.0001.

10. Use L'Hôpital's Rule and any formulas from elementary calculus you wish, to evaluate the following limits:

    (a)
    $$\lim_{x \to 0} \frac{\cos x - 1 + x^2/2 - x^4/24}{x^6}.$$

    (b) $\lim_{x \to 0} (e^{2x} - x)^{1/x}$.

    (c) $\lim_{x \to \infty} x^{2/x}$.

    (d)
    $$\lim_{x \to 0} \frac{2^x - 1}{3^x - 1}.$$

    (e) $\lim_{x \to \pi/2} \left(\frac{\pi}{2} - x\right) \tan x$.

11. Let $f(x) = (ax + b)/(cx + d)$ with $ad - bc \neq 0$.

    (a) Show that f is a 1-1 function and find $f^{-1}(x)$.

    (b) For what values of a, b, c and d (if any) does $f^{-1} = f$?

12. Let $f(x) = x^7 + x^5 + x^3 + x$. Find $(f^{-1})'(4)$.

13. For $-\pi/2 \leq x \leq \pi/2$, $\sin x$ is a 1-1 function. Define $f(x) = \sin x$ on $[-\pi/2, \pi/2]$. Show that

    $$(f^{-1})'(x) = \frac{1}{\sqrt{1 - x^2}}.$$

14. Suppose one knew that if f is a 1-1 and differentible function with $f'(x) \neq 0$ on (a, b), then f^{-1} is differentiable on its domain. Use the Chain Rule to find a formula for $(f^{-1})'(f(x_0))$.

15. Suppose f is differentiable on (a, b) and $c \in (a, b)$.

(a) Show that
$$\lim_{h \to 0} \frac{f(c + h) - f(c - h)}{2h} = f'(c).$$

(b) Show that
$$f(x) - f(c) = f'(c)(x - c) + g(x)(x - c) \text{ where } \lim_{x \to c} g(x) = 0.$$

16. Suppose f' exists on (a, b) and $c \in (a, b)$. Show that there is a sequence $\{x_n\} \subset (a, b)$, $x_n \neq c$, where $\{f'(x_n)\} \to f'(c)$.

17. Suppose that f' exists and is increasing on $(0, \infty)$ and that f is continuous on $[0, \infty)$ with $f(0) = 0$. Show that $g(x) = [f(x)]/x$ is increasing on $(0, \infty)$.

18. Here we show that the derivative of a function must satisfy the intermediate value property. Suppose f is differentiable on an open interval containing $[a, b]$ and $f'(a) < f'(b)$. Let θ be between $f'(a)$ and $f'(b)$. We shall show that there is a $c \in (a, b)$ where $f'(c) = \theta$.

(a) Define a function g on $[a, b]$ by
$$g(x) = \begin{cases} \frac{f(x) - f(a)}{x - a} & \text{if } x \neq a \\ f'(a) & \text{if } x = a. \end{cases}$$

Show that $g(x)$ attains every value between $f'(a)$ and $[f(b) - f(a)]/(b - a)$.

(b) Define a function h on $[a, b]$ by
$$h(x) = \begin{cases} \frac{f(b) - f(x)}{b - x} & \text{if } x \neq b \\ f'(b) & \text{if } x = b. \end{cases}$$

Show that $h(x)$ attains every value between $f'(b)$ and $[f(b) - f(a)]/(b - a)$.

(c) Let $\gamma = [f(b) - f(a)]/(b - a)$. Our proof will assume $f'(a) < \gamma < f'(b)$, but the idea of the proof is valid for any relationship for γ, $f'(a)$, and $f'(b)$. Choose $\theta \in (f'(a), f'(b))$. Show that if $\theta \in [\gamma, f'(b)]$, there is a number $x_0 \in [a, b]$ where $h(x_0) = \theta$.

(d) Show that if $\theta \in [f'(a), \gamma]$, there is a number $x_0 \in [a, b]$ where $g(x_0) = \theta$.

(e) Use the Mean Value Theorem to show there is a number $c \in [x_0, b]$ where $f'(c) = h(x_0) = \theta$ or a number $c \in [a, x_0]$ where $f'(c) = g(x_0) = \theta$.

19. Show that if $f(x)$ is function whose derivative $f'(x)$ is monotonic, then $f'(x)$ is continuous.

6

Integration

In this chapter we study the theory of integration. In the first section we define the Riemann integral of a bounded function on a closed bounded interval. We shall see that the integral of such a function does not always exist. One of the main results of the first section is the determination of a condition that characterizes Riemann integrable functions. That is, we shall find a condition such that a function is Riemann integrable if and only if it satisfies this condition.

In Section 6.2 we derive some of the properties of the Riemann integral and prove some of the devices of elementary calculus, including the Fundamental Theorem of Calculus and the integration by parts formula.

There are many kinds of integrals. The Riemann integral was historically the first one invented and is perhaps the simplest. In Section 6.3 we study a more general integral, the Riemann-Stieltjes integral. We say this is a more general integral because the Riemann integral is a special case of the Riemann-Stieltjes integral. Also, there are applications where the Riemann-Stieltjes integral provides an acceptable tool and the Riemann integral is inappropriate.

6.1   The Riemann Integral

Throughout this chapter we shall assume that f is a bounded function whose domain includes the interval $[a, b]$ unless otherwise stated. We begin our study of integration with some results about sets of numbers.

Theorem 6.1: If A and B are bounded sets of real numbers with $A \subset B$ then

$$\sup A \leq \sup B \text{ and } \inf A \geq \inf B.$$

The proof is left for Exercise 1, Section 6.1.   ■

Theorem 6.2: (a) Suppose A and B are nonempty sets of real numbers such that if $x \in A$ and $y \in B$, then $x \leq y$. Then $\sup A$ and $\inf B$ are finite and $\sup A \leq \inf B$.

(b) Suppose that A and B are as in part (a). Then $\sup A = \inf B$ if and only if, for any $\epsilon > 0$, there exist an $x(\epsilon) \in A$ and $y(\epsilon) \in B$ such that $y(\epsilon) - x(\epsilon) < \epsilon$.

Proof:

(a) We first observe that for any $x \in A$, x is a lower bound for the set B, since if $y \in B$, then $x \leq y$. Thus $x \leq \inf B$ for any $x \in A$ but this means that $\inf B$ is an upper bound for A, and thus $\sup A \leq \inf B$.

(b) To show $\sup A = \inf B$, we need to show that for any $\epsilon > 0$,

$$\inf B - \sup A < \epsilon$$

since $\inf B \geq \sup A$ by part (a).

Let $\epsilon > 0$ be given. Suppose there are numbers $x(\epsilon) \in A$ and $y(\epsilon) \in B$ such that $y(\epsilon) - x(\epsilon) < \epsilon$. But $y(\epsilon) \geq \inf B$ and $x(\epsilon) \leq \sup A$, so $\inf B - \sup A \leq y(\epsilon) - x(\epsilon) < \epsilon$.

Conversely, suppose $\inf B = \sup A$. Let $\epsilon > 0$ be given. There is a number $x(\epsilon) \in A$ with $\sup A \geq x(\epsilon) > \sup A - \epsilon/2$, and there is a number $y(\epsilon) \in B$ with $\inf(B) \leq y(\epsilon) < \inf(B) + \epsilon/2$. Then, if $\alpha = \inf B = \sup(A)$, we have

$$\alpha \geq x(\epsilon) > \alpha - \epsilon/2, \text{ and } \alpha \leq y(\epsilon) < \alpha + \epsilon/2$$

so $0 < y(\epsilon) - x(\epsilon) < (\alpha + \epsilon/2) - (\alpha - \epsilon/2) = \epsilon.$ ∎

Definition: A *partition* $P = \{x_0, x_1, ..., x_n\}$ of the interval $[a, b]$ is a finite set of numbers $x_0, x_1, ..., x_n$ such that

$$a = x_0 < x_1 < \cdots < x_n = b.$$

Definition: Let P and Q be partitions of $[a, b]$. We say that Q is a refinement of P if $P \subset Q$.

We use partitions to divide the interval $[a, b]$ into closed subintervals, the subintervals having the points of the partition as their endpoints. If Q is a refinement of P, then the subintervals determined by Q are smaller than the subintervals by P in the sense that any subinterval determined by P is the union of subintervals determined by Q.

Example 6.1

Let $[a, b] = [0, 2]$. Let $P = \{0, 1, 2\}$ and $Q = \{0, 1/2, 1, 2\}$. Then Q is a refinement of P. The subintervals determined by P are $[0, 1]$ and $[1, 2]$. The subintervals determined by Q are $[0, 1/2], [1/2, 1]$, and $[1, 2]$. Note that $[0, 1] = [0, 1/2] \cup [1/2, 1]$.

We now develop some notation. Let $P = \{x_0, x_1, ..., x_n\}$ be a partition of $[a, b]$. For each subinterval $[x_{i-1}, x_i]$ determined by P, let

$$m_i(f) = \inf\{f(x) | x \in [x_{i-1}, x_i]\}$$

$$M_i(f) = \sup\{f(x)|x \in [x_{i-1}, x_i]\}$$

and

$$\Delta x = x_i - x_{i-1}, i = 1, ..., n.$$

Since we are assuming f is bounded on $[a, b]$, $m_i(f)$ and $M_i(f)$ are finite. Also $\Delta x_i > 0, i = 1, ..., n$. Let

$$\overline{S}(f; P) = \sum_{i=1}^{n} M_i(f)\Delta x_i \text{ and } \underline{S}(f; P) = \sum_{i=1}^{n} m_i(f)\Delta x_i.$$

Definition: With the preceding notation, $\overline{S}(f; P)$ and $\underline{S}(f; P)$ are called the *upper Riemann sum* and *lower Riemann sum*, respectively, of f on $[a, b]$ with respect to the partition P.

Definition: Let $P = \{x_0, x_1, ..., x_n\}$ be a partition of $[a, b]$. For each subinterval $[x_{i-1}, x_i]$, choose $\overline{x}_i \in [x_{i-1}, x_i]$. Then

$$\sum_{i=1}^{n} f(\overline{x}_i)\Delta x_i = S(f; P)$$

is called a *Riemann sum* of f on $[a, b]$ with respect to the partition P.

Since $m_i(f) \leq f(\overline{x}_i) \leq M_i(f)$ and $\Delta x_i > 0$ for $i - 1, ..., n$, then

$$\underline{S}(f; P) \leq \sum f(\overline{x}_i)\Delta x_i \leq \overline{S}(f; P).$$

Our first major task is to show that if P and Q are partitions of $[a, b]$ then $\underline{S}(f; P) \leq \overline{S}(f; Q)$. We can then use Theorem 6.2 to conclude

$$\sup S(f; P)|P \text{ is a partition of } [a, b]\}$$

$$\leq \inf \overline{S}(f; P)|P \text{ is a partition of } [a, b]\}.$$

Theorem 6.3:

(a) Suppose P and Q are partitions of $[a, b]$, and Q is a refinement of P. Then

$$\underline{S}(f; P) \leq \underline{S}(f; Q) \text{ and } \overline{S}(f; Q) \leq \overline{S}(f; P).$$

(b) If P and Q are any partitions of $[a, b]$, then

$$\underline{S}(f; P) \leq \overline{S}(f; Q).$$

(c) Let

$$\underline{S}(f) = \sup\{\underline{S}(f;P)|P \text{ is a partition of } [a,b]\}$$

and

$$\overline{S}(f) = \inf\{\overline{S}(f;P)|P \text{ is a partition of } [a,b]\}.$$

Then $\underline{S}(f)$ and $\overline{S}(f)$ are finite and $\underline{S}(f) \le \overline{S}(f)$.

Proof:

(a) We show that $\underline{S}(f,P) \le \underline{S}(f;Q)$ and leave the proof that $\overline{S}(f;P) \ge \overline{S}(f;Q)$ for Exercise 2, Section 6.1. Let $P = \{x_0, x_1, ..., x_n\}$ be a partition of $[a,b]$ and let P_1 be a refinement of P obtained by adding exactly one point to P. Suppose

$$P_1 = \{x_0, x_1, ..., x_{j-1}, x', x_j, ..., x_n\}.$$

Then

$$\underline{S}(f;P) = \sum_{i=1}^{n} m_i(f)\Delta x_i = \sum_{i=1}^{j-1} m_i(f)\Delta x_i + m_j(f)\Delta x_j + \sum_{i=j+1}^{n} m_i(f)\Delta x_i$$

and

$$\underline{S}(f;P_1) = \sum_{i=1}^{j-1} m_i(f)\Delta x_i + m_j'(f)(x' - x_{j-1})$$

$$+ m_j''(f)(x_j - x') + \sum_{i=j+1}^{n} m_i(f)\Delta x_i$$

where

$$m_j'(f) = \inf\{f(x)|x \in [x_{j-1}, x']\}$$

and

$$m_j''(f) = \inf\{f(x)|x \in [x', x_j]\}.$$

Thus, to compare $\underline{S}(f;P)$ and $\underline{S}(f;P_1)$, we only need to compare $m_j(f)\Delta x_j$ with $m_j'(f)(x' - x_{j-1}) + m_j''(f)(x_j - x')$.

By Theorem 6.1

$$m_j'(f) \ge m_j(f) \text{ and } m_j''(f) \ge m_j(f)$$

so

$$m_j(f)\Delta x_j = m_j(f)(x_j - x_{j-1})$$
$$= m_j(f)(x_j - x' + x' - x_{j-1}$$
$$= m_j(f)(x_j - x') + m_j(f)(x' - x_{j-1})$$
$$\le m_j''(f)(x_j - x') + m_j'(f)(x' - x_{j-1}).$$

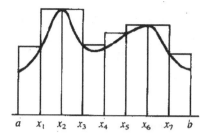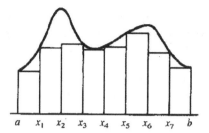

Figure 6.1
Riemann sums

Thus
$$\underline{S}(f;P) \leq \underline{S}(f;P_1).$$

Now let Q be any refinement of P. There is a set of partitions of $[a,b], P_1, ..., P_k$ such that

$$P \subset P_1 \subset \cdots \subset P_k \subset Q$$

and each partition is obtained from the preceding one by adding exactly one point. Then

$$\underline{S}(f;P) \leq S(f;P_1) \leq \cdots \leq \underline{S}(f;P_k) \leq \underline{S}(f;Q).$$

(b) Let P and Q be partitions of $[a,b]$. The $P \cup Q$ is a refinement of both P and Q. By part (a)

$$\underline{S}(f;P) \leq \underline{S}(f;P \cup Q) \text{ and } \overline{S}(f;P \cup Q) \leq \overline{S}(f;Q).$$

But $\underline{S}(f;P \cup Q) \leq \overline{S}(f;P \cup Q)$, and so the result follows.
(c) Let

$$m = \inf\{f(x)|x \in [a,b]\} \text{ and } M = \sup\{f(x)|x \in [a,b]\}.$$

Since f is bounded, m and M are finite. Then for any partition P

$$m(b-a) \leq \underline{S}(f;P) \leq S(f;P) \leq \overline{S}(f;P) \leq M(b-a).$$

Thus $\underline{S}(f)$ and $\overline{S}(f)$ are finite. By Theorem 6.2,

$$\underline{S}(f) \leq \overline{S}(f). \quad \blacksquare$$

If $f(x) \geq 0$ for $x \in [a,b]$, the upper Riemann sum is the area of circumscribed rectangles on the graph of f, and the lower Riemann sum is the area of inscribed rectangles on the graph of f. (See Figure 6.1.) One intuitively believes the area under f is smaller than any upper Riemann sum and larger than any lower Riemann sum. If it is possible to choose a partition that makes the difference between the inscribed rectangles and the circumscribed rectangles arbitrarily small, then $\overline{S}(f) = \underline{S}(f)$ and this common value is a reasonable way to define the area under $y = f(x)$ between $x = a$ and $x = b$. The next example shows that it is not always possible to do this.

Example 6.2:
Let f be defined on $[0, 1]$ by

$$f(x) = \begin{cases} 1 & \text{if } x \text{ is rational} \\ 0 & \text{if } x \text{ is irrational.} \end{cases}$$

We show $\underline{S}(f) = 0$ and $\overline{S}(f) = 1$.
Let $P = \{x_0, x_1, ..., x_n\}$ be a partition of $[0, 1]$. Any interval $[x_{i-1}, x_i]$ contains both rational and irrational points, so

$$m_i(f) = 0 \text{ and } M_i(f) = 1, \quad i = 1, ..., n.$$

Thus

$$\underline{S}(f; P) = \sum_{i=1}^{n} m_i(f)\Delta x_i = \sum_{i=1}^{n} 0\,\Delta x_i = 0$$

and

$$\overline{S}(f; P) = \sum_{i=1}^{n} M_i(f)\Delta x_i = \sum_{i=1}^{n} 1\,\Delta x_i = 1$$

since the sum of the subintervals, $\sum \Delta x_i$, is equal to the length of the interval $[0, 1]$. Thus for any partition P, $\underline{S}(f; P) = 0$ and $\overline{S}(f; P) = 1$ so that $\underline{S}(f) = 0$ and $\overline{S}(f) = 1$.

Definition: Let f be a bounded function on $[a, b]$. We say that f is Riemann integrable on $[a, b]$ if $\underline{S}(f) = \overline{S}(f)$ in the notation of Theorem 6.3(c). If this is the case, then the Riemann integral of f on $[a, b]$ is the common value of $\underline{S}(f)$ and $\overline{S}(f)$, which is denoted $\int_a^b f(x)\,dx$ or $\int_a^b f$.

Because the infimum and supremum of a nonempty set of numbers are unique, the Riemann integral of a Riemann integrable function is unique.
The next theorem will be an extremely useful tool for proving the existence of the Riemann integral of functions.

Theorem 6.4 (Riemann Condition for Integrability): A bounded function f defined on $[a, b]$ is Riemann integrable on $[a, b]$ if and only if, given $\epsilon > 0$, there is a partition $P(\epsilon)$ of $[a, b]$ such that

$$\overline{S}(f; P(\epsilon)) - \underline{S}(f; P(\epsilon)) < \epsilon.$$

The proof is an application of Theorem 6.2 and is left for Exercise 6, Section 6.1. ∎

The next theorem is sometimes useful in proving certain properties of the Riemann integral.

Theorem 6.5: Suppose that f is a Riemann integrable function on $[a,b]$. If I is a number such that

$$\underline{S}(f;P) \leq I \leq \overline{S}(f;P)$$

for *every* partition P, then $I = \int_a^b f$.

The proof is left for Exercise 7, Section 6.1. ∎

Definition: Let $P = \{x_0, x_1, .., x_n\}$ be a partition of $[a,b]$. The *norm* (or *mesh*) of P, denoted $||P||$ is $\max\{x_i - x_{i-1}|, i = 1, ..., n\}$.

Thus the norm of P is the length of the largest subinterval defined by P. For example, if $P = \{1, 3/2, 2, 3\}$ is a partition of $[1,3]$ then $||P|| = 1$.

The next theorem say that a function is Riemann integrable if and only if, given $\epsilon > 0$, there is a number $\delta(\epsilon) > 0$ such that $\overline{S}(f;P) - \underline{S}(f,P) < \epsilon$ for every partition P with $||P|| < \delta(\epsilon)$. This characteristic distinguishes the Riemann integral from the other integral we shall study, the Riemann-Stieltjes integral. It says that the particular points chosen for the partition are inconsequential, and the important property (as far as Riemann integrability is concerned) is the maximum length of the intervals that the partition determines.

Theorem 6.6: A bounded function f is Riemann integrable on $[a,b]$ if and only if, given $\epsilon > 0$ there is a $\delta(\epsilon) > 0$ such that if P is any partition with $||P|| < \delta(\epsilon)$ then

$$\overline{S}(f;P) - \underline{S}(f;P) < \epsilon.$$

Proof: We first suppose that f is Riemann integrable and show that the condition holds. To do this, we apply the Riemann condition for integrability. Let $\epsilon > 0$ be given and let $P(\epsilon) = \{x_0, x_1, ..., x_N\}$ be a partition with $\overline{S}(f; P(\epsilon)) - \underline{S}(f; P(\epsilon)) < \epsilon/2$. We shall show that there is a number $\delta(\epsilon)$ such that if P is any partition with $||P|| < \delta(\epsilon)$ then $\overline{S}(f;P) - \underline{S}(f;P) < \epsilon$.

Let

$$M = \sup\{f(x)|x \in [a,b]\} \text{ and } m = \inf\{f(x)|x \in [a,b]\}.$$

If f is not constant on $[a,b]$, then $M - m > 0$. We shall assume $M - m > 0$. Let

$$\delta(\epsilon) = \frac{\epsilon}{2(M-m)N}.$$

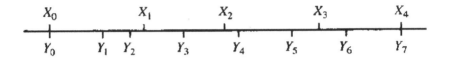

Figure 6.2
Two partitions of an interval

(Remember $P(\epsilon)$ has $N + 1$ points and so determines N subintervals of $[a, b]$.) Suppose P is any partition with $||P|| < \delta(\epsilon)$. Let $P = \{y_0, y_1, ..., y_s\}$. Now some of the subintervals determined by P may overlap parts of more than one of the subintervals determined by $P(\epsilon)$, and some of the subintervals determined by P may be contained entirely in the subintervals contained by $P(\epsilon)$. Figure 6.2 may clarify this.

We split the subintervals determined by P into two disjoint groups. Let G_1 consist of those subintervals determined by P that are contained in one of the subintervals of $P(\epsilon)$ and let G_2 consist of the other intervals of P. In Figure 6.2,

$$G_1 = \{[Y_0, Y_1], [Y_1, Y_2], [Y_4, Y_5], [Y_6, Y_7]\}$$

$$G_2 = \{[Y_2, Y_3], [Y_3, Y_4], [Y_5, Y_6]\}.$$

Then

$$\overline{S}(f; P) - \underline{S}(f; P) = \sum_{G_1}[M_j(f) - m_j(f)]\Delta y_j + \sum_{G_2}[M_j(f) - m_j(f)]\Delta y_j.$$

Consider the first summand. Suppose the subintervals $[y_{j-1}, y_j], ..., [y_{k-1}, y_k]$ are all contained in the subinterval $[x_{h-1}, x_h]$. Then

$$M_j(f), ..., M_k(f) \leq M_h(f)$$

and

$$m_j(f), ..., m_k(f) \geq m_h(f).$$

Also

$$\sum_{i=j}^{k} \Delta y_i \leq \Delta x_h$$

so that

$$\sum_{i=j}^{k}[M_i(f) - m_i(f)]\Delta y_i \leq [M_h(f) - m_h(f)]\sum_{i=j}^{k} \Delta y_i$$

$$\leq [M_h(f) - m_h(f)]\Delta x_h.$$

Thus

$$\sum_{G_1}[M_i(f) - m_i(f)]\Delta y_i \le \sum_{i=1}^{N}[M_i(f) - m_i(f)]\Delta x_i$$

$$= \overline{S}(f; P(\epsilon)) - \underline{S}(f; P(\epsilon)) < \frac{\epsilon}{2}.$$

Now consider the $\sum_{G_2}$ summand. There can be no more than $(N-1)$ terms in this summand, since each subinterval in G_2 contains parts of more than one subinterval of those subintervals that $P(\epsilon)$ determines, and $P(\epsilon)$ determines N subintervals. Thus

$$\sum_{G_2}[M_j(f) - m_j(f)]\Delta y_j \le \sum_{G_2}(M-m)\Delta y_j = (M-m)\sum_{G_2}\Delta y_j$$

$$< (M-m)N\frac{\epsilon}{2(M-m)N} = \frac{\epsilon}{2}.$$

The last inequality holds because $\sum_{G_2}\Delta y_j$ consists of no more than N terms, each of which is smaller than $\epsilon/[2(M-m)N]$. Thus

$$\overline{S}(f; P) - \underline{S}(f; P) < \frac{\epsilon}{2} + \frac{\epsilon}{2} = \epsilon \text{ if } ||P|| < \delta(\epsilon) = \frac{\epsilon}{2(M-m)N}.$$

Conversely, given $\epsilon > 0$, suppose there is a $\delta(\epsilon) > 0$ such that if $||P|| < \delta(\epsilon)$, then $\overline{S}(f; P) - \underline{S}(f; P) < \epsilon$. Then for our partition $P(\epsilon)$ in the Riemann condition for integrability, we can take any partition whose norm is less than $\delta(\epsilon)$. ∎

So far we haven't determined which functions are Riemann integrable. An exact classification of Riemann integrable functions will be somewhat involved. However, part of the answer is easily obtained. We do that now.

Theorem 6.7: If f is a continuous real-valued function on $[a, b]$, then f is Riemann integrable.

Proof: Since f is continuous on a closed bounded interval, it is bounded and *uniformly* continuous. The *uniform* continuity means that, given $\epsilon > 0$ there is a number $\delta(\epsilon) > 0$ such that if $x, y \in [a, b]$ and $|x - y| < \delta(\epsilon)$, then

$$|f(x) - f(y)| < \frac{\epsilon}{b - a}.$$

Let $\epsilon > 0$ be given, and let $P = \{x_0, x_1, ..., x_n\}$ be a partition of $[a, b]$ with $||P|| < \delta(\epsilon)$. On every subinterval $[x_{i-1}, x_1]$, f assumes its maximum value at some point, say x_i', and its minimum value at some point, say x_i''. This is because a continuous function on a closed, bounded interval assumes its maximum and minimum values on that interval. Thus

$$\overline{S}(P; f) - \underline{S}(P; f) = \sum[f(x_i' - f(x_i'')]\Delta x_i$$

$$< \frac{\epsilon}{b-a} \sum \Delta x_i = \frac{\epsilon}{b-a}(b-a) = \epsilon.$$

The first inequality follows because if $||P|| < \delta(\epsilon)$, then $|x_i' - x_i''| < \delta(\epsilon)$ if they are in the same subinterval, so that

$$f(x_i') - f(x_i'') < \frac{\epsilon}{b-a}.$$

Thus, by the Riemann condition, f is Riemann integrable. ∎

Now we determine a condition that characterizes Riemann integrable functions. To do this, we have to define the (Lebesgue) measure of a set; at least we have to define what it means for a set to have measure zero.

Definition: The (Lebesgue) *measure* of an open interval (a, b) is $b - a$. The measure of $(-\infty, a), (a, \infty)$, or $(-\infty, \infty)$ is infinite. We denote the measure of an open interval I by $m(I)$.

Clearly, the measure of an open interval is the length of the interval.

Definition: A set E has measure zero if, for any $\epsilon > 0$, there is a countable collection of open intervals $\{I_1, I_2, \ldots\}$ such that

$$E \subset \cup I_i \text{ and } \sum_{i=1}^{\infty} m(I_i) < \epsilon.$$

Example 6.3:
We show that a finite set has measure zero. Let $\{x_1, \ldots, x_N\}$ be a finite set and let $\epsilon > 0$ be given. Then

$$\left\{ (x_1 - \frac{\epsilon}{4N}, x_1 + \frac{\epsilon}{4N}), \ldots, (x_N - \frac{\epsilon}{4N}, x_N + \frac{\epsilon}{4N}) \right\}$$

is an open cover of $\{x_1, .., x_N\}$. There are N intervals each of measure $\frac{\epsilon}{2N}$ so that the open cover has measure $N \frac{\epsilon}{2N} = \epsilon/2$.

Theorem 6.8: A subset of a set of measure 0 has measure 0.
We leave the proof for Exercise 9, Section 6.1. ∎

Theorem 6.9: The union of a countable collection of sets of measure zero is a set of measure zero.
We defer the proof to Exercise 11, Section 7.2, since the proof will require a discussion of double series. ∎

Corollary 6.9: A countable set has measure zero. ■

One might wonder whether there are uncountable sets of measure zero. There are, the most famous of which is the Cantor set that we will study in a subsequent section.

This is all the measure theory we shall need to prove the culminating result of this section. Before doing this we need one more preliminary result.

Definition: The *oscillation* of a function f on a set A is

$$\sup\{|f(x) - f(y)| \mid x, y \in A \cap \mathcal{D}(f)\}.$$

The oscillation of f at X is

$$\lim_{h \downarrow 0}(\sup\{|f(x') - f(x'')| \mid x', x'' \in (x - h, x + h) \cap \mathcal{D}(f)\}.$$

We denote the oscillation of f at x by $\operatorname{osc}(f; x)$.

Example 6.4:
Let
$$f(x) = \begin{cases} 1 & \text{if } x \text{ is irrational} \\ -1 & \text{if } x = 0 \\ 0 & \text{if } x \text{ is rational and nonzero.} \end{cases}$$

Then $\operatorname{osc}(f; 0) = 2$ since any interval containing 0 contains irrational numbers. If $x \neq 0$ then $\operatorname{osc}(f; x) = 1$.

Theorem 6.10: A function f is continuous at $x \in \mathcal{D}(f)$ if and only if $\operatorname{osc}(f; x) = 0$.

Proof: Suppose $\operatorname{osc}(f; x) = 0$. We show that f is continuous at x. Let $\epsilon > 0$ be given. Since

$$\lim_{h \downarrow 0}(\sup\{|f(x') - f(x'')| \mid x', x'' \in (x - h, x + h) \cap \mathcal{D}(f)\}) = 0$$

there is a $\delta > 0$ such that if $0 < h < \delta$, then

$$\sup\{|f(x') - f(x'')| \mid x', x'' \in (x - h, x + h) \cap \mathcal{D}(f)\} < \epsilon.$$

Thus, if $|x - y| < \delta$ and $y \in \mathcal{D}(f)$, then $|f(x) - f(y)| < \epsilon$, so f is continuous at x.

Conversely, suppose f is continuous at x. We show $\operatorname{osc}(f; x) = 0$ by showing that for any $\epsilon > 0$, there is a $\delta > 0$ such that

$$\sup\{|f(x') - f(x'')| \mid x', x'' \in (x - h, x + h) \cap \mathcal{D}(f)\} < \epsilon.$$

Since f is continuous at x, there is a $\delta > 0$ such that if $|x - y| < \delta$ and $y \in \mathcal{D}(f)$ then $|f(x) - f(y)| < \epsilon/2$. Thus if $x', x'' \in (x - \delta, x + \delta) \cap \mathcal{D}(f)$, then

$$|f(x') - f(x'')| \leq |f(x') - f(x)| + |f(x) - f(x'')| < \epsilon/2 + \epsilon/2 = \epsilon. \quad \blacksquare$$

We now state and prove the Riemann-Lebesgue Theorem. The proof of the Riemann-Lebesgue Theorem is substantially more difficult than any we have encountered thus far. However, one does not have to understand the proof of a theorem in order to understand the statement of the theorem and be able to apply it. Keep this fact in mind if the following proof seems overwhelming.

Theorem 6.11 (Riemann-Lebesgue Theorem): A bounded function f defined on $[a, b]$ is Riemann integrable on $[a, b]$ if and only if the set of discontinuities of f on $[a, b]$ has measure zero.

Proof: Suppose f is a bounded function defined on $[a, b]$ that is continuous except on a set of measure zero. We show that f is Riemann integrable. Let

$$M = \sup\{f(x) | x \in [a, b]\} \text{ and } m = \inf\{f(x) | x \in [a, b]\}.$$

(Assume $M - m > 0$, which is the case unless f is constant.) Let

$$A = \{x \in [a, b] | f \text{ is not continuous at } x\}.$$

Recall that $x \in A$ if and only if $\operatorname{osc}(f; x) > 0$. Thus, for any $s > 0$, the set

$$A_s = \{x \in [a, b] | \operatorname{osc}(f; x) \geq s\}$$

is a subset of A. Since A has measure zero, so does A_s.

Let $\epsilon > 0$ be given. There is a countable collection of open intervals $\{I_1, I_2, \dots\}$ such that

$$A_{\epsilon/(2(b-a))} \subset \bigcup_{i=1}^{\infty} I_i \text{ and } \sum_{i=1}^{\infty} m(I_i) < \frac{\epsilon}{2(M - m)}.$$

In Exercise 8, Section 6.1, we show that $A_{\epsilon/(2(b-a))}$ is compact. Thus there is a finite subcover of $A_{\epsilon/(2(b-a))}$ by $\{I_1, I_2, \dots\}$, call it $\{I_1, \dots, I_N\}$.

Now for any

$$x \in [a, b] \setminus \left(\bigcup_{i=1}^{N} I_i \right) \subset [a, b] \setminus A_{\epsilon/(2(b-a))}, \; \operatorname{osc}(f; x) < \frac{\epsilon}{2(b - a)}.$$

Thus for

$$x \in [a, b] \setminus \left(\bigcup_{i=1}^{N} I_i \right)$$

there is an interval $(x - \delta_x, x + \delta_x)$ such that if $x', x'' \in (x - \delta_x, x + \delta_x)$, then

$$|f(x') - f(x'')| < \frac{\epsilon}{2(b - a)}.$$

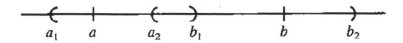

Figure 6.3
Overlapping intervals in an open cover of $[a, b]$.

The set

$$[a, b] \setminus \left(\bigcup_{i=1}^{N} I_i \right)$$

is closed (since $[a, b]$ is closed and $\cup_{i=1}^{N} I_i$ is open) and bounded and so is compact. Thus from the open cover

$$\left\{ (x - \delta_x, x + \delta_x) | x \in [a, b] \setminus \left(\bigcup_{i=1}^{N} I_i \right) \right\}$$

we may extract a finite subcover, call it

$$\{ (\overline{x}_1 - \delta_1, \overline{x}_1 + \delta_1), \ldots, (\overline{x}_k - \delta_k, \overline{x}_k + \delta_k) \}.$$

Now we are ready to construct our partition P of $[a, b]$ for which $\overline{S}(f; P) - \underline{S}(f; P) < \epsilon$. Notice that we have a *finite* subcover of $[a, b]$ by the open intervals

$$(\overline{x}_1 - \delta_1, \overline{x}_1 + \delta_1), \ldots, (\overline{x}_k - \delta_k, \overline{x}_k + \delta_k), I_1, \ldots, I_N.$$

Let $P = \{x_0, x_1, \ldots, x_j\}$ be any partition of $[a, b]$ such that any subinterval $[x_{i-1}, x_i]$ is contained in one of the intervals in the open subcover. We illustrate how this can be done for two subintervals. Consider (a_1, b_1) and (a_2, b_2) as shown in Figure 6.3.

Then

$$\left[a, \frac{a + a_2}{2} \right], \quad \left[\frac{a + a_2}{2}, \frac{a_2 + b_1}{2} \right], \quad \left[\frac{a_2 + b_1}{2}, b \right]$$

are each contained in either (a_1, b_1) or (a_2, b_2).

Now group the subintervals determined by P into two classes. Let c_1 consist of those subintervals contained in some I_i, and let c_2 consist of the remaining subintervals. Then

$$\overline{S}(f; P) - \underline{S}(f; P) = \sum_{c_1} [M_i(f) - m_i(f)] \Delta x_i + \sum_{c_2} [M_i(f) - m_i(f)] \Delta x_i$$

where $\sum_{c_i}$ indicates we are summing over those subintervals in c_i. Consider the first summand:

$$\sum_{c_1} [M_i(f) - m_i(f)] \Delta x_i \leq (M - m) \sum_{c_1} \Delta x_i \leq (M - m) \sum_{i=1}^{N} m(I_i)$$

$$< (M - m)\frac{\epsilon}{2(M - m)} = \frac{\epsilon}{2}.$$

For the second summand, if $[x_{i-1}, x_i] \subset (\bar{x}_r - \delta_r, \bar{x}_r + \delta_r)$, then

$$|f(x') - f(x'')| < \frac{\epsilon}{2(b - a)} \text{ for any } x', x'' \in [x_{i-1}, x_i].$$

Then

$$\sum_{c_2}[M_i(f) - m_i(f)]\Delta x_i \leq \frac{\epsilon}{2(b - a)}\sum_{c_2}\Delta x_i \leq \frac{\epsilon}{2(b - a)}(b - a) = \frac{\epsilon}{2}.$$

Thus $\overline{S}(f; P) - \underline{S}(f; P) < \epsilon$, so f is Riemann integrable on $[a, b]$.

Conversely, suppose f is a bounded function on $[a, b]$, which is Riemann integrable. We shall show that the set of points at which f is not continuous on $[a, b]$ has measure zero. Let

$$A = \{x \in [a, b] | f \text{ is not continuous at } x\}$$

$$= \{x \in [a.b] \mid \text{osc}(f; x) > 0\}.$$

For each positive integer k, let

$$A_k = \left\{x \in [a, b] \mid \text{osc}(f; x) \geq \frac{1}{k}\right\}.$$

Then $A = \cup_{k=1}^{\infty} A_k$, so the measure of A will be zero if the measure of each A_k is zero.

Let $\epsilon > 0$ be given, and suppose that k is a fixed positive integer. Since f is Riemann integrable on $[a, b]$, there is a partition $P = \{x_0, x_1, \ldots, x_n\}$ of $[a, b]$ for which

$$\overline{S}(f; P) - \underline{S}(f; P) < \frac{\epsilon}{2k}.$$

Consider the interval (x_{i-1}, x_i). If $x \in A_k$ and $x \in (x_{i-1}, x_i)$, then

$$[M_i(f) - m_i(f)]\Delta x_i \geq \frac{1}{k}\Delta x_i.$$

Now

$$\frac{\epsilon}{2k} > \overline{S}(f; P) - \underline{S}(f; P) \geq \Sigma'[M_i(f) - m_i(f)]\Delta x_i$$

where Σ' indicates we are summing only over those subintervals where $(x_{i-1}, x_i) \cap A_k \neq \emptyset$. Now

$$\Sigma'[M_i(f) - m_i(f)]\Delta x_i \geq \frac{1}{k}(\Sigma'\Delta x_i)$$

so that

$$\frac{\epsilon}{2} > \Sigma'\Delta x_i.$$

Thus $A_k \setminus \{x_0, x_1, \ldots, x_n\}$ is contained in a collection of open intervals whose measure is less than $\epsilon/2$. Since $\{x_0, x_1, \ldots, x_n\}$ can be contained in open intervals, the sum of whose lengths is less than $\epsilon/2$, it follows that $m(A_k) = 0$ for every k. Thus

$$m(A) = m\left(\cup_{k=1}^{\infty} A_k\right) = 0. \quad \blacksquare$$

While the proof of the Riemann-Lebesgue Theorem is difficult, the importance of the result cannot be overstated. We shall use it often in the next section. We can use it immediately to obtain some important facts.

Corollary 6.11:

(a) A bounded function that has at most countably many points of discontinuity is Riemann integrable.

(b) A monotone function is Riemann integrable.

(c) The sum or product of Riemann integrable functions is Riemann integrable. $\quad \blacksquare$

Exercises 6.1

1. Show that if A and B are nonempty bounded sets of real numbers with $A \subset B$ then $\sup A \le \sup B$ and $\inf A \ge \inf B$.

2. Show that if P and Q are partitions of $[a, b]$ such that Q is a refinement of P, then $\overline{S}(f; P) \ge \overline{S}(f; Q)$.

3. Let $f(x) = x^2 - x$, and let $P = \{0, 1/2, 1, 3/2, 2\}$. Form the upper and lower Riemann sums of f with respect to the partition P.

4. Use the definition of Riemann integral to evaluate $\int_1^2 f(x)\, dx$ where
   (a) $f(x) = x^2$. \qquad (b) $f(x) = 3x - 1$.

5. Let f be a function that is continuous on $[0, 1]$.
   (a) Show that

$$\lim_{n \to \infty} \frac{1}{n} \sum_{k=1}^{n} f\left(\frac{k}{n}\right) = \int_0^1 f(x)\, dx.$$

   (b) Find $\lim_{n \to \infty} \frac{1}{n} \sum_{k=1}^{n} \frac{k}{n}$.

6. Prove Theorem 6.4.

7. Prove Theorem 6.5.

8. Let f be a function with $\mathcal{D}(f) = [a, b]$. Show that for any positive number S, $A_S = \{x \in [a, b] \mid \operatorname{osc}(f; x) \ge S\}$ is compact.

9. Prove that a subset of a set of measure zero has measure zero.

10. Prove that a monotonic function on $[a, b]$ is Riemann integrable.

11. **(a)** Show that the union of two sets, each of measure zero, has measure zero.
    (b) Prove that if f and g are Riemann integrable on $[a, b]$, then $f \cdot g$ and $f + g$ are Riemann integrable on $[a, b]$.

12. Let
$$f(x) = \begin{cases} 0 & \text{if } x \text{ is irrational or } 0 \\ 1/n & \text{if } x = m/n \text{ in lowest terms, } n > 0. \end{cases}$$

    (a) Show that f is Riemann integrable on $[0, 1]$.
    (b) Find $\int_0^1 f(x)\, dx$.

13. Show that a bounded function f is Riemann integrable on $[a, b]$ if and only if there is a number I such that, given $\epsilon > 0$, there is a partition P_ϵ such that for any Riemann sum $S(f; P_\epsilon)$ of f with respect to P_ϵ, $|S(f; P_\epsilon) - I| < \epsilon$.

14. Give an example where the function f is not Riemann integrable, but $|f|$ is Riemann integrable.

15. **(a)** Give an example of a set of measure zero that is not compact.
    (b) Give an example of a set A that has measure zero and such that $\overline{A} = \mathbb{R}$.

16. Let
$$f(x) = \begin{cases} \sin(1/x) & \text{if } x \neq 0 \\ 0 & \text{if } x = 0. \end{cases}$$

    (a) Find $\mathrm{osc}(f; 0)$.
    (b) Find $\mathrm{osc}(f; x)$ where x is any nonzero number.
    (c) Find the oscillation of f on the set $[2/\pi, 4/\pi]$.

6.2   Some Properties and Applications of the Riemann Integral

Properties of the Riemann integral

Most of our effort in this section will be directed toward deriving the properties of the Riemann integral that one uses in elementary calculus. The first theorems that we prove establish the linearity properties of Riemann integrable functions. That is

$$\int_a^b (\alpha f + \beta g)\, dx = \alpha \int_a^b f\, dx + \beta \int_a^b g\, dx$$

for any real numbers α and β and any functions f and g that are Riemann integrable on $[a, b]$. The following result was proven earlier and will facilitate our work.

Theorem 6.12: Let f and g be bounded functions on a set A. Then

(a) $\sup\{(f + g)(x)|x \in A\} \le \sup\{f(x)|x \in A\} + \sup\{(g)(x)|x \in A\}.$

$$\inf\{(f + g)(x)|x \in A\} \ge \inf\{f(x)|x \in A\} + \inf\{g(x)|x \in A\}.$$

(b) If $c \ge 0, \sup\{(cf)(x)|x \in A\} = c\sup\{f(x)|x \in A\}$ and

$$\inf\{cf(x)|x \in A\} = c\inf\{f(x)|x \in A\}.$$

If $c < 0, \sup\{(cf)(x)|x \in A\} = c\inf\{f(x)|x \in A\}$ and

$$\inf\{(cf)(x)|x \in A\} = c\sup\{f(x)|x \in A\}. \quad \blacksquare$$

In the notation of Section 6.1, as the result of this theorem we have

$$M_i(f + g) \le M_i(f) + M_i(g) \text{ and } m_i(f + g) \ge m_i(f) + m_i(g).$$

If $c \ge 0$, we also have

$$M_i(cf) = cM_i(f) \text{ and } m_i(cf) = cm_i(f)$$

and if $c < 0$,

$$M_i(cf) = cm_i(f) \text{ and } m_i(cf) = cM_i(f).$$

Theorem 6.13: Suppose f and g are Riemann integrable functions on $[a, b]$. Then

(a) $f + g$ is Riemann integrable on $[a, b]$ and

$$\int_a^b (f + g) \, dx = \int_a^b f \, dx + \int_a^b g \, dx$$

(b) For any number c, cf is Riemann integrable on $[a, b]$ and

$$\int_a^b cf \, dx = c \int_a^b f \, dx.$$

Proof:
(a) From Corollary 6.11(c) to the Riemann-Lebesgue Theorem, we know that $f + g$ is Riemann integrable. We first show that

$$\int_a^b (f + g) \le \int_a^b f + \int_a^b g.$$

Let P_1 and P_2 be partitions of $[a, b]$. Then

$$\int_a^b (f + g) \leq \overline{S}(f + g; P_1 \cup P_2)$$

$$\leq \overline{S}(f; P_1 \cup P_2) + \overline{S}(g; P_1 \cup P_2)$$

$$\leq \overline{S}(f; P_1) + \overline{S}(g; P_2). \qquad (6.1)$$

The inequality $\overline{S}(f + g; P_1 \cup P_2) \leq \overline{S}(f; P_1 \cup P_2) + \overline{S}(g; P_1 \cup P_2)$ follows from $M_i(f + g) \leq M_i(f) + M_i(g)$. Also, $\overline{S}(f; P_1 \cup P_2 \leq \overline{S}(f; P_1)$, since $P_1 \cup P_2$ is a refinement of P_1, and similarly $\overline{S}(g; P_1 \cup P_2) \leq \overline{S}(g; P_2)$. Since

$$\int_a^b f = \inf\{\overline{S}(f; P)|P \text{ is a partition of } [a, b]\}$$

and

$$\int_a^b (f + g) \leq \overline{S}(f; P_1) + \overline{S}(g; P_2)$$

for any partitions P_1 and P_2 of $[a, b]$, it follows that

$$\int_a^b (f + g) \leq \int_a^b f + \overline{S}(g; P_2). \qquad (6.2)$$

Since Equation (6.2) holds for any partition P_2 of $[a, b]$ and since

$$\int_a^b g = \inf\{\overline{S}(g; P)||P \text{ is a partition of } [a, b]\}$$

it follows that

$$\int_a^b f + g \leq \int_a^b f + \int_a^b g.$$

In a similar manner, but using lower Riemann sums, we can show that

$$\int_a^b (f + g) \geq \int_a^b f + \int_a^b g.$$

Thus

$$\int_a^b (f + g) = \int_a^b f + \int_a^b g.$$

(b) Since f is Riemann integrable on $[a, b]$, the set of discontinuities of f on $[a, b]$ has measure zero. Thus, for any constant c, the function cf is continuous except on a set of measure zero, and so cf is Riemann integrable on $[a, b]$.

If $c > 0$, then

$$\int_a^b cf = \inf\{\overline{S}(cf; P)|P \text{ is a partition of } [a, b]\}$$

$$= \inf\{c\overline{S}(f; P)|P \text{ is a partition of } [a, b]\}$$

$$= c\inf\{\overline{S}(f; P)|P \text{ is a partition of } [a, b] = c\int_a^b f.$$

If $c < 0$, then

$$\int_a^b cf = \sup\{\underline{S}(cf; P)|P \text{ is a partition of } [a, b]\}$$

$$= \sup\{c\overline{S}(f; P)|P \text{ is a partition of } [a, b]\}$$

$$= c\inf\{\overline{S}(f; P)|P \text{ is a partition of } [a, b]\} = c\int_a^b f.$$

For $c = 0$, $\overline{S}(cf; P) = \underline{S}(cf; P) = 0$ for any partition P. Thus, for any real number c, we have that cf is Riemann integrable on $[a, b]$ and

$$\int_a^b cf = c\int_a^b f. \quad \blacksquare$$

Theorem 6.14: A function f is Riemann integrable on $[a, b]$ if and only if f is Riemann integrable on $[a, c]$ and $[c, b]$ for every $c \in (a, b)$. If this is the case, then

$$\int_a^b f = \int_a^c f + \int_c^b f. \tag{6.3}$$

Proof: The function f is Riemann integrable on $[a, b]$ if and only if the sets of discontinuities of f in $[a, c]$ and in $[c, b]$ each has measure zero for all $c \in [a, b]$.

To show Equation (6.3), let P_1 be a partition of $[a, c]$ and let P_2 be a partition of $[c, b]$. Then $P_1 \cup P_2$ is a partition of $[a, b]$. Now

$$\overline{S}(f; P_1 \cup P_2) = \overline{S}(f; P_1) + \overline{S}(f; P_2) \tag{6.4}$$

and

$$\underline{S}(f; P_1 \cup P_2) = \underline{S}(f; P_1) + \underline{S}(f; P_2) \tag{6.5}$$

where $\overline{S}(f; P_1 \cup P_2)$ is the upper Riemann sum for f on $[a, b]$ with respect to the partition $P_1 \cup P_2$ and $\overline{S}(f; P_1)$ and $\overline{S}(f; P_2)$ are the upper Riemann sums of f on $[a, c]$ and $[c, b]$ with respect to the partitions P_1 and P_2, respectively. Thus

$$\inf\{\overline{S}(f; P)|P \text{ is a partition of } [a, b]\}$$

$$\leq \inf\{\overline{S}(f; P_1|P_1 \text{ is a partition of } [a, c]\}$$

$$+ \inf\{\overline{S}(f; P_2|P_2 \text{ is a partition of } [c, b]\}. \text{ (Why?)}$$

Therefore

$$\int_a^b f \leq \int_a^c f + \int_c^b f.$$

Similarly,

$$\sup\{\underline{S}(f; P)|P \text{ is a partition of } [a, b]\}$$

$$\geq \sup\{\underline{S}(f; P_1)|P_1 \text{ is a partition of } [a, c]\}$$

$$+ \sup\{\underline{S}(f; P_2)|P_2 \text{ is a partition of } [c, b]\}$$

so

$$\int_a^c f + \int_c^b f \leq \int_a^b f.$$

Thus

$$\int_a^c f + \int_c^b f = \int_a^b f. \quad \blacksquare$$

Definition: If f is Riemann integrable on $[a, b]$, then we define $\int_b^a f$ by

$$\int_b^a f = -\int_a^b f.$$

Definition: If $a \in \mathcal{D}(f)$ then $\int_a^a f = 0$.

Corollary 6.14: If a, b, c are numbers and f is a function such that $\int_a^b f, \int_c^b f$, and $\int_a^c f$ exist, then

$$\int_a^c f + \int_c^b f = \int_a^b f$$

regardless of the relative sizes of a, b and c.

The next group of theorems we prove establishes some inequalities of integrals and a mean value property.

Theorem 6.15: Suppose f is Riemann integrable on $[a, b]$ and $f \geq 0$. Then

$$\int_a^b f \geq 0.$$

Proof: Let P be any partition of $[a, b]$. Then

$$\underline{S}(f; P) = \sum m_i(f) \Delta x_i.$$

But $m_i(f) \geq 0$, since $f \geq 0$, so

$$0 \leq \underline{S}(f; P) \leq \int_a^b f. \quad \blacksquare$$

Corollary 6.15(a): If f and g are Riemann integrable functions on $[a, b]$ and $f \geq g$, then

$$\int_a^b f \geq \int_a^b g.$$

Proof: We have $f - g \geq 0$, so

$$\int_a^b f - \int_a^b g = \int_a^b (f - g) \geq 0. \quad \blacksquare$$

Corollary 6.15(b): If f is Riemann integrable on $[a, b]$, then $|f|$ is Riemann integrable on $[a, b]$ and

$$\left| \int_a^b f \right| \leq \int_a^b |f|.$$

Proof: We first show that if f is Riemann integrable on $[a, b]$, then so is $|f|$. It was shown in Exercise 11, Section 4.1, that if f is continuous at x, then so is $|f|$. Thus

$$\{x \in [a, b] \mid |f| \text{ is not continuous at } x\} \subset \{x \in [a, b] \mid f \text{ is not continuous at } x\}.$$

Since f is Riemann integrable on $[a, b]$, the set of discontinuities of f on $[a, b]$ has measure zero. Therefore, $|f|$ is continuous on $[a, b]$ except on a set of measure zero. Thus $|f|$ is Riemann integrable on $[a, b]$.

Now

$$-|f| \leq f \leq |f|$$

so

$$-\int_a^b |f| \leq \int_a^b f \leq \int_a^b |f|$$

and thus

$$\left| \int_a^b f \right| \le \int_a^b |f|. \quad \blacksquare$$

Corollary 15(c): If f is a bounded Riemann integrable function on $[a, b]$ and $m \le f(x) \le M$ for $x \in [a, b]$, then

$$m(b - a) \le \int_a^b f \le M(b - a). \quad \blacksquare$$

Theorem 6.16 (Mean Value Theorem for Integrals): Suppose f is a continuous function on $[a, b]$. Then there is a number $c \in [a, b]$ such that

$$\int_a^b f = f(c)(b - a).$$

Proof: Since f is continuous on the closed bounded interval $[a, b]$, it is bounded and assumes its maximum and minimum values somewhere on the interval. Let

$$m = \min\{f(x)|x \in [a, b]\} \text{ and } M = \max\{f(x)|x \in [a, b]\}.$$

Suppose x_0 and x_1 are points in $[a, b]$ where $f(x_0) = m$ and $f(x_1) = M$. By Corollary 6.15(c) we have

$$m \le \frac{1}{b - a} \int_a^b f \le M.$$

The Intermediate Value Theorem tells us that since f is a continuous function, it assumes all values in the interval $[m, M]$ between x_0 and x_1. Thus there is a value c between x_0 and x_1 where

$$f(c) = \frac{1}{b - a} \int_a^b f. \quad \blacksquare$$

Thus far, we have done nothing to indicate how an integral may be evaluated. The main device for evaluating integrals is the Fundamental Theorem of Calculus. Next we prove a group of theorems culminating in that result. This is done by beginning with a function f that is Riemann integrable on $[a, b]$ then defining a function $F(x) = \int_a^x f$ for $x \in [a, b]$. We first show that F is continuous. We next prove that if f is continuous, then F is differentiable. Thus the integral of a function is more "well behaved" than the original function. This principle is exploited in several ways in applications.

Theorem 6.17: Let f be a bounded Riemann integrable function on $[a,b]$ and let $F(x) = \int_a^x f$ for $x \in [a,b]$. Then F is continuous on $[a,b]$.

Proof: We first notice that if $f(x) = 0$ on $[a,b]$, then $F(x) = 0$ on $[a,b]$, and we are done. If this is not the case, let

$$M = \sup\{|f(x)| \mid x \in [a,b]\}.$$

Let $\epsilon > 0$ be given, and fix $x_0 \in [a,b]$. Then if $x \in [a,b]$,

$$|F(x) - F(x_0)| = \left| \int_a^x f - \int_a^{x_0} f \right| = \left| \int_x^{x_0} f \right| \leq \int_x^{x_0} |f|$$

by Corollary 6.14 and Corollary 6.15(b). Now

$$\int_x^{x_0} |f| \leq M|x - x_0|$$

by Corollary 6.15(c). Thus if $|x - x_0| < \epsilon/M$ and $x \in [a,b]$, then

$$|F(x) - F(x_0)| \leq M|x - x_0| < M \cdot \frac{\epsilon}{M} = \epsilon$$

so F is continuous at $x_0 \in [a,b]$. ∎

Theorem 6.18: Let f be a continuous function on $[a,b]$ and let $F(x) = \int_a^x f$ For $x \in (a,b)$. Then F is differentiable at x and $F'(x) = f(x)$.

Proof: Let x_0 be a point in (a,b). Let $\{x_n\}$ be a sequence of numbers in (a,b) such that $x_n \neq x_0$ for all n and $\lim x_n = x_0$. Then, as in the proof of Theorem 6.17,

$$F(x_n) - F(x_0) = \int_{x_0}^{x_n} f.$$

By Theorem 6.16, since f is continuous, there is a number c_n between x_0 and x_n for which $\int_{x_0}^{x_n} f = f(c_n)(x_n - x_0)$. Thus

$$\frac{F(x_n) - F(x_0)}{x_n - x_0} = f(c_n)$$

where c_n is between x_n and x_0. Since c_n is between x_n and x_0, and since $\lim_{n\to\infty} x_n = x_0$, it follows that $\lim_{n\to\infty} c_n = x_0$. Because f is continuous,

$$\lim_{n\to\infty} f(c_n) = f\left(\lim_{n\to\infty} c_n\right) = f(x_0)$$

so that

$$\lim_{n\to\infty} \frac{F(x_n) - F(x_0)}{x_n - x_0} = \lim_{n\to\infty} f(c_n) = f(x_0).$$

Thus $F'(x_0) = f(x_0)$. ∎

Theorem 6.19 (Fundamental Theorem of Calculus): Suppose f is a Riemann integrable function on $[a, b]$, and suppose F is a function defined on $[a, b]$ such that

(i) F is continuous on $[a, b]$ and differentiable on (a, b).

(ii) $F'(x) = f(x)$ for $x \in (a, b)$.

Then

$$\int_a^b f = F(b) - F(a).$$

Note: If f is a function that has properties (i) and (ii), then F is called an *antiderivative* of f on $[a, b]$. By Corollary 5.8(b), we know that any two antiderivatives of f differ by a constant.

Proof: Let $P = \{x_0, x_1, ..., x_n\}$ be any partition of $[a, b]$. Then we can write

$$F(b) - F(a)$$

$$= (F(x_1) - F(a)) + (F(x_2) - F(x_1)) + (F(x_3) - F(x_2))$$

$$+ ... + (F(x_{n-1}) - F(x_{n-2})) + (F(b) - F(x_{n-1}))$$

$$= \sum_{i=1}^n [F(x_i) - F(x_{i-1})].$$

(Remember $x_0 = a$ and $x_n = b$.)

Now F is continuous on $[a, b]$ and differentiable on (a, b). Thus for any subinterval $[x_{k-1}, x_k]$, we have

$$F(x_k) - F(x_{k-1}) = F'(\overline{x}_k)(x_k - x_{k-1}) = f(\overline{x}_k)(x_k - x_{k-1})$$

for some $\overline{x}_k \in (x_{k-1}, x_k)$ by the Mean Value Theorem for Derivatives. But

$$m_k(f)\Delta x_k \le f(\overline{x}_k)(x_k - x_{k-1}) \le M_k(f)\Delta x_k.$$

Thus

$$\underline{S}(f; P) = \sum_{i=1}^n m_i(f)\Delta x_i \le \sum_{i=1}^n f(\overline{x}_i)(x_i - x_{i-1})$$

$$= \sum_{i=1}^n [F(x_i) - F(x_{i-1})]$$

$$= F(b) - F(a) \le \sum_{i=1}^n M_i(f)\Delta x_i = \overline{S}(f; P).$$

This means that for any Partition P

$$\underline{S}(f;P) \le F(b) - F(a) \le \overline{S}(f;P).$$

Thus, since f is Riemann integrable, we have $\int_a^b f = F(b) - F(a)$. ∎

Corollary 6.19(Integration by Parts Formula): Suppose f and g are functions that have continuous derivatives on $[a,b]$. Then $\int_a^b fg'$ and $\int_a^b gf'$ exist and

$$\int_a^b fg' = f(b)g(b) - f(a)g(a) - \int_a^b gf'.$$

Proof: The hypotheses imply that gf' and fg' are continuous on $[a,b]$, so both integrals exist. Now

$$(gf)' = fg' + gf'$$

so

$$\int_a^b fg' + \int_a^b gf' = \int_a^b (fg)' = f(b)g(b) - f(a)g(a). \quad ∎$$

In elementary calculus one often performs integrations by using a "u-substitution." This is an application of the Fundamental Theorem and the Chain Rule. The next theorem justifies the validity of this technique.

Theorem 6.20 (Change of Variables Theorem): Suppose g is a function whose derivative g' is continuous on $[a,b]$. Suppose that f is a function that is continuous on $\{g(x)|a \le x \le b\}$. Then

$$\int_a^b (f(g(x))g'(x)dx = \int_{g(a)}^{g(b)} f(u)du.$$

Proof: Since g is continuous, then $g([a,b])$ is a compact connected set, that is, either a point or a closed bounded interval. (See Exercise 26, Section 4.1.) We leave it for Exercise 9, Section 6.2 to show that both integrals are zero in the case that $g([a,b])$ is a point. Otherwise, for $x \in g([a,b])$, define F by

$$F(x) = \int_{g(a)}^x f. \qquad (6.6)$$

Then $F'(x) = f(x)$. Notice we are not saying that $g([a,b]) = [g(a),g(b)]$, and we have not said that $x \ge g(a)$. The statement is valid without those assumptions.

Since g' is continuous on $[a,b]$, $f(g(x)g'(x)$ is continuous on $[a,b]$, and

$$(f \circ g)'(x) = F'(g(x))g'(x) = f(g(x))g'(x)$$

by the Chain Rule, so that $F \circ g$ is an antiderivative of $f(g(x))g'(x)$ on $[a, b]$. Thus

$$\int_a^b f(g(x))g'(x)\,dx = (F \circ g)(b) - (F \circ g)(a) = F(g(b)) - F(g(a)).$$

Also, from Equation (6.6), $F(g(a)) = 0$ and so

$$\int_{g(a)}^{g(b)} f = F(g(b)) = F(g(b)) - F(g(a))$$

from which the result follows.   ■

Improper Riemann Integrals

Thus far we have restricted our attention to the integral of bounded functions on bounded intervals. If we allow either the interval or the function to be unbounded, the integral is said to be improper. Improper Riemann integrals are defined as the limit of Riemann integrals.

Definition: Suppose f is a real valued function defined on $[a, \infty)$ that is Riemann integrable on $[a, t]$ for all $t \in (a, \infty)$. Then the *improper Riemann integral of f on* $[a, \infty)$, denoted $\int_a^\infty f$, is defined to be

$$\int_a^\infty f = \lim_{t \to \infty} \int_a^t f \qquad (6.7)$$

provided this limit exists. If this limit exists, the improper integral is said to *converge* to the value of the limit. Otherwise, the integral is said to be *divergent*. If the limit in Equation (6.7) is $\pm\infty$, the integral is said to diverge to that value.

If the interval is of the form $(-\infty, a]$, the improper integral $\int_{-\infty}^a f$ is defined to be

$$\int_{-\infty}^a f = \lim_{t \to -\infty} \int_t^a f.$$

For the integral of f on $(-\infty, \infty)$ to converge, we require that both the integrals $\int_{-\infty}^a f$ and $\int_a^\infty f$ converge for some number a. In this case,

$$\int_{-\infty}^\infty f = \int_{-\infty}^a f + \int_a^\infty f.$$

Example 6.21:
Let $f(x) = x$. Then

$$\int_0^\infty x = \lim_{t\to\infty} \int_0^t x = \lim_{t\to\infty} \left.\frac{x^2}{2}\right|_0^t = \infty$$

so that $\int_0^\infty x$ diverges, and thus so does $\int_{-\infty}^\infty x$. However, if we were to let

$$\int_{-\infty}^\infty x = \lim_{t\to\infty} \int_{-t}^c x = \lim_{t\to\infty} \left.\frac{x^2}{2}\right|_{-t}^t = \lim_{t\to\infty} 0 = 0$$

we would conclude that $\int_{-\infty}^\infty x$ converges, which is not correct.

The other reason that an integral may be classified as improper is if the function being integrated is unbounded on the interval of integration.

Definition: Suppose f is a function defined on $(a, b]$ that is Riemann integrable on $[t, b]$ for all $t \in (a, b)$, and suppose $\lim_{t\downarrow a} f(t) = \pm\infty$. Then the *improper Riemann integral of f on $[a, b]$*, denoted $\int_a^b f$, is defined to be

$$\int_a^b f = \lim_{t\downarrow a} \int_t^b f \qquad (6.8)$$

provided the limit exists. If this limit does exist, the improper integral is said to be *convergent* and converges to the value of the limit. Otherwise the integral is said to be *divergent*. If the limit in Equation (6.8) is $\pm\infty$, the integral is said to diverge to that value.

A similar definition can be made if the function becomes unbounded at b. If the function becomes unbounded at $c \in (a, b)$, then we say $\int_a^b f$ converges if both $\int_a^c f$ and $\int_c^b f$ converge. If this is the case, then

$$\int_a^b f = \int_a^c f + \int_c^b f.$$

The Natural Logarithm Function

As an application of some of our earlier theorems, we shall develop the natural logarithm function $\ln x$ and the exponential function e^x. There are several points at which one can begin this development. We begin by defining

$$\ln x = \int_1^x \frac{dt}{t} \text{ for } x > 0.$$

Thus the domain of $\ln x$ is the set of positive real numbers.

Since $f(t) = 1/t$ is a continuous function for $t > 0$, by Theorem 6.18 $\ln x$ is a differentiable function and

$$\frac{d}{dx}\ln x = \frac{1}{x} \text{ for } x > 0.$$

Now $1/x > 0$ for $x > 0$, so $\ln x$ is an increasing function. In particular, by the Mean Value Theorem, if $x_2 > x_1 > 0$, then

$$\ln x_2 - \ln x_1 = \frac{1}{x_0}(x_2 - x_1) > 0 \text{ for some } x_0 \in (x_1, x_2).$$

Thus $\ln x$ is strictly increasing and is a 1-1 function.

Next we show that $\ln x$ has the properties that define a logarithm function. Namely,

(i) $\ln 1 = 0$

(ii) $\ln(ab) = \ln a + \ln b$ for $a, b > 0$

(iii) $\ln(a/b) = \ln a - \ln b$ for $a, b > 0$

(iv) $\ln(a^r) = r \ln a$ for $a > 0$ and r any real number.

Property (i) is immediate by definition of $\ln x$. We exhibit two ways to prove properties (ii),(iii), and (iv). The first method is to appeal directly to the definition of $\ln x$. We use this approach to prove (ii). Now

$$\ln(ab) = \int_1^{ab} \frac{1}{t}dt = \int_1^a \frac{1}{t}dt + \int_a^{ab} \frac{1}{t}dt.$$

In the integral $\int_a^{ab}(dt/t)$, we make the change of variables $u = t/a$. Then

$$\int_a^{ab} \frac{1}{t}\,dt = \int_1^b \frac{1}{u}\,du$$

and so

$$\ln(ab) = \int_1^{ab} \frac{1}{t}\,dt + \int_1^b \frac{1}{u}\,du = \ln a + \ln b.$$

We shall prove (iv) using the following ideas. Suppose $f(x)$ and $g(x)$ are differentiable functions on an interval (c, d) and $f'(x) = g'(x)$ for all $x \in (c, d)$. Then $f(x) = g(x) + k$ on (c, d) for some constant k. If there is a point $x_0 \in (c, d)$ at which we know $f(x_0)$ and $g(x_0)$, then we can solve for k; that is, $k = f(x_0) - g(x_0)$.

Let $f(x) = \ln(x^r)$ and $g(x) = r \ln x$ for $x > 0$. Then[1]

$$f'(x) = \frac{1}{x^r}rx^{r-1} = \frac{r}{x} \text{ and } g'(x) = \frac{r}{x}.$$

---

[1] We have not proven that $(d/dx)(x^r) = rx^{r-1}$ if r is irrational, but we shall in Chapter 8.

Thus $f(x) = g(x) + k$. At $x_0 = 1$, $f(x_0) = \ln(1^r) = 0$ and $g(x_0) = r\ln(1) = 0$, so $k = 0$. Hence $f(x) = g(x)$.

We leave the proof of (iii) for Exercise 15, Section 6.2.

Next we obtain an estimate for $\ln n$ for n a positive integer. We do this by estimating $\int_1^n \frac{dx}{x}$ by upper and lower Riemann sums. Fix a positive integer n and let $P = \{1, 2, 3, ..., n\}$ be partition of $[1, n]$. Now on the interval $[k, k+1]$,

$$\frac{1}{k+1} \le \frac{1}{x} \le \frac{1}{k}$$

so if $f(x) = 1/x$, then

$$\underline{S}(f;P) = \frac{1}{2} + \frac{1}{3} + \cdots + \frac{1}{n} \le \int_1^n \frac{dx}{x} \le 1 + \frac{1}{2} + \frac{1}{3} + \cdots + \frac{1}{n-1} = \overline{S}(f;P).$$

We shall show that $\frac{1}{2} + \frac{1}{3} + \cdots + \frac{1}{n}$ can be made arbitrarily large by taking n sufficiently large. Notice that

$$\frac{1}{2} + \frac{1}{3} > 2 \cdot \frac{1}{4} = \frac{1}{2}$$

$$\left(\frac{1}{2} + \frac{1}{3}\right) + \left(\frac{1}{4} + \frac{1}{5} + \frac{1}{6} + \frac{1}{7}\right) > 2 \cdot \frac{1}{4} + 4 \cdot \frac{1}{8} = 2 \cdot \frac{1}{2}$$

and one can show that

$$\frac{1}{2} + \cdots + \frac{1}{2^n - 1} > (n-1)\frac{1}{2}.$$

Thus

$$\frac{n-1}{2} < \frac{1}{2} + \frac{1}{3} + \cdots + \frac{1}{2^n - 1} \le \int_1^{2^n - 1} \frac{dx}{x}.$$

Since $\ln x$ is an increasing function, it follows that $\lim_{x \to \infty}(\ln x) = \infty$.

We now show $\lim_{x \downarrow 0} \ln x = -\infty$. For n a positive integer,

$$\ln(1/n) = \ln 1 - \ln n = -\ln n$$

by properties (i) and (iii). Thus, since $\ln x$ is continuous on $(0, \infty)$,

$$\lim_{x \downarrow 0} \ln x = \lim_{n \to \infty} \ln(1/n) = \lim_{n \to \infty}(-\ln n) = -\infty.$$

Since $\ln x$ is continuous, it follows that the range of $\ln x$ is all the real numbers by the Mean Value Theorem. Thus, there is a number, that we shall call e, at which $\ln e = 1$.

Our next project is to determine an expression for e that can be used to estimate its size. Since $(d/dx)(\ln x) = 1/x$, it follows that $(d/dx)(\ln x)|_{x=1} = 1$. By definition, the derivative of $\ln x$ at $x = 1$ is

$$\lim_{h \to 0} \frac{\ln(1+h) - \ln(1)}{h} = \lim_{h \to 0} \frac{1}{h}\ln(1+h) = \lim_{h \to 0} \ln(1+h)^{1/h}$$

since $\ln 1 = 0$ and $r \ln x = \ln(x^r)$. Thus $\lim_{h \to 0} \ln(1 + h)^{1/h} = 1$. Since $\ln x$ is a continuous function,

$$\lim_{h \to 0} \ln(1 + h)^{1/h} = \ln\left(\lim_{h \to 0}(1 + h)^{1/h}\right) = 1.$$

But $\ln e = 1$ and since $\ln x$ is a 1-1 function, it follows that

$$e = \lim_{h \to 0}(1 + h)^{1/h}.$$

This can also be written

$$e = \lim_{n \to \infty}(1 + 1/n)^n$$

with the change of variables $n = 1/h$. Later we shall show that e is irrational.

We now develop the exponential function e^x. Since $\ln x$ is a differentiable, increasing 1-1 function, then it has an inverse that is also increasing and differentiable. Taking a clue from other logarithmic functions, we expect that the inverse is an exponential function with base e. Thus we denote the inverse of $\ln x$ by e^x. We note that the domain of e^x is the range of $\ln x$, that is, the real numbers, and the range of e^x is the domain of $\ln x$, that is, the positive real numbers. We show that e^x has the properties that determine an exponential function. Namely,

(i) $e^0 = 1$

(ii) $e^{a+b} = e^a \cdot e^b$

(iii) $e^{a-b} = e^a / e^b$

(iv) $e^{ax} = (e^a)^x$.

To prove (i), we have $\ln(e^0) = 0$, since $\ln(e^x) = x$ because e^x is the inverse of $\ln x$. But $\ln 0 = 1$ and $\ln x$ is a 1-1 function, so $e^0 = 1$.

To prove (ii), let $x = e^a$ and $y = e^b$. Then $\ln x = \ln e^a = a$ and $\ln y = \ln e^b = b$, so

$$\ln e^{a+b} = a + b = \ln x + \ln y = \ln(xy) = \ln(e^a \cdot e^b).$$

Therefore $e^{a+b} = e^a \cdot e^b$, since $\ln x$ is a 1-1 function. The proofs of (iii) and (iv) are similar and are left for Exercise 16, Section 6.2.

To calculate the derivative of e^x, we note that

$$1 = \frac{d}{dx}(x) = \frac{d}{dx}(\ln(e^x)) = \frac{1}{e^x}\frac{d}{dx}(e^x)$$

so that $e^x = (d/dx)(e^x)$.

Finally, we note that $\lim_{x \to \infty} e^x = \infty$ and $\lim_{x \to -\infty} e^x = 0$, since e^x is an increasing function whose domain is the real numbers and whose range is $(0, \infty)$.

Exercises 6.2

1. **(a)** Suppose that f and g are Riemann integrable functions on $[a, b]$ and $f(x) = g(x)$ if $x \neq c$. Show that $\int_a^b f = \int_a^b g$.
   (b) Suppose that f and g are Riemann integrable functions on $[a, b]$ and $f(x) = g(x)$ if $x \neq c_1, ..., c_n$. Show that $\int_a^b f = \int_a^b g$.
   (c) Suppose that f is Riemann integrable function on $[a, b]$. If g is bounded and $f = g$ except for countably many points, must it be true that g is Riemann integrable on $[a, b]$?

2. A step function f on $[a, b]$ is a function for which there are a finite number of disjoint intervals $I_1, ..., I_n$ with $[a, b] = I_1 \cup ... \cup I_n$ and for which f is constant on each of the intervals.
   (a) Let I_i be an interval with endpoints a_i and b_i. Suppose $f(x) = c_i$ on I_i. Show that f is Riemann integrable on $[a, b]$ and that

$$\int_a^b f = \sum_{i=1}^n c_i(b_i - a_i).$$

   (b) Let f be a continuous (and thus uniformly continuous) function on $[a, b]$. Show that, given $\epsilon > 0$, there is a step function g_ϵ defined on $[a, b]$ such that

$$|g_\epsilon(x) - f(x)| < \epsilon \text{ for every } x \in [a, b]$$

   and

$$\left| \int_a^b g_\epsilon - \int_a^b f \right| < \epsilon(b - a).$$

3. Prove Corollary 6.14.

4. Suppose that f and g are Riemann integrable functions on $[a, b]$.
   (a) Show that

$$\left| \int_a^b f(x)g(x) \, dx \right| \leq \left\{ \int_a^b f^2(x) \, dx \right\}^{1/2} \left\{ \int_a^b g^2(x) \, dx \right\}^{1/2}.$$

   This is the Schwarz inequality for integrals. [*Hint:* for α and β real numbers $\int_a^b (\alpha f(x) - \beta g(x))^2 \, dx \geq 0$.]
   (b) Show that

$$\left\{ \int_a^b [f(x) + g(x)]^2 \, dx \right\}^{1/2} \leq \left\{ \int_a^b f^2(x) \, dx \right\}^{1/2} \left\{ \int_a^b g^2(x) \, dx \right\}^{1/2}.$$

   This is the Minkowski inequality for integrals.

5. Let f be a bounded and continuous function on $[0,1]$. For n a positive integer, define

$$||f||_n = \left\{ \int_0^1 |f(x)|^n \, dx \right\}^{1/n}.$$

Show that $\lim_{n\to\infty} ||f||_n = \max\{|f(x)| \mid x \in [0,1]\}$.

6. **(a)** If f is continuous on $[a,b]$ and $\int_a^x f = 0$ for every $x \in [a,b]$, show that $f(x) = 0$ for all $x \in [a,b]$.
   (b) If f is continuous on $[a,b]$ and $\int_a^x f = \int_x^b f$ for all $x \in [a,b]$, show that $f(x) = 0$ for all $x \in [a,b]$.

7. Suppose f is a continuous function on $[a,b]$ and $\int_a^b f(x)g(x) \, dx = 0$ for every continuous function g. Prove that $f(x) = 0$ on $[a,b]$.

8. **(a)** Suppose f and g are continuous functions on $[a,b]$ and $g(x) \geq 0$ on $[a,b]$. Show there is a number $c \in [a,b]$ where

$$\int_a^b f(x)g(x) \, dx = f(c) \int_a^b g(x) \, dx.$$

   (b) Give an example where the result in part (a) is not true if it is not required that $g(x) \geq 0$. [*Hint:* Consider $g(x) = \cos x$ on $[0, \pi]$.]

9. Prove Theorem 6.20 in the case that $g([a,b])$ is a point.

10. Give an example where Theorem 6.18 is not true if $f(x)$ is not continuous.

11. Suppose f and g are continuous functions on $[a,b]$ with $g(x) \geq 0$ and

$$m = \inf\{f(x) \mid x \in [a,b]\}, \quad M = \sup\{f(x) \mid x \in [a,b]\}.$$

Show that there is a number $c \in [a,b]$ where

$$\int_a^b f(x)g(x) \, dx = m \int_a^c g(x) \, dx + M \int_c^b g(x) \, dx.$$

[*Hint:* For $x \in [a,b]$, let $F(x) = m \int_a^x g(t) \, dt + M \int_x^b g(t) \, dt$. Show $\min F(x) \leq F(b) = m \int_a^b g(t) \, dt \leq M \int_a^b g(t) \, dt = F(a) \leq \max F(x)$.]

12. **(a)** Suppose f, g and h are differentiable functions on $\mathbb{R}$. Show that

$$\frac{d}{dx} \int_{g(x)}^{h(x)} f(t) \, dt = f(h(x))h'(x) - f(g(x))g'(x).$$

   (b) Find

$$\frac{d}{dx} \int_{\sin x}^{\cos x} t^2 \, dt.$$

13. **(a)** Find the value of c where

$$\int_2^6 f(x)\, dx = f(c)(6 - 2) \text{ if } f(x) = x^2 + 1.$$

(b) Give an example where the Mean Value Theorem for Integrals fails if f is not a continuous function.

14. **(a)** If f and g are continuous functions on $[a, b]$ with $\int_a^x f \geq \int_a^x g$ for every $x \in [a, b]$, must it be true that $f(x) \geq g(x)$ on $[a, b]$?
    (b) If f and g are continuous functions on $[a, b]$ and $\int_\alpha^\beta f \geq \int_\alpha^\beta g$ for every interval $[\alpha, \beta] \subset [a, b]$, show that $f(x) \geq g(x)$ on $[a, b]$.

15. Show that $\ln(x/y) = \ln x - \ln y$ for $x, y > 0$.

16. Prove that $e^{a-b} = e^a/e^b$ and $e^{ab} = (e^a)^b$.

17. Show that if $f'(x) = f(x)$ for $x \in \mathbb{R}$, then $f(x) = Ce^x$.

18. Find $\lim_{n\to\infty}(1 + 2/n)^n$.

19. **(a)** Show that $e^x \geq 1 + x$ for $x \geq 0$. Conclude $e \geq 2$.
    (b) Show that $e^x \geq 1 + x + x^2/2$ for $x \geq 0$.
    (c) Show that $e^x \geq 1 + x + x^2/2 + \cdots + x^n/n!$ for $x \geq 0$ and any positive integer n.

20. Let $F(x) = \int_0^x f$, where f is a bounded function.
    (a) Show that if f has a removable discontinuity at $x = c$, then F is differentiable at c and $F'(c) = \lim_{x\to c} f(x)$.
    (b) Show that if f has a jump discontinuity at $x = c$ then F is not diffentiable at $x = c$.

21. An important function that is defined as an improper integral is the gamma function. Here we define this function and develop some of its properties. Let

$$\Gamma(x) = \int_0^\infty e^{-t} t^{x-1}\, dt.$$

(a) Show that the integral that defines $\Gamma(x)$ converges for $x > 0$.
(b) Use integration by parts to show that $\Gamma(x + 1) = x\Gamma(x)$ for $x > 0$.
(c) Show that $\Gamma(1) = 1$.
(d) Use induction to show that $\Gamma(n) = (n - 1)!$ for every positive integer n. (Recall that $0! = 1$.)
(e) Show that $\lim_{x\downarrow 0} \Gamma(x) = \infty$.

22. For n a positive integer, define

$$a_n = \left(\sum_{k=1}^n \frac{1}{k}\right) - \ln n.$$

Show that $\{a_n\}$ is a convergent sequence by showing it is a bounded monotone sequence. The limit of this sequence is called Euler's constant.

6.3   The Riemann-Stieltjes Integral

This section may be omitted without loss of continuity.

In this section we develop a second type of integral called the Riemann-Stieltjes integral. In our discussion, we shall be dealing with two functions, an integrand function (typically denoted by f) and a integrator function (typically denoted by g). Throughout this section we assume that each function is defined on an interval $[a, b]$ and is bounded. We shall begin by assuming that the integrator function is nondecreasing but later allow for a more general function. We shall use the notation $M_i(f)$ and $m_i(f)$ as defined in Section 6.1.

Definition: Let f be a bounded function on $[a, b]$, let g be a nondecreasing function on $[a, b]$, and let $P = \{x_0, x_1, ..., x_n\}$ be a partition of $[a, b]$. The *upper (lower) Riemann-Stieltjes sum of f with respect to g with respect to the partition P*, denoted $\overline{S}(f; g; P)$ $(\underline{S}(f; g; P))$, is given by

$$\overline{S}(f; g; P) = \sum_{i=1}^{n} M_i(f)[g(x_i) - g(x_{i-1})] = \sum_{i=1}^{n} M_i(f)\Delta g(x_i)$$

$$\underline{S}(f; g; P) = \sum_{i=1}^{n} m_i(f)[g(x_i) - g(x_{i-1}] = \sum_{i=1}^{n} m_i(f)\Delta g(x_i).$$

The hypothesis that g is nondecreasing means that $\Delta g(x_i) \geq 0$, so that $\overline{S}(f; g; P) \geq \underline{S}(f; g; P)$ for any partition P of $[a, b]$. This will enable us to prove several of the theorems about the Riemann-Stieltjes integral with an argument that is virtually identical to the one given for the corresponding theorem with the Riemann integral. When this is the case, we shall make reference to the appropriate theorem rather than rewriting the proof.

Notice that if we let $g(x) = x$, then the Riemann-Stieltjes integral becomes the Riemann integral. The advantage in using a more general integrator function g is that it enables us to work with a class of problems other than those where Δx_i is interpreted as the length of the interval $[x_{i-1}, x_i]$. For example we may want to consider a variable "mass density" on the x-axis. Then $\Delta g(x_i)$ could be the mass between x_{i-1} and x_i. A possibly familiar application occurs in probability theory where the integrator function g is the probability distribution function.

Because of our familiarity with the Riemann integral and its similarities with the Riemann-Stieltjes integral, much of our effort will be directed toward highlighting the differences and similarities of the two integrals.

Theorem 6.21:

(a) If Q is a refinement of the partition P, then

$$\overline{S}(f;g;Q) \leq \overline{S}(f;g;P) \text{ and } \underline{S}(f;g;Q) \geq \underline{S}(f;g;P).$$

(b) For any partitions P and Q of $[a,b]$

$$\underline{S}(f;g;P) \leq \overline{S}(f;g;Q).$$

(c) Let

$$\overline{S}(f;g) = \inf\{\overline{S}(f;g;P)|P \text{ is a partition of } [a,b]\}$$

and

$$\underline{S}(f;g) = \sup\{\underline{S}(f;g;P)|P \text{ is a partition of } [a,b]\}.$$

Then

$$\underline{S}(f;g) \leq \overline{S}(f;g).$$

Proof: See Theorem 6.3. ∎

Definition: If $\overline{S}(f;g) = \underline{S}(f;g)$, then f is said to be integrable with respect to g on $[a,b]$. In this case we let

$$\int_a^b f\, dg = \overline{S}(f;g) = \underline{S}(f;g)$$

and say $\int_a^b f\, dg$ is the *Riemann-Stieltjes integral of f with respect to g on $[a,b]$.*

Theorem 6.22 (Riemann Condition for Integrability): The function f is integrable with respect to g on $[a,b]$ if and only if, given $\epsilon > 0$, there is a partition $P(\epsilon)$ for which

$$\overline{S}(f;g;P(\epsilon)) - \underline{S}(f;g;P(\epsilon)) \leq \epsilon.$$

Proof: See Theorem 6.4. ∎

Example 6.6:

Let

$$f(x) = \begin{cases} 0 & \text{if } 0 \leq x \leq 1 \\ 1 & \text{if } 1 < x \leq 2, \end{cases}$$

$$g(x) = \begin{cases} 0 & \text{if } 0 \leq x < 1 \\ 1 & \text{if } 1 \leq x \leq 2. \end{cases}$$

We first show that f is integrable with respect to g on $[0,2]$. Let $P = \{x_0, x_1, ..., x_n\}$ be any partition of $[0,2]$ *that includes the number 1.*

Suppose $x_j = 1$. Then

$$\overline{S}(f; g; P) = \sum_{i=1}^{n} M_i(f)[g(x_i) - g(x_{i-1})]$$

$$= M_j(f)[g(x_i) - g(x_{i-1})].$$

This is true because $g(x_i) - g(x_{i-1}) = 0$ unless $i = j$. Now

$$g(x_j) - g(x_{j-1}) = g(1) - g(x_{j-1}) = 1 - 0$$

since $g(x) = 0$ if $x < 1$ and $x_{j-1} < 1$. Thus

$$\overline{S}(f; g; P) = M_j(f)[g(x_j) - g(x_{j-1})] = M_j(f).$$

But $M_j(f) = \sup\{f(x) \mid x \in [x_{j-1}, 1]\} = 0$, since $f(x) = 0$ if $x \leq 1$ and $x_{j-1} < x_j = 1$. Thus

$$\overline{S}(f; g; P) = M_j(f)[g(x_j) - g(x_{j-1})] = M_j(f) = 0.$$

Likewise

$$\underline{S}(f; g; P) = \sum_{i=1}^{n} m_i(f)[g(x_i) - g(x_{i-1})] = m_j(f) = 0$$

so that $\underline{S}(f; g; P) = \overline{S}(f; g; P) = 0$, and thus f is integrable with respect to g and $\int_0^2 f \, dg = 0$.

We want to use this example to point out a difference between the Riemann integral and Riemann-Stieltjes integral. Suppose $Q = \{x_0, x_1, ..., x_n\}$ is any partition of $[0, 2]$ that does *not* include 1 as a partition point. Suppose $x_{j-1} < 1 < x_j$. Then

$$\overline{S}(f; g; Q) = \sum_{i=1}^{n} M_i(f)[g(x_i) - g(x_{i-1})]$$

$$= M_j(f)[g(x_j) - g(x_{j-1}] = M_j(f)$$

and

$$\underline{S}(f; g; Q) = m_j(f).$$

Now the interval $[x_{j-1}, x_j]$ contains values of x smaller than 1 and values of x larger than 1 so that $M_j(f) = 1$ and $m_j(f) = 0$. Thus

$$\overline{S}(f; g; Q) = 1 \text{ and } \underline{S}(f; g; Q) = 0$$

no matter how small $||Q||$ is. In this example, it is the points in the partition rather than the norm of the partition that governs the behavior of the upper and lower sums. This is in contrast to the Riemann integral (see Theorem 6.6). The difficulty is caused by f and g having a common point of discontinuity.

In fact, the following is true. Let

$$g(x) = \begin{cases} 0 & \text{if } a \leq x \leq c \\ 1 & \text{if } c < x \leq b. \end{cases}$$

Then f is integrable with respect to g on $[a, b]$ if and only if f is continuous from the right at $x = c$, as will be shown in Exercise 1, Section 6.3.

Next we determine some conditions that ensure f is integrable with respect to g. This is a more difficult problem than the Riemann integral problem, and we shall not be able to obtain a complete answer. Part of the problem is that two functions are involved, and sometimes ill behavior of one can be overcome by sufficiently nice behavior of the other.

Theorem 6.23: If f is continuous on $[a, b]$, then f is integrable with respect to g on $[a, b]$.
Proof: See Theorem 6.7. ∎

Next we prove a result that relates the Riemann integral and Riemann-Stieltjes integral.

Theorem 6.24: Suppose f is Riemann integrable on $[a, b]$ and g is an increasing function on $[a, b]$ such that g' is defined and Riemann integrable on $[a, b]$. Then f is integrable with respect to g on $[a, b]$, fg' is Riemann integrable on $[a, b]$ and

$$\int_a^b f \, dg = \int_a^b f(x) g'(x) \, dx.$$

Note: The integral on the left is a Riemann-Stieltjes integral and the integral on the right is a Riemann integral.
Proof: Since f and g' are Riemann integrable, then fg' is Riemann integrable as was shown in Exercise 11, Section 6.1.

Let $P = \{x_0, x_1, ..., x_n\}$ be any partition of $[a, b]$. Then

$$\overline{S}(f; g; P) - \underline{S}(f; g; P) = \sum [M_i(f) - m_i(f)][g(x_i) - g(x_{i-1})].$$

Since g' exists on $[a, b]$, by the Mean Value Theorem we have

$$g(x_i) - g(x_{i-1}) = g'(t_i)(x_i - x_{i-1}) \text{ where } t_i \in (x_{i-1}, x_i).$$

Thus

$$\overline{S}(f; g; P) - \underline{S}(f; g; P) = \sum [M_i(f) - m_i(f)] g'(t_i)(x_i - x_{i-1}).$$

Since g' is Riemann integrable on $[a, b]$, it is bounded. Thus there is a positive number k such that $g'(x) \leq k$ for $x \in [a, b]$. Therefore

$$\overline{S}(f; g; P) - \underline{S}(f; g; P) \leq k \sum [M_i(f) - m_i(f)](x_i - x_{i-1})$$

$$= k[\overline{S}(f; P) - \underline{S}(f; P)] \qquad (6.9)$$

where $\overline{S}(f; P)$ and $\underline{S}(f; P)$ are the upper and lower Riemann sums, respectively, of f with respect to the partition P.

Let $\epsilon > 0$ be given. Since f is Riemann integrable on $[a, b]$, there is a partition $P(\epsilon)$ such that $\overline{S}(f; P(\epsilon)) - \underline{S}(f; P(\epsilon)) < \epsilon/k$. From Equation (6.8), we have

$$\overline{S}(f; g; P(\epsilon)) - \underline{S}(f; g; P(\epsilon)) \leq k[\overline{S}(f; P(\epsilon)) - \underline{S}(f; P(\epsilon))] < k\frac{\epsilon}{k} = \epsilon.$$

Thus f is integrable with respect to g. We now know that $\int_a^b f \, dg$ and $\int_a^b f(x)g'(x) \, dx$ exist.

We now show that $\int_a^b f \, dg = \int_a^b f(x)g'(x) \, dx$. Let $\epsilon > 0$ be given. The we can choose a partition $P = \{x_0, x_1, ..., x_n\}$ of $[a, b]$ such that

$$\int_a^b f(x)g'(x) \, dx - \epsilon < \sum_{i=1}^n (fg')(\psi_i)\Delta x_i < \int_a^b f(x)g'(x) \, dx + \epsilon$$

for any choice of ψ_i in $[x_i - x_{i-1}]$.

Now

$$\overline{S}(f; g; P) = \sum M_i(f)[g(x_i) - g(x_{i-1})]$$

$$= \sum M_i(f)g'(t_i)(x_i - x_{i-1}) \geq \sum f(t_i)g'(t_i)\Delta x_i$$

since $g'(t) \geq 0$ if g is increasing. Thus

$$\overline{S}(f; g; P) > \int_a^b f(x)g'(x) \, dx - \epsilon$$

so

$$\int_a^b f \, dg \geq \int_a^b f(x)g'(x) \, dx. \qquad (6.10)$$

Likewise

$$\underline{S}(f; g; P) = \sum m_i(f)[g(x_i) - g(x_{i-1})]$$

$$= \sum m_i(f)g'(t_i)(x_i - x_{i-1}) \leq \sum f(t_i)g'(t_i)\Delta x_i.$$

Thus

$$\underline{S}(f; g; P) < \int_a^b f(x)g'(x) \, dx + \epsilon$$

so

$$\int_a^b f\, dg \leq \int_a^b f(x)g'(x)\, dx. \tag{6.11}$$

Together, Equations (6.10) and (6.11) show that

$$\int_a^b f\, dg = \int_a^b f(x)g'(x)dx. \quad \blacksquare$$

Next we prove some of the linearity properties of the Riemann-Stieljtes integral. There are two functions that must be considered, the integrand and the integrator, and we show that the Riemann-Stieljtes integral is linear in both.

Theorem 6.25:

(a) Suppose f_1 and f_2 are integrable with respect to g on $[a, b]$. Then for any real numbers α and β, $\alpha f_1 + \beta f_2$ is integrable with respect to g on $[a, b]$ and

$$\int_a^b (\alpha f_1 + \beta f_2)\, dg = \alpha \int_a^b f_1\, dg + \beta \int_a^b f_2\, dg.$$

(b) Suppose f is integrable with respect to the increasing functions g_1 and g_2. Then for any nonnegative numbers α and β, f is integrable with respect to $\alpha g_1 + \beta g_2$ and

$$\int_a^b f\, d(\alpha g_1 + \beta g_2) = \alpha \int_a^b f\, dg_1 + \beta \int_a^b f\, dg_2.$$

(The condition that α and β be nonnegative is to ensure that $\alpha g_1 + \beta g_2$ is increasing.)

Proof:

(a) The proof is virtually identical to the proof of Theorem 6.13.

(b) Let P be a partition of $[a, b]$. Then

$$\overline{S}(f; \alpha g_1 + \beta g_2; P) = \sum M_i(f)[(\alpha g_1 + \beta g_2)(x_i) - (\alpha g_1 + \beta g_2)(x_{i-1})]$$

$$= \sum M_i(f)[\alpha g_1(x_i) - \alpha g_1(x_{i-1})] + M_i(f)[\beta g_2(x_i) - \beta g(x_{i-1})]$$

$$= \alpha \sum M_i(f)[g_1(x_i) - g_1(x_{i-1})] + \beta \sum M_i(f)[g_2(x_i) - g(2(x_{i-1})]$$

$$= \alpha \overline{S}(f; g_1; P) + \beta \overline{S}(f; g_2; P).$$

Likewise,

$$\underline{S}(f; \alpha g_1 + \beta g_2; P) = \alpha \underline{S}(f; g_1; P) + \beta \underline{S}(f; g_2; P).$$

Let $\epsilon > 0$ be given. Since f is integrable with respect to g_1, there is a partition P_1 such that

$$\overline{S}(f; g_1; P_1) - \underline{S}(f; g_1; P_1) < \frac{\epsilon}{2(\alpha + 1)} \qquad (6.12)$$

and since f is integrable with respect to g_2, there is a partition P_2 such that

$$\overline{S}(f; g_2; P_2) - \underline{S}(f; g_2; P_2) < \frac{\epsilon}{2(\beta + 1)}. \qquad (6.13)$$

Let $Q = P_1 \cup P_2$. Then Q is a refinement of both P_1 and P_2, so both Equations (6.12) and (6.13) hold. Thus

$$\overline{S}(f; \alpha g_1 + \beta g_2; Q) - \underline{S}(f; \alpha g_1 + \beta g_2; Q)$$

$$= [\alpha \overline{S}(f; g_1; Q) + \beta \overline{S}(f; g_2; Q)] - [\alpha \underline{S}(f; g_1; Q) + \beta \underline{S}(f; g_2; Q)]$$

$$= \alpha [\overline{S}(f; g_1; Q) - \underline{S}(f; g_1; Q)] + \beta [\overline{S}(f; g_2; Q) - \underline{S}(f; g_2; Q)]$$

$$\leq \frac{\alpha \epsilon}{2(\alpha + 1)} + \frac{\beta \epsilon}{2(\beta + 1)} < \epsilon$$

which shows that f is integrable with respect to $\alpha g_1 + \beta g_2$. Since

$$\overline{S}(f; \alpha g_1 + \beta g_2; P) = \alpha \overline{S}(f; g_1; P) + \beta \overline{S}(f; g_2; P)$$

and

$$\underline{S}(f; \alpha g_1 + \beta g_2; P) = \alpha \underline{S}(f; g_1; P) + \beta \underline{S}(f; g_2; P)$$

for any partition P, it follows that

$$\int_a^b f\, d(\alpha g_1 + \beta g_2) = \alpha \int_a^b f\, dg_1 + \beta \int_a^b f\, dg_2. \qquad \blacksquare$$

The next theorem and its corollary give a useful device for computing Riemann-Stieltjes integrals. It is not the most general possible result, but it will suffice for most applications. We do not give the proof of the theorem, but describe the ideas that would be involved in a proof.

Theorem 6.26: Let f and g be functions defined on $[a, b]$, with g increasing. Suppose for some $c \in (a, b)$,

(i) fg' has an antiderivative $F(x)$ on $[a, c)$ and an antiderivative $G(x)$ on $(c, b]$ and

$$\lim_{x \uparrow c} F(x) \text{ and } \lim_{x \downarrow c} G(x)$$

exist.

(ii) g has a jump discontinuity at c, and f is continuous at c.

Then

$$\int_a^b f \, dg = \lim_{x \uparrow c}[F(x) - F(a)] + f(c) \left[\lim_{x \downarrow c} g(x) - \lim_{x \uparrow c} g(x)\right] + \lim_{x \downarrow c}[G(b) - G(x)].$$

Idea of the Proof: For $\delta > 0$ sufficiently small,

$$\int_a^b f \, dg = \int_a^{c-\delta} f \, dg + \int_{c-\delta}^{c+\delta} f \, dg + \int_{c+\delta}^b f \, dg.$$

The hypotheses imply that g is differentiable on $[a, c - \delta)$ and $(c + \delta, b]$ for any $\delta > 0$ so

$$\int_a^{c-\delta} f \, dg = \int_a^{c-\delta} f(x) g'(x) \, dx = F(c - \delta) - F(a)$$

and

$$\int_{c+\delta}^b f \, dg = \int_{c+\delta}^b f(x) g'(x) \, dx = G(b) - G(c + \delta).$$

Now consider $\int_{c-\delta}^{c+\delta} f \, dg$. Since f is continuous at c, $f(x)$ can be made arbitrarily close to $f(c)$ on $[c - \delta, c + \delta]$ by making δ sufficiently small. Thus $\int_{c-\delta}^{c+\delta} f \, dg$ can be made arbitrarily close to $f(c) \int_{c-\delta}^{c+\delta} dg$ by making δ sufficiently small. But

$$\int_{c-\delta}^{c+\delta} dg = g(c + \delta) - g(c - \delta)$$

so $\int_{c-\delta}^{c+\delta} f \, dg$ can be made arbitrarily close to $f(c)[g(c + \delta) - g(c - \delta)]$ if δ is sufficiently small. Thus $\int_a^b f \, dg$ may be made close to

$$F(c - \delta) - F(a) + f(c)[g(c + \delta) - g(c - \delta)] + G(b) - G(c + \delta)$$

by taking δ sufficiently small. Now take the limit as $\delta \to 0$ to get the result. ∎

Corollary 6.26: Theorem 6.26 remains true if g has a finite number of discontinuities and f is continuous at points where g is discontinuous. ∎

Example 6.7:
Let $f(x) = e^x$ and

$$g(x) = \begin{cases} x^2 & \text{if } 0 \le x < 1 \\ 1 + 2x & \text{if } 1 \le x \le 2 \\ 5 + x & \text{if } 2 < x \le 3. \end{cases}$$

Then f is continuous, and g has jump discontinuities at $x = 1$ and $x = 2$. Then according to Corollary 6.26,

$$\int_0^3 f \, dg \int_0^1 e^x (2x) \, dx + e \left[\lim_{x \downarrow 1}(1 + 2x) - \lim_{x \uparrow 1}(x^2) \right]$$

$$+ \int_1^2 e^x \cdot 2 \, dx + e^2 \left[\lim_{x \downarrow 2}(5 + x) - \lim_{x \uparrow 2}(1 + 2x) \right] + \int_2^3 e^x \, dx$$

$$= \left[2 \left(xe^x \Big|_0^1 - \int_0^1 e^x \, dx \right) \right]$$

$$+ e(3 - 1) + 2(e^2 - e) + e^2(7 - 5) + (e^3 - e^2)$$

$$= 2 + 3e^2 + e^3.$$

Example 6.8:

Let $f(x) = x^2 + 3$ and

$$g(x) = \begin{cases} 0 & \text{if } x = 0 \\ 1 + x^3 & \text{if } 0 < x \leq 1. \end{cases}$$

Then

$$\int_0^1 f \, dg = \lim_{x \downarrow 0}[g(x) - g(0)]f(0) + \int_0^1 f(x)g'(x) \, dx$$

$$= \lim_{x \downarrow 0}[(1 + x^3) - 0] \cdot 3 + \int_0^1 (x^2 + 3)(3x^2) \, dx$$

$$= 3 + \int_0^1 (3x^4 + 9x^2) \, dx = 3 + \left(\frac{3}{5} + 3 \right).$$

Theorem 6.27: Suppose f is a bounded function on $[a, b]$. Then f is Riemann-Stieltjes integrable with respect to the increasing function g on $[a, b]$ if and only if, for every $c \in (a, b)$, f is Riemann-Stieltjes integrable with respect to g on $[a, c]$ and $[c, b]$. If this is the case, then

$$\int_a^b f \, dg = \int_a^c f \, dg + \int_c^b f \, dg.$$

Proof: See Theorem 6.14.   ■

Thus far we have restricted our attention to the case where the integrator function g is an increasing function. We now enlarge the class of functions that may be considered as integrator functions. First notice that if g is a nonincreasing function, then $-g$ is a nondecreasing function.

Definition: Suppose g is a nonincreasing function, and suppose that f is Riemann-Stieltjes integrable with respect to $-g$. Then we define the Riemann-Stieltjes integral of f with respect to g, denoted $\int_a^b f \, dg$, to be

$$\int_a^b f \, dg = -\int_a^b f \, d(-g).$$

In a similar manner, we can allow our integrator function to be a linear combination of increasing functions. In fact, it suffices to consider functions of the type $g = g_1 - g_2$ where g_1 and g_2 are nondecreasing functions. A function that may be so expressed is said to be of *bounded variation*.

Definition: Let g_1 and g_2 be nondecreasing functions, and suppose f is Riemann-Stieltjes integrable with respect to g_1 and g_2 on $[a, b]$. Then we define the Riemann-Stieltjes integral of f with respect to $g = g_1 - g_2$, denoted $\int_a^b f \, dg$, by

$$\int_a^b f \, dg = \int_a^b f \, dg_1 - \int_a^b f \, dg_2.$$

With this definition, Theorem 6.26(b) can be extended to include the case where α or β is negative.

Every theorem we have proven in this section is valid for the integrator functions g of bounded variation where $g = g_1 - g_2$, if g_1 and g_2 satisfy the hypotheses of the theorem. Where one must be careful is in trying to prove theorems involving inequalities similar to Theorem 6.15 and its corollaries. Often an analogous result will hold if the integrator function is nondecreasing, but one should be extremely cautious in trying to prove such theorems when the integrator function is an arbitrary function of bounded variation.

As our final result in this section, we want to prove an "integration by parts" theorem. We first prove the following preliminary result.

Theorem 6.28: If f and g are nondecreasing functions on $[a, b]$ and $P = \{x_0, x_1, ..., x_n\}$ is any partition of $[a, b]$, then

$$\overline{S}(f; g; P) = f(b)g(b) - f(a)g(a) - \underline{S}(f; g; P)$$

and

$$\underline{S}(f; g; P) = f(b)g(b) - f(a)g(a) - \overline{S}(f; g; P).$$

Proof: Let $P = \{x_0, x_1, ..., x_n\}$ be a partition of $[a, b]$. Then

$$\overline{S}(f; g; P) = \sum_{i=1}^{n} f(x_i)[g(x_i) - g(x_{i-1})]$$

since $M_i = f(x_i)$ if f is nondecreasing. Now

$$\sum_{i=1}^{n} f(x_i)[g(x_i) - g(x_{i-1})]$$

$$= f(x_1)[g(x_1) - g(x_0)] + f(x_2)[g(x_2) - g(x_1)]$$

$$+ f(x_3)[g(x_3) - g(x_2)] + \cdots + f(x_{n-1})[g(x_{n-1} - g(x_{n-2})]$$

$$+ f(x_n)[g(x_n) - g(x_{n-1})]$$

$$= f(x_n)g(x_n) + g(x_{n-1})[f(x_{n-1} - f(x_n)]$$

$$+ g(x_{n-2})[f(x_{n-2}) - f(x_{n-1})]$$

$$+ \cdots + g(x_1)[f(x_1) - f(x_2)] - f(x_1)g(x_0)$$

$$= -f(x_1)g(x_0) + \sum_{i=1}^{n-1} g(x_i)[f(x_i) - f(x_{i+1})] + f(x_n)g(x_n). \qquad (6.14)$$

Now adding and subtracting $f(x_0)g(x_0)$ to Equation (6.14) gives

$$\overline{S}(f; g; P) = -f(x_0)g(x_0) + f(x_0)g(x_0) - f(x_1)g(x_0)$$

$$- \sum_{i=1}^{n-1} g(x_i)[f(x_{i+1}) - f(x_i)] + f(x_n)g(x_n)$$

$$= -f(x_0)g(x_0) + f(x_n)g(x_n) - \sum_{i=0}^{n-1} g(x_i)[f(x_{i+1}) - f(x_i)]$$

$$= f(b)g(b) - f(a)g(a) - \sum_{i=1}^{n} g(x_{i-1})[f(x_i) - f(x_{i-1})]$$

(since $x_0 = a$ and $x_n = b$)

$$= f(b)g(b) - f(a)g(a) - \underline{S}(g; f; P)$$

since $m_i(g) = g(x_{i-1})$ if g is increasing.

The second equality follows immediately from the first. ■

Theorem 6.29 (Integration by Parts): Suppose f and g are increasing functions on $[a, b]$. Then f is Riemann-Stieltjes integrable with respect to g if and only if g is Riemann-Stieltjes integrable with respect to f. If this is the case, then

$$\int_a^b f\ dg = f(b)g(b) - f(a)g(a) - \int_a^b g\ df.$$

Proof: Suppose f is Riemann integrable with respect to g. Let $\epsilon > 0$ be given. There is a partition P of $[a, b]$ such that

$$\overline{S}(f; g; P) - \underline{S}(f; g; P) < \epsilon.$$

Now

$$\overline{S}(f; g; P) - \underline{S}(f; g; P) = [f(b)g(b) - f(a)g(a) - \underline{S}(f; g; P)]$$
$$-[f(b)g(b) - f(a)g(a) - \overline{S}(f; g; P)]$$
$$= \overline{S}(g; f; P) - \underline{S}(g; f; P)$$

by Theorem 6.28. Thus g is Riemann-Stieltjes integrable with respect to f. Likewise, if g is Riemann-Stieltjes integrable with respect to f, then f is Riemann-Stieltjes integrable with respect to g.

Now, if both f and g are Riemann-Stieltjes integrable with respect to each other, there are partitions P_1 and P_2 of $[a, b]$ such that

$$\overline{S}(f; g; P_1) - \int_a^b f\ dg < \epsilon/2$$

and

$$\int_a^b g\ df - \underline{S}(g; f; P_2) < \epsilon/2.$$

Then if $P = P_1 \cup P_2$

$$\left| \int_a^b f\ dg - \left(f(b)g(b) - f(a)g(a) - \int_a^b g\ df \right) \right|$$

$$= \left| \left(\int_a^b f\ dg - \overline{S}(f; g; P) \right) + \left\{ \overline{S}(f; g; P) - \left[f(b)g(b) - f(a)g(a) - \int_a^b g\ df \right] \right\} \right|$$

$$= \left| \left(\int_a^b f\ dg - \overline{S}(f; g; P) \right) + [f(b)g(b) - f(a)g(a) - \underline{S}(g; f; P)] \right.$$

$$\left. - \left[f(b)g(b) - f(a)g(a) - \int_a^b g\ df \right] \right|$$

$$\leq \left| \int_a^b f\ dg - \overline{S}(f; g; P) \right| + \left| \int_a^b g\ df - \underline{S}(f; g; P) \right|$$

$$< \epsilon/2 + \epsilon/2 = \epsilon$$

which shows $\int_a^b f dg = f(b)g(b) - f(a)g(a) - \int_a^b g\ df$. ∎

Corollary 6.29(a): If $f = f_1 - f_2$ and $g = g_1 - g_2$ where f_1, f_2, g_1, and g_2 are nondecreasing functions and f_i is Riemann-Stieltjes integrable with respect to g_j $(i, j = 1, 2)$, then f is Riemann-Stieltjes integrable with respect to g and

$$\int_a^b f \, dg = f(b)g(b) - f(a)g(a) - \int_a^b g \, df.$$

The proof is left for Exercise 6, Section 6.3. ∎

Also, as a corollary to Theorems 6.23 and 6.29, we have

Corollary 6.29(b): If f is monotone and g is continuous on $[a, b]$, then f is Riemann-Stieltjes integrable with respect to g on $[a, b]$. ∎

Exercises 6.3

1. Let
$$g(x) = \begin{cases} 0 & \text{if } a \leq x \leq c \\ 1 & \text{if } c < x \leq b. \end{cases}$$

Show that f is Riemann-Stieltjes integrable with respect to g on $[a, b]$ if and only if f is continuous from the right at $x = c$.

2. Calculate $\int_0^3 f \, dg$ for the following functions:
   (a)
$$f(x) = \begin{cases} 2x & \text{if } 0 \leq x < 1 \\ 2x^3 & \text{if } 1 < x \leq 3/2 \\ e^x & \text{if } 3/2 < x \leq 3, \end{cases}$$

$$g(x) = \begin{cases} 0 & \text{if } 0 \leq x < 1 \\ 1 & \text{if } 1 \leq x \leq 2 \\ 3 & \text{if } 2 < x \leq 3. \end{cases}$$

   (b)
$$f(x) = \begin{cases} \sin x & \text{if } 0 \leq x < \pi/2 \\ x^2 & \text{if } \pi/2 \leq x \leq 3, \end{cases}$$

$$g(x) = \begin{cases} 2x & \text{if } 0 \leq x < 1 \\ 4x^3 & \text{if } 1 \leq x \leq 3. \end{cases}$$

   (c)
$$f(x) = \begin{cases} [x] & \text{if } 0 \leq x < 3/2 \\ e^x & \text{if } 3/2 \leq x \leq 3, \end{cases}$$

$$g(x) = x + [x] \text{ on } [0, 3].$$

3. Suppose f is continuous and g is increasing on $[a, b]$. Show that

$$\int_a^b f \, dg \leq \int_a^b |f| \, dg \leq \left(\int_a^b |f| \, dx \right) (g(b) - g(a)).$$

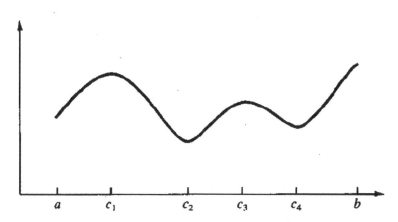

Figure 6.4
Graph for Exercise 5(b).

4. During the course of the proof of Theorem 6.26, we showed that under the hypotheses of the theorem

$$\overline{S}(f; \alpha g_1 + \beta g_2; P) = \alpha \overline{S}(f; g_1; P) + \beta \overline{S}(f; g_2; P)$$

and

$$\underline{S}(f; \alpha g_1 + \beta g_2; P) = \alpha \underline{S}(f; g_1; P) + \beta \underline{S}(f; g_2; P).$$

Show how one can conclude from this that

$$\int_a^b f\, d(\alpha g_1 + \beta g_2) = \alpha \int_a^b f\, dg_1 + \beta \int_a^b f\, dg_2.$$

5. An important class of continuous functions on a closed, bounded interval can be expressed as the difference of two continuous nondecreasing functions. Show how this can be done in the following examples.
   (a) $f(x) = x^2$ on [-1,1].
   (b) The graph of $f(x)$ is as in Figure 6.4.

6. Prove Corollary 6.29(a).

7

Series of Real Numbers

7.1 Tests for Convergence of Series

In this chapter we consider series of real numbers, a topic that is intimately related to sequences of real numbers. What one would like to do in a series is to add an infinite number of numbers. We begin with a sequence $\{a_n\}$, that consists of the numbers we want to sum. From $\{a_n\}$ we construct a new sequence $\{S_n\}$ as follows:

Let

$$S_1 = a_1$$

$$S_2 = a_1 + a_2$$

$$\vdots$$

$$S_k = a_1 + a_2 + \cdots + a_k.$$

Thus $\{S_k\}$ is the sum of the first k terms of the sequence $\{a_n\}$. Notice that $S_k - S_{k-1} = a_k$ for $k > 1$.

Definition: A series $\sum_{n=1}^{\infty} a_n$ is defined as the ordered pair of sequences $(\{a_n\}, \{S_n\})$ where the sequences are related as above.

Definition: Let $\{a_n\}$ and $\{S_n\}$ be the sequences above. Then $\{S_n\}$ is called the *sequence of partial sums* for the series $\sum_{n=1}^{\infty} a_n$. The numbers a_n are called the *terms* of the series $\sum_{n=1}^{\infty} a_n$.

Definition: We say *the series* $\sum_{n=1}^{\infty} a_n$ *converges to* L if the sequence of partial sums converges to L and we say the series $\sum_{n=1}^{\infty} a_n$ *diverges* if the sequence of partial sums diverges. We say *the series* $\sum_{n=1}^{\infty} a_n$ *diverges to* ∞ or $-\infty$ if the sequence of partial sums does so.

Notice that since

$$S_n = \sum_{i=1}^{n} a_i$$

then

$$\lim_{n \to \infty} S_n = \lim_{n \to \infty} \sum_{i=1}^{n} a_i$$

and we shall sometimes write

$$\sum_{i=1}^{\infty} a_i = \lim_{n \to \infty} \sum_{i=1}^{n} a_i.$$

For compactness of notation, we shall often write $\sum a_n$ for $\sum_{n=1}^{\infty} a_n$.

It is very important to understand that the question of convergence or divergence of a series is exactly the same question as convergence or divergence of the sequence of partial sums for the series. This means we can apply the results of Chapter 2 in determining properties of series. An illustration of this is given by the proof of Theorem 7.1.

Theorem 7.1: Suppose that the series $\sum a_n$ converges to a and the series $\sum b_n$ converges to b. Then for any numbers α and β the series $\sum (\alpha a_n + \beta b_n)$ converges to $\alpha a + \beta b$. That is,

$$\sum (\alpha a_n + \beta b_n) = \alpha a + \beta b.$$

Note: Just as with sequences, we can combine series to make a new series. The series $\sum (\alpha a_n + \beta b_n)$ is the series whose terms are the terms in the sequence $\{\alpha a_n + \beta b_n\}$.

Proof: Let $\{S_n\}$ and $\{T_n\}$ be the sequences of partial sums for $\sum a_n$ and $\sum b_n$, respectively. By Theorem 2.4

$$\lim(\alpha S_n + \beta T_n) = \alpha \lim S_n + \beta \lim T_n = \alpha a + \beta b. \quad \blacksquare$$

Theorem 7.2: The convergence or divergence of a series is not affected by the addition of a finite number of terms to the series.

Proof: For simplicity, we first demonstrate the proof of the theorem when only one number is added to the series. Let $\sum a_n$ be a series and t be a given number. Let $\{S_n\}$ be the sequence of partial sums associated with the series $\sum a_n$. By definition, we know the series $\sum a_n$ converges exactly when the sequence $\{S_n\}$ converges. We also know that the sequence $\{S_n\}$ converges if and only if it is a Cauchy sequence. Suppose that we add the number t between the k and $k+1$ terms of the original series. Let the sequence of partial sums be denoted by $\{S'_n\}$. Suppose that $m > n > k+1$. Then

$$S'_m = a_1 + \cdots + a_k + t + a_{k+1} + \cdots + a_{n-1} + a_n + \cdots + a_{m-1}$$

$$S_n = a_1 + \cdots + a_k + t + a_{k+1} + \cdots + a_{n-1}$$

so that
$$S'_m - S'_n = a_n + \cdots + a_{m-1} = S_{m-1} - S_{n-1}.$$

Thus $\{S'_n\}$ is a Cauchy sequence if and only if $\{S_n\}$ is a Cauchy sequence. To complete the proof for an arbitrary finite number of additional terms, we observe that these numbers can be added one at a time, and that convergence is not affected by the addition of a single term. ∎

This is the first of several places where the Cauchy criterion for convergence will be used. Notice that to apply the criterion, we do not need to have a candidate for the number to which the sequence might converge. In Theorem 7.2, if a series does converge, adding additional terms to the series will affect the number to which the series converges.

In dealing with series, one usually does not try to find an explicit formula for the nth term in the sequence of partial sums. This task is often impossible. Most often, one must be satisfied with determining whether the series converges. That is, even if we are able to determine that a series converges, often we shall be unable to find the number to which it converges. An exception to this viewpoint is geometric series, which will be discussed shortly. But first, we present some more preliminary results.

Theorem 7.3: If $\sum a_n$ converges, then $\lim a_n = 0$.

Proof: Let $\sum a_n$ be a convergent series, and let $\{S_n\}$ be the sequence of partial sums associated with $\sum a_n$. To prove the result, we need to show that given any $\epsilon > 0$, $|a_n| < \epsilon$ for n sufficiently large.

Let $\epsilon > 0$ be given. Since $\sum a_n$ is a convergent series, $\{S_n\}$ is a convergent sequence and is thus a Cauchy sequence. Then there is a number N such that if $m > n > N$, then $|S_m - S_n| < \epsilon$. So if $m = n + 1$ and $n > N$, we have $|S_{n+1} - S_n| = |a_{n+1}| < \epsilon$, proving the theorem. ∎

In the proof of Theorem 7.3, N depends on ϵ and perhaps this should be emphasized by writing $N(\epsilon)$. However, we shall often use the less cumbersome notation for simplicity.

It is important to realize that Theorem 7.3 does *not* say that if $\lim a_n = 0$ then the series $\sum a_n$ converges. There are, in fact, many series where $\lim a_n = 0$ and the series diverges. It does say that if $\lim a_n \neq 0$, then the series must diverge. The dominant problem in this section will be, given a series, find whether it converges or diverges. We shall develop several theorems that deal with this problem, but not every theorem will be applicable to every series. In practice, the difficulty is often in deciding which theorem to apply. Usually one checks whether $\lim a_n = 0$ first, because it is often an easy question to answer. However, if it is true that $\lim a_n = 0$, we must do further analysis to determine whether the series converges or diverges.

Definition: A geometric series is a series of the form

$$a + ar + ar^2 + ar^3 + ar^4 + \cdots.$$

Thus each term in a geometric series is obtained by multiplying the previous term by some fixed number.

Geometric series are particularly nice because the partial sums can be explicitly calculated. They are also about the only type of series that we shall actually sum. The next theorem tells us how to do this.

Theorem 7-4: Let

$$a + ar + ar^2 + ar^3 + ar^4 + \cdots$$

be a geometric series. If $a = 0$ the series converges to 0. If $a \neq 0$, the series converges to $a/(1-r)$ if $|r| < 1$ and diverges if $|r| \geq 1$.

Proof: If $a = 0$, then every term in the series is 0, so the series converges to 0. If $a \neq 0$, then the nth partial sum can be written

$$S_n = a + ar + ar^2 + ar^3 + \cdots + ar^{n-1}$$

so

$$r S_n = ar + r^2 + ar^3 + \cdots + ar^n.$$

Subtracting gives

$$S_n - r S_n = (1-r)S_n = a - ar^n.$$

Thus, if $r \neq 1$,

$$S_n = \frac{a - ar^n}{1-r}.$$

It has been shown in Exercise 2, Section 2.2, that if $|r| < 1$, then $\lim r^n = 0$. So in this case

$$\lim S_n = \frac{a}{1-r}$$

as claimed.

If $|r| \geq 1$, then $\lim a_n \neq 0$, so the series diverges.   ∎

Example 7.1:

The series

$$40 + 4 + .4 + .04 + \cdots$$

is a geometric series with $a = 40$ and $r = .1$ (a is always the first term). Since $|r| < 1$, the series converges and, in fact, converges to $40/(1 - .1) = 400/9$.

Example 7.2:
The series

$$1 - 4 + 16 - 64 + \cdots$$

is a geometric series with $a = 1$ and $r = -4$. Since $|r| \geq 1$, the series diverges.

Example 7.3:
We show that a countable set has measure zero. Let $\{x_1, x_2, ...\}$ be a countable set, and let $\epsilon > 0$ be given. The set of open intervals

$$\left\{ \left(x_i - \frac{\epsilon}{2^{i+1}}, x_i + \frac{\epsilon}{2^{i+1}} \right) \middle| i = 1, 2, ... \right\}$$

contains the countable set. The sum of the lengths of the intervals is

$$\sum_{i=1}^{\infty} \frac{\epsilon}{2^i} = \epsilon.$$

In Chapter 2 we saw that monotonic sequences are relatively simple to analyze because monotonic sequences converge if and only if they are bounded. We want to exploit this property in our analysis of series. Notice that if we have a series $\sum a_n$ whose terms are all nonnegative, then this series gives rise to a monotonic increasing sequence of partial sums because

$$S_k - S_{k-1} = a_k \geq 0$$

Thus, if we restrict our attention to series whose terms are nonnegative, our analysis may be simplified. This is what we shall do for the next several theorems.

Theorem 7.5: Let $\sum a_n$ be a series of nonnegative terms. Then $\sum a_n$ either converges or it diverges to ∞.
Proof: Let $\{S_n\}$ be the sequence of partial sums for $\sum a_n$. Since

$$S_k - S_{k-1} = a_k \geq 0,$$

$\{S_n\}$ is a monotonic increasing sequence. Thus $\{S_n\}$ (and thus also $\sum a_n$) converges if it is bounded and diverges to ∞ otherwise. ∎

Theorem 7.6 (Comparison Test): Let $\sum a_n$ and $\sum b_n$ be series of nonnegative terms with $a_n \leq b_n$ for every n. Then
(a) if $\sum b_n$ converges, then so does $\sum a_n$;
(b) if $\sum a_n$ diverges, then so does $\sum b_n$.
Proof:
(a) Let $\{S_n\}$ and $\{T_n\}$ be the sequence of partial sums for $\sum a_n$ and $\sum b_n$, respectively. By Theorem 7.5, we only need to show that $\{S_n\}$ is bounded.

By the same result, we know that $\{T_n\}$ is bounded, since $\sum b_n$ converges. Since $a_i \le b_i$ for every i, we have

$$S_n = a_1 + \cdots + a_n \le b_1 + \cdots + b_n = T_n \text{ for every } n.$$

Thus, if $T_n \le L$ for every n, then $S_n \le L$ for every n, and so $\{S_n\}$ is a bounded monotone increasing sequence, which must converge. That is, $\sum a_n$ converges.

(b) We use the same notation as in part (a). To prove the result, we must show that $\{T_n\}$ diverges to ∞ if $\{S_n\}$ diverges to ∞. That is, given a number M, we must show that there is a number N such that if $n > N$ then $T_n > M$. Let M be given. Since $\sum a_n$ diverges to ∞, so does $\{S_n\}$, and thus there is a number N such that $S_n > M$ if $n > N$. Then

$$T_n \ge S_n > M \text{ for } n > N. \quad \blacksquare$$

Corollary 7.6(a): Let $\sum a_n$ and $\sum b_n$ be series such that there is a number N with $b_n \ge a_n \ge 0$ for $n > N$. Then the same conclusions as in Theorem 7.6 are valid.

The proof is left for Exercise 6, Section 7.1. $\quad \blacksquare$

Corollary 7.6(b): Let $\sum a_k$ be a series of the form

$$a_k = \frac{k^n + b_{n-1}k^{n-1} + \cdots + b_0}{k^m + c_{m-1}k^{m-1} + \cdots + c_0}.$$

Then $\sum a_k$ converges if and only if $\sum(k^n/k^m)$ converges.
Proof: We have

$$k^n + b_{n-1}k^{n-1} + \cdots + b_0 = k^n\left(1 + \frac{b_{n-1}}{k} + \cdots + \frac{b_0}{k^n}\right).$$

For k sufficiently large,

$$\frac{1}{2} < 1 + \frac{b_{n-1}}{k} + \cdots + \frac{b_0}{k^n} < \frac{3}{2}$$

so that

$$\frac{1}{2}k^n < k^n + b_{n-1}k^{n-1} + \cdots + b_0 < \frac{3}{2}k^n$$

for k sufficiently large. Similarly

$$\frac{1}{2}k^m < k^m + c_{m-1}k^{m-1} + \cdots + c_0 < \frac{3}{2}k^m$$

for k sufficiently large.

Thus

$$\frac{\frac{1}{2}k^n}{\frac{3}{2}k^m} < a_k < \frac{\frac{3}{2}k^n}{\frac{1}{2}k^m}$$

for k sufficiently large. Thus $\sum a_n$ converges if and only if $\sum(k^n/k^m)$ converges. ∎

If we have a series of positive terms, $\sum a_n$ and $\lim a_n = 0$, then the series *may* converge. (It may also diverge.) Whether it converges depends on how "rapidly" the terms of $\sum a_n$ go to zero. For example, the terms in the series $\sum(1/n^2)$ go to zero more rapidly than the terms of the series $\sum(1/n)$ because for a given positive integer n, $1/n^2$ is smaller than $1/n$. The more rapidly the terms go to zero, the more likely it is that the series will converge. Our next two theorems give tests for convergence of series with positive terms. These two tests, the ratio test and the root test, will normally work only for series whose terms go to zero very rapidly. The reason is that series to which these tests apply to show convergence must be comparable to geometric series whose terms go to zero very rapidly. Notice this, as the proofs of these theorems are discussed. This is not to minimize their importance – they are among our most useful tests.

Theorem 7.7 (Ratio Test): Let $\sum a_n$ be a series of positive terms. Let

$$R = \overline{\lim}\frac{a_{n+1}}{a_n} \text{ and } r = \underline{\lim}\frac{a_{n+1}}{a_n}.$$

Then the series $\sum a_n$ converges if $R < 1$ and diverges if $r > 1$.

Note: The test is inconclusive if $r \le 1 \le R$.

Before writing a formal proof, we give the idea of the proof, which is to compare the series with a geometric series. Suppose that $a_{n+1}/a_n = r$ for every n. Then $a_{n+1} = ra_n$ for every n. Thus

$$a_2 = ra_1$$

$$a_3 = ra_2 = r(ra_1) = r^2a_1$$

$$a_4 = ra_3 = r(r^2a_1) = r^3a_1$$

$$\vdots$$

$$a_n = r^{n-1}a_1$$

so that

$$a_1 + a_2 + \cdots + a_n = a_1 + ra_1 + r^2a_1 + \cdots + r^{n-1}a_1.$$

That is, we have a geometric series that converges if $r < 1$ and diverges if $r > 1$.

Proof: Let

$$R = \overline{\lim}(a_{n+1}/a_n)$$

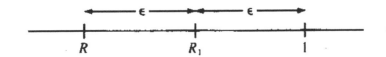

Figure 7.1
See proof of Theorem 7.7.

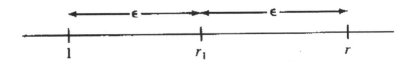

Figure 7.2
See proof of Theorem 7.7.

and suppose that $R < 1$. We shall show that the series converges. In Theorem 2.17 we showed that there are only finitely many terms a_{n+1}/a_n larger than $R + \epsilon$ for any $\epsilon > 0$. Choose $\epsilon = (1 - R)/2$, and let $R_1 = R + \epsilon$. (See Figure 7.1.)

Then $R_1 < 1$ and $a_{n+1}/a_n < R_1$ for all but finitely many n. Thus there is a number N such that if $n \geq N$ then $a_{n+1} < R_1 a_n$. So

$$a_N + a_{N+1} + a_{N+2} + \cdots < a_N + a_N R_1 + a_N (R_1)^2 + \cdots.$$

The series on the right is a convergent geometric series, since $R_1 < 1$. Thus, by the comparison test, the series on the left converges. Since adding a finite number of terms to a series does not affect its convergence, the original series converges.

We now show that if $r = \underline{\lim}(a_{n+1}/a_n) > 1$, the series $\sum a_n$ diverges. Suppose $r > 1$. By Theorem 2.17, for any $\epsilon > 0$ there are at most finitely many terms a_{n+1}/a_n that are less than $r - \epsilon$. Choose $\epsilon = (r - 1)/2$ and let $r_1 = r - \epsilon > 1$. (See Figure 7.2.) Then there is a number N such that if $n > N$, $a_{n+1}/a_n > r_1$. Thus

$$a_{N+k} > a_N (r_1)^k > a_N \text{ and } \lim a_n \neq 0.$$

Therefore the series diverges. ■

Example 7.4:
Consider the series whose nth term is given by $a_n = n/2^n$. Then

$$\frac{a_{n=1}}{a_n} = \left(\frac{n+1}{2^{n+1}}\right) \div \left(\frac{n}{2^n}\right) = \frac{n+1}{n} \cdot \frac{1}{2}$$

so that $\lim(a_{n+1}/a_n) = 1/2$, and the series converges by the ratio test.

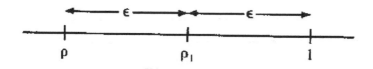

Figure 7.3
See proof of Theorem 7.8.

Example 7.5:
Consider the series

$$\frac{1}{2} + \frac{1}{3} + \left(\frac{1}{2}\right)^2 + \left(\frac{1}{3}\right)^2 + \left(\frac{1}{2}\right)^3 + \left(\frac{1}{3}\right)^3 + \cdots$$

that is, the series $\sum a_n$ where

$$a_{2n-1} = (1/2)^n \text{ and } a_{2n} = (1/3)^n.$$

Then $a_{2n+1}/a_{2n} = (3/2)^n(1/2)$ and $a_{2n}/a_{2n-1} = (2/3)^n$. Thus

$$\overline{\lim}\frac{a_{n+1}}{a_n} = \infty \text{ and } \underline{\lim}\frac{a_{n+1}}{a_n} = 0$$

so the test is inconclusive.

Theorem 7.8(Root Test): Let $\sum a_n$ be a series of nonnegative terms, and let

$$\rho = \overline{\lim}(a_n)^{1/n}.$$

Then the series $\sum a_n$ converges if $\rho < 1$ and diverges if $\rho > 1$.
Note: The test is inconclusive if $\rho = 1$.
We again discuss the idea behind the proof before giving a formal argument. Suppose $\rho = \lim(a_n)^{1/n}$. Then ρ^n is approximately equal to a_n for n sufficiently large. Thus the series is approximately a geometric series for n large enough. The series $\sum \rho^n$ converges if $\rho < 1$ and diverges if $\rho > 1$.
Proof: Let $\rho = \overline{\lim}(a_n)^{1/n}$, and suppose that $\rho < 1$. Then by Theorem 2.17, for any $\epsilon > 0$ there are at most finitely many terms that exceed $\rho + \epsilon$. Choose $\epsilon = (1 - \rho)/2$, and let $\rho_1 = \rho + \epsilon < 1$. (See Figure 7.3.)
Thus there is a number N such that if $n > N$, then $(a_n)^{1/n} < \rho_1$; that is, $a_n < (\rho_1)^n$. Thus

$$a_N + a_{N+1} + a_{N+2} + \cdot < (\rho_1)^N + (\rho_1)^{N+1} + (\rho_1)^{N+2} + \cdots$$

so that beyond the Nth term the series is dominated by a convergent geometric series; and so the series on the left must converge by the comparison test. Since

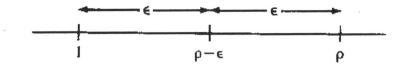

Figure 7.4
See proof of Theorem 7.8.

adding a finite number of terms to a series does not affect its convergence, the original series converges.

Now suppose that $\rho > 1$. We show that the series diverges. By Theorem 2.17, we know that for any $\epsilon > 0$, there are infinitely many terms of the series $\sum a_n$ for which $(a_n)^{1/n} > \rho - \epsilon$. Choose $\epsilon = (\rho - 1)/2$. (See Figure 7.4.)

Then, for any number N, there is a number $n > N$ with $a_n > (\rho - \epsilon)^n > 1$. Thus $\lim a_n \neq 0$, so the series $\sum a_n$ diverges. ■

Example 7.6:
Again consider the series

$$\frac{1}{2} + \frac{1}{3} + \left(\frac{1}{2}\right)^2 + \left(\frac{1}{3}\right)^2 + \left(\frac{1}{2}\right)^3 + \left(\frac{1}{3}\right)^3 + \cdots.$$

Earlier we saw that the ratio test was not conclusive for this series. However, applying the root test gives

$$\lim(a_{2n})^{1/2n} = \lim\left(\frac{1}{3n}\right)^{1/2n} = \frac{1}{\sqrt{3}} < 1$$

and

$$\lim(a_{2n-1})^{1/(2n-1)} = \lim\left(\frac{1}{2^n}\right)^{1/(2n-1)} = \frac{1}{\sqrt{2}} < 1$$

so the series converges by the root test.

If the ratio test is conclusive for a given series, then the root test will also be conclusive, as we prove in Exercise 19, Section 7.1. Example 7.6 shows that the converse does not hold. However the root test is often more difficult to apply.

As we said, series that can be shown to converge using the ratio test or the root test converge very rapidly. To illustrate this, we show in an exercise (Exercise 10, Section 7.1) that if $\sum a_n$ is a series that can be shown to converge by either of these tests, then the series $\sum n^k a_n$ can also be shown to converge by the same test. Several other tests are available when the ratio test is inconclusive in that $\lim(a_{n+1}/a_n) = 1$, including Kummer's Test, Raabe's test and Gauss's test, which may be found in many other texts. (See Olmstead, pp. 395–397 for example.)

   Theorem 7.9 (Integral Test): Let $\sum a_n$ be a series of positive numbers with

$$a_1 \geq a_2 \geq a_3 \geq \cdots .$$

Let $f(x)$ be a nonincreasing continuous function on $[0, \infty)$ such that $f(n) = a_n$ for each positive integer n. Then

$$S_n = a_1 + \cdots + a_n \geq \int_1^n f(x)\, dx$$

and

$$S_n - a_1 = a_2 + \cdots + a_n \leq \int_1^n f(x)\, dx$$

so that the series $\sum a_n$ converges if and only if the improper integral $\int_1^\infty f(x)\, dx$ converges. If the series converges, then

$$\sum a_n \geq \int_1^\infty f(x)\, dx \geq \sum a_n - a_1 .$$

Thus, if the integral test applies, we have bounds on the number to which the series converges.

   Proof: Suppose the hypotheses of the theorem hold. On the real number line between the positive integers n and $n + 1$ draw a rectangle of height a_n, for each positive integer n. Then $a_1 + \cdots + a_n = S_n =$ (area of the rectangles above the interval $[1, n + 1]$). (See Figure 7.5.)

   If $a_1 \geq a_2 \geq a_3 \geq \cdots$ as shown in Figure 7.5, and if $f(x)$ is a continuous nonincreasing function such that $f(n) = a_n$ for each positive integer n, then it is clear that

$$S_n = a_1 + \cdots + a_n \geq \int_1^n f(x)\, dx .$$

Now if we keep $f(x)$ fixed but slide the rectangles one unit to the left as shown in Figure 7.6, then it is again clear that

$$a_2 + \cdots + a_n \leq \int_1^n f(x)\, dx$$

so that $S_n - a_1 \leq \int_1^n f(x)dx \leq S_n$
for every positive integer n. Taking the limit as $n \to \infty$ gives the result. ∎

   Corollary 7.9(p-series Test): The series $\sum 1/n^p$ converges if $p > 1$ and diverges if $p \leq 1$.

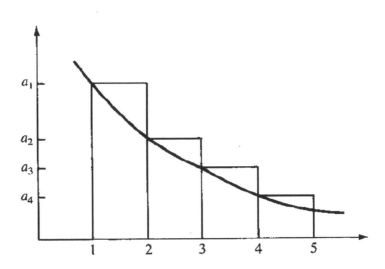

Figure 7.5
A sum that exceeds the area under the curve.

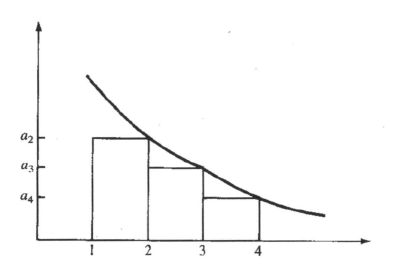

Figure 7.6
A sum that is less than the area under the curve.

Proof: Let $f(x) = x^{-p}$. Then the hypotheses of Theorem 7.9 are satisfied, so the series $\sum 1/n^p$ converges if and only if the integral $\int_1^\infty x^{-p}\, dx$ converges. This occurs if and only if $p > 1$. ∎

Note: For $p = 1$ the series $\sum 1/n^p = \sum 1/n$ is called the *harmonic series*.

Example 7.7:
The series
$$\sum \frac{k^3 - 2k^2 + 3k - 1}{k^5 - 5k^4 + 6}$$
converges because, by the proof of Corollary 7.6(b),
$$\frac{1}{3k^2} < \frac{k^3 - 2k^2 + 3k - 1}{k^5 + 5k^4 + 6} < \frac{3}{k^2}$$
for k sufficiently large. But by the p-series test $\sum 1/k^2$ converges.

Example 7.8:
Determine the convergence or divergence of the series whose nth term is given by

(a) $1/(n \ln n)$. Using the integral test gives
$$\int_2^\infty \frac{1}{x \ln x}\, dx = \ln(\ln x)\Big|_2^\infty \to \infty$$
so the series diverges. An interesting aspect of this series is the slow rate at which it diverges. (Notice also that we must begin summing at $n = 2$ to avoid dividing by 0.) To exceed the number n, we need e^{e^n} terms of the series; so to exceed 10, for example, we need $e^{e^{10}}$ or about 10^{9500} terms.

(b) $n^n/n!$ The ratio test will give $\lim(a_{n+1}/a_n) = \lim(1 + 1/n)^n = e$, so the series diverges. However, it is much easier to observe that $\lim a_n \neq 0$.

(c) $2^n/n!$ Here $\lim a_n = 0$, so the series may converge or diverge. Since a factorial is involved, we try the ratio test. This gives
$$\lim \frac{a_{n+1}}{a_n} = \lim \frac{2^{n+1}/(n+1)!}{2^n/n!} = \lim \frac{2}{n+1} = 0$$
so the series converges.

(d) e^n/n^n. This may be written $(e/n)^n$, so the root test seems an obvious choice. It gives
$$\lim(a_n)^{1/n} = \lim \frac{e}{n} = 0$$
so the series converges.

(e) $(\ln n)/(\exp\sqrt{n})$. Now $\sqrt{n} > 3\ln n$ so that

$$\exp\sqrt{n} > \exp(3\ln n) = n^3$$

and $\ln n < n$. Thus

$$\frac{\ln n}{\exp\sqrt{n}} < \frac{n}{n^3} = \frac{1}{n^2}$$

and the series $\sum 1/n^2$ converges by the p-series test.

Exercises 7.1

1. Determine the convergence or divergence of a series whose nth term is given by
   (a) $n^4/(3n^2 + 4n - 1)$.
   (b) $n!/n^n$.
   (c) $1/(n + 4\sqrt{n})$.
   (d) $1/[n(\ln n)(\ln(\ln(n)))]$, $n \geq 10$.
   (e) $n!/e^n$.
   (f) $n!/e^{n^2}$.
   (g) $\sqrt{n+1} - \sqrt{n}$.
   (h) $1/(\ln n)^{\ln n}$, $n \geq 2$.
   (i) $1/[n(\ln n)^2]$, $n \geq 2$.
   (j) $(n!)^2/(2n)!$
   (k) $1/(\ln n)^{\sqrt{n}}$, $n \geq 2$.

2. (a) Show that $\sum_{k=2}^{\infty} 1/(\ln k)^n$ diverges for any value of n. (b) Show that $\sum_{k=2}^{\infty}(\ln k)^n/k^{1+\epsilon}$ converges for any value of n and any $\epsilon > 0$.

3. Show that the series $\sum a_k$ converges if and only if, given any $\epsilon > 0$, there is a number N (which depends on ϵ) such that

$$\left| \sum_{k=n}^{m} a_k \right| < \epsilon \text{ for any } m > n > N.$$

4. This is a somewhat different type of comparison test. Suppose that $\sum a_n$ and $\sum b_n$ are series of positive terms and that there is a positive number c such that $a_n \leq cb_n$ for every n. Show that if $\sum b_n$ converges, then $\sum a_n$ converges, and if $\sum a_n$ diverges, then $\sum b_n$ diverges.

5. Use Exercise 4 to show that if $\sum a_n$ and $\sum b_n$ are series of positive numbers and $\lim a_n/b_n = 1$, then $\sum a_n$ converges if and only if $\sum b_n$ converges. [*Hint:* for n large enough, $\frac{1}{2} < a_n/b_n < 2$.]

6. Prove Corollary 7.6(a).

7. Show that if $b_0, b_1, \ldots, b_{n-1}$ are real numbers, then for k sufficiently large

$$\frac{1}{2} < 1 + \frac{b_{n-1}}{k} + \cdots + \frac{b_0}{k^n} < \frac{3}{2}.$$

8. Show that for any $A > 1$ and any positive integer k,

$$\lim_{n \to \infty} \frac{n^k}{A^n} = 0.$$

9. Earlier we showed that $\{(1 + 1/n)^n\}$ is an increasing sequence and

$$\left(1 + \frac{1}{n}\right)^n \le \frac{1}{0!} + \frac{1}{1!} + \frac{1}{2!} + \cdots + \frac{1}{n!}.$$

(a) Show that if $n \ge 2$, then $n! \ge 2^{n-1}$.
(b) Show that $\{(1+1/n)^n\}$ is bounded above by 3. Conclude that $\lim(1 + 1/n)^n$ exists.

10. (a) Suppose that the series of positive numbers $\sum a_n$ can be shown to converge using either the ratio test or the root test. Show that the series $\sum n^k a_n$ converges for any positive integer k.
(b) Show that the series $\sum n^k r^n$ converges if $0 < r < 1$ for any positive integer k.

11. Suppose that the series of nonnegative terms $\sum a_n$ converges. Use the Schwarz inequality (see Exercise 18, Section 1.2) to show that $\sum \sqrt{a_n}/n^\alpha$ converges for any $\alpha > 1/2$.

12. Suppose that $\sum a_k$ is a convergent series whose terms are positive. Let $\{S_n\}$ be the sequence of partial sums for $\sum a_k$. Let $S = \lim S_n$ and let $\rho_k = a_{k+1}/a_k$.
(a) Suppose there is a number N for which $\rho_k \ge \rho_{k+1}$ if $k \ge N$ and $\rho_N < 1$. Show that $S - S_N \le a_{N+1}/(1 - \rho_N)$.
(b) Suppose there is a number N for which $\rho_k \le \rho_{k+1}$ if $k \le N$ and that there is a number $\rho < 1$ for which $\rho = \lim \rho_n$. Show that $S - S_N \le a_{N+1}/(1 - \rho)$.

13. Suppose $\{a_n\}$ and $\{b_n\}$ are series of positive terms.
(a) Suppose $a_{n+1}/a_n \le b_{n+1}/b_n$ and $\sum b_n$ converges. Show that $\sum a_n$ converges.
(b) Show that if $a_{n+1}/a_n \ge b_{n+1}/b_n$ and $\sum a_n$ diverges, then $\sum b_n$ diverges.

14. (a) Use the Binomial Theorem to show that if k is a positive integer, then for n sufficiently large, $(1 - k/n) \le (1 - 1/n)^k$.
(b) Suppose k is a positive integer larger than 1, and $\sum a_n$ is a series of positive terms with

$$\frac{a_{n+1}}{a_n} \le \left(1 - \frac{k}{n}\right).$$

Show that $\sum a_n$ converges. [*Hint:* Use Exercise 13 to compare a_{n+1}/a_n with $(1/n^{k+1})/(1/n^k)$.]

(c) Show that the series whose nth term is

$$\frac{(2)(4)\cdots(2n)}{(5)(7)\cdots(2n+5)}$$

converges.

15. Later we shall show that $\sum_{k=0}^{\infty} 1/k! = e$. For now, assume this is true. Let $S_n = \sum_{k=0}^{n} 1/k!$, so that $\lim S_n = e$. We want to estimate $e - S_n$.
    (a) Show that $\sum_{k=1}^{\infty} 1/(n+1)^k = 1/n$.
    (b) Show that $\sum_{k=n+1}^{\infty} 1/k! < 1/(n!n)$ so that $e - S_n < 1/(n!n)$.
    (c) How large must n be to ensure $\sum_{k=1}^{\infty} 1/k!$ approximates e to within 10^{-10}?

16. Suppose $\sum a_n$ is a series of positive terms with $\lim(a_{n+1}/a_n) = L$. Show that $\lim(a_n)^{1/n} = L$. This may be done by the following steps:
    (a) First suppose $L > 0$. Then choose $\epsilon > 0$ with $0 < \epsilon < L$. Then there is an N such that if $n > N$,

$$L - \epsilon < \frac{a_{n+1}}{a_n} < L + \epsilon.$$

(b) Show that $(L - \epsilon)^k a_N < a_{N+k} < (L + \epsilon)^k a_N$.
(c) Now

$$(L - \epsilon)^{N+k} \frac{a_N}{(L - \epsilon)^N} < a_{N+k} \tag{1}$$

and

$$a_{N+k} < \frac{a_N}{(L + \epsilon)^N}(L + \epsilon)^{N+k}. \tag{2}$$

(d) Take the $N + k$ roots of inequalities (1) and (2), and let $k \to \infty$. Conclude

$$L - \epsilon \le \lim_{k \to \infty}(a_{N+k})^{1/(N+k)} \le L + \epsilon.$$

(e) For the case $L = 0$, we have $\lim_{k \to \infty}(a_{N+k})^{1/(N+k)} \le \epsilon$. But $a_n \ge 0$, so $(a_n)^{1/n} \ge 0$ for all n.

17. Suppose that $\sum a_k$ is a divergent series of positive terms, and $S_n = a_1 + \cdots + a_n$ is the nth partial sum of the series. In this exercise we show that $\sum a_k/S_k$ diverges and $\sum a_k/(S_k)^2$ converges.
    (a) Let $b_k = a_k/S_k$. Show that

$$\sum_{k=m}^{n} b_k \ge \frac{1}{S_n}(a_m + \cdots + a_n) = \frac{1}{S_n}(S_n - S_{m-1}).$$

(b) Use part (a) to show that the sequence of partial sums for the series $\sum b_k$ is not a Cauchy sequence. Conclude that $\sum b_k = \sum a_k/S_k$ diverges.

(c) Let $c_k = a_k/(S_k)^2$. Use partial fractions to show that

$$c_k < \frac{a_k}{S_k S_{k-1}} = \left(\frac{1}{S_{k-1}} - \frac{1}{S_k}\right).$$

(d) Show that $\sum_{k=m}^n c_k =< 1/S_m$. Conclude that the sequence of partial sums for the series $c_k = \sum a_k/(S_k)^2$ is a Cauchy sequence and thus the series converges.

18. **(a)** Give an example of a series of nonnegative terms $\sum a_n$ for which $\sum a_n$ diverges, but $\sum a_n^2$ converges.
    (b) Show that if $\sum a_n$ is a series of nonnegative terms for which $\sum a_n^2$ converges, then $\sum a_n/n$ converges.

19. Suppose $\sum a_n$ is a series of positive terms. Here we show that

$$\underline{\lim}\frac{a_{n+1}}{a_n} \leq \underline{\lim}(a_n)^{1/n} \text{ and } \overline{\lim}(a_n)^{1/n} \leq \overline{\lim}\frac{a_{n+1}}{a_n}.$$

(Note that the limits need not exist as they must in Exercise 16.) This shows that if the ratio test can be used to illustrate the convergence or divergence of a series, then the root test will do the same for that particular series.

(a) Show $\overline{\lim}(a_{n+1}/a_n) \geq \overline{\lim}(a_n)^{1/n}$. This may be done by the following steps, which mimic Exercise 16.

(i) Let $L = \overline{\lim}(a_{n+1}/a_n)$. Let $\epsilon > 0$ be given. Show that there is an N such that
$$a_{N+k} < (L+\epsilon)^k a_N = (L+\epsilon)^{N+k}\frac{a_N}{(L+\epsilon)^N}.$$

(ii) Take the $(N+k)$th roots of both sides, and let $k \to \infty$.

(b) Show $\underline{\lim}(a_{n+1}/a_n) \leq \underline{\lim} a_n^{1/n}$. The proof requires minor modifications to the proof in part (a).

7.2   Operations Involving Series

Absolute and Conditional Convergence

In Section 7.1 we developed some tests to determine whether a series converges. Usually these tests (except for geometric series) applied to series with positive terms. We begin this section by discussing more general series. We shall see that there are two ways in which a series may converge, absolutely and conditionally. Then we shall investigate the questions of whether certain manipulations with series, in particular grouping and rearranging the terms, are valid. The answers depend on whether a series converges and the way in which it converges. Next, we discuss series in which the sign of each term differs from that of its predecessor (alternating series). Finally, we discuss the multiplication of two series.

Definition: A series $\sum a_n$ is said to be *absolutely convergent* if $\sum |a_n|$ converges. If $\sum a_n$ converges but $\sum |a_n|$ diverges, the series is said to be *conditionally convergent*.

Example 7.9:
Let $\sum a_n = 1 - 1 + 1/2 - 1/2 + 1/3 - 1/3 + \cdots$. If S_n is the sequence of partial sums for the series, then

$$S_n = \begin{cases} 0 & \text{if } n \text{ is even} \\ \frac{2}{n+1} & \text{if } n \text{ is odd.} \end{cases}$$

Thus $\lim S_n = 0$, so the series converges to 0. However,

$$\sum |a_n| = 1 + 1 + 1/2 + 1/2 + 1/3 + 1/3 \cdots$$

diverges, so the series is conditionally convergent.

Theorem 7.10: If a series is absolutely convergent, then it is convergent.
Proof: Let $\sum a_n$ be an absolutely convergent series. Let $\{S_n\}$ be the sequence of partial sums associated with $\sum a_n$, and let $\{T_n\}$ be the sequence of partial sums associated with $\sum |a_n|$. We use the Cauchy criterion to prove the theorem.
Let $\epsilon > 0$ be given. Since $\sum |a_n|$ converges, $\{T_n\}$ converges. So there is a number N such that if $m > n > N$, then $|T_m - T_n| < \epsilon$. Now

$$|S_m - S_n| = |a_m + \cdots + a_{n+1}| \leq |a_m| + \cdots + |a_{n+1}| = |T_m - T_n|$$

by the triangle inequality.
Therefore, if $m > n > N$, then $|S_m - S_n| < \epsilon$; that is, $\{S_n\}$ is a Cauchy sequence. Thus $\{S_n\}$ converges, and so does the series $\sum a_n$. ∎

Corollary 7.7: Let $\sum a_n$ be a series of nonzero, but not necessarily positive, terms. Let

$$R = \overline{\lim} \frac{|a_{n+1}|}{|a_n|} \quad \text{and} \quad r = \underline{\lim} \frac{|a_{n+1}|}{|a_n|}.$$

Then the series $\sum a_n$ converges absolutely if $R < 1$ and diverges if $r > 1$. ∎

Corollary 7.8: Let $\sum a_n$ be a series, not necessarily of positive terms. Let

$$\rho = \overline{\lim} \, |a_n|^{1/n}.$$

Then the series converges absolutely if $\rho < 1$ and diverges if $\rho > 1$. ■

Remarks on the Proof of the Corollaries: The conditions for the series to converge absolutely are identical to those of the theorems that dealt with series whose terms were nonnegative. To see why the series diverge under the given conditions, recall that in the proofs of the theorems, we showed that $\lim a_n \neq 0$ if $r > 1$ or $\rho > 1$.

Theorem 7.11: Let $\sum a_n$ be a conditionally convergent series. Choose the positive terms of the series $\sum a_n$ and form the series $\sum b_n$. Choose the negative terms of the series $\sum a_n$ and form the series $\sum c_n$. Then the series $\sum b_n$ diverges to ∞ and the series $\sum c_n$ diverges to $-\infty$.

Idea of the Proof: We first argue that at least one of the series $\sum b_n$ or $\sum c_n$ must diverge. If $\sum b_n$ converges to L and $\sum c_n$ converges to $-K$, then it is not hard to show that the sequence of partial sums for $\sum |a_n|$ is bounded by $L + K$ so the series $\sum a_n$ would then be absolutely convergent.

Now suppose that exactly one of the series $\sum b_n$ or $\sum c_n$ diverges. We shall argue that the series $\sum a_n$ must diverge. Suppose that $\sum b_n$ diverges to ∞ and $\sum c_n$ converges to $-K$. Then one can show that the series $\sum a_n$ diverges to ∞, so that $\sum a_n$ was not conditionally convergent. ■

Rearrangement and Regrouping of Series

One often would like to do certain operations with series that are valid arithmetic operations, such as grouping the terms. The problem is that the usual rules of arithmetic do not always apply to series. For example, if we have a finite collection of numbers that we want to add, say $1 + 2 + 3 + \cdots + 10$, then we can group the terms any way we like by inserting parentheses, say $(1+2) + (3+4) + \cdots + (9+10)$ and get the same answer. Consider, however, if the series

$$1 - 1 + 1 - 1 + 1 - 1 + \cdots$$

is grouped as

$$1 + (-1+1) + (-1+1) + (-1+1) \cdots = 1,$$

then we get a different answer than if the terms are grouped as

$$(1-1) + (1-1) + (1-1) + (1-1) + \cdots = 0.$$

Thus, one cannot blindly apply the associative law to series. The same holds for the commutative law. With a finite number of terms, we are free to move

the numbers around any way we want and the answer will always be the same. For example,
$$1 + 2 + 3 = 2 + 1 + 3 = 3 + 1 + 2 = 6.$$

With series, we shall see this is not always the case. That is, one is not always free to rearrange the terms of a series. By a rearrangement of a series $\sum a_n$, we mean a new series $\sum b_n$ whose terms consist of the same numbers as those of $\sum a_n$ but which appear in a different order. Thus if

$$\sum a_n = 1 + 1/2 + 1/3 + 1/4 + 1/5 + \cdots$$

and

$$\sum b_n = 1/2 + 1 + 1/3 + 1/4 + 1/5 + \cdots,$$

then $\sum b_n$ is a rearrangement of $\sum a_n$. More precisely, we give the following definition.

Definition: Let f be a 1-1 onto function from the positive integers to the positive integers. Let $\sum a_n$ be a series and define a new series $\sum b_n$ by

$$b_n = a_{f(n)}$$

for n a positive integer. Then $\sum b_n$ is a *rearrangement* of $\sum a_n$.

In our preceding example $f(1) = 2$, $f(2) = 1$, and $f(n) = n$, otherwise.

The next theorem shows that by rearranging a series, it may be possible to get a series that behaves radically different from the original series.

Theorem 7.12: Let $\sum a_n$ be a conditionally convergent series. Then given any number L (which may be finite or infinite), it is possible to construct a rearrangement of $\sum a_n$ that converges to L.

Proof: We demonstrate the proof in the case that L is positive and finite. Choose L and let $\epsilon > 0$ be given.

Let b_k denote the kth nonnegative term of the series $\sum a_n$ and c_k denote the kth negative term of the series $\sum a_n$. Since $\sum a_n$ converges, there is a number N such that $|a_n| < \epsilon$ if $n \geq N$. Notice that this implies $|b_k| < \epsilon$ and $|c_k| < \epsilon$ if $k \geq N$. By Theorem 7.11, $\sum b_n$ diverges to ∞ and $\sum c_n$ diverges to $-\infty$. We inductively construct a rearrangement of $\sum a_n$ that converges to L. First add $b_1 + \cdots + b_{n_1}$ so that

$$b_1 + \cdots + b_{n_1} > L$$

but

$$b_1 + \cdots + b_{n_1 - 1} \leq L.$$

That is, we choose the first numbers in the rearrangement by adding the first positive terms of $\sum a_n$ until L is exceeded.

Next we add the first negative terms of $\sum a_n$ until our sum falls below L. That is

$$b_1 + \cdots + b_{n_1} + c_1 + \cdots + c_{n_2} < L$$

but

$$b_1 + \cdots + b_{n_1} + c_1 + \cdots + c_{n_2-1} \geq L.$$

Now we add the next positive terms until we just exceed L, and so on.

We claim this rearrangement of $\sum a_n$ converges to L.

Suppose that $\{S_n\}$ is the sequence of partial sums for the rearranged series. Suppose that n is large enough so that we have included at least N of the b_i's and at least N of the c_i's. Then

$$|S_n - L| \leq \sup\{b_k, |c_k| \mid k \geq N\} \leq \epsilon$$

so $\lim S_n = L$ and the rearranged series converges to L. ∎

We now know some caution is in order when trying to apply the rules of arithmetic to series. A reasonable question is: When can we regroup or rearrange series without affecting the convergence of the original series?

First we further investigate the effect of inserting parentheses into a series. Suppose we begin with a series $a_1 + a_2 + \cdots$ and then add parentheses, grouping some of the terms. The new series looks like

$$(a_1 + \cdots + a_{n_1}) + (a_{n_1+1} + \cdots + a_{n_2}) + \cdots .$$

To sum the new series, we first sum the terms within parentheses and then add the parenthetical terms. In effect, we have made a new series $\sum b_n$ where

$$b_1 = a_1 + \cdots + a_{n_1}$$

$$b_2 = a_{n_1+1} + \cdots + a_{n_2}$$

$$\vdots$$

We say that the series $\sum b_n$ is obtained from the series $\sum a_n$ by inserting parentheses. We want to know when is $\sum a_n = \sum b_n$. We know by our earlier example that this is not always the case.

Theorem 7.13: Suppose $\sum a_n$ converges to L and $\sum b_n$ is obtained from $\sum a_n$ by inserting parentheses. Then $\sum b_n$ also converges to L. That is, inserting parentheses has no effect on a convergent series.

Proof: Let $\{S_n\}$ be the sequence of partial sums associated with $\sum a_n$. The hypotheses imply that $\lim S_n = L$. Let $\{T_n\}$ be the sequence of partial sums associated with $\sum b_n$. That is,

$$T_1 = b_1 = (a_1 + \cdots + a_{n_1}) = S_{n_1}$$

$$T_2 = b_1 + b_2 = (a_1 + \cdots + a_{n_1}) + (a_{n_1+1} + \cdots + a_{n_2}) = S_{n_2}$$

$$\vdots$$

$$T_k = b_1 + \cdots + b_k = a_1 + \cdots + a_{n_k} = S_{n_k}$$

where $n_1 < n_2 < \cdots < n_k < \cdots$. That is, $\{T_n\}$ is a subsequence of $\{S_n\}$. But a subsequence of a convergent sequence must converge to the same number as the original sequence. Thus

$$\sum b_n = \lim T_n = \lim S_n = L. \quad \blacksquare$$

For a rearrangement of a series to exhibit the same convergence as the original series, we need more restrictive hypotheses, as Theorem 7.12 shows. What is needed is absolute convergence.

Theorem 7.14: Let $\sum a_n$ be an absolutely convergent series. Let $\sum a'_n$ be the series whose terms are defined by

$$a'_n = \begin{cases} a_n & \text{if } a_n \geq 0 \\ 0 & \text{if } a_n < 0 \end{cases}$$

and $\sum a''_n$ be the series whose terms are defined by

$$a''_n = \begin{cases} a_n & \text{if } a_n < 0 \\ 0 & \text{if } a_n \geq 0. \end{cases}$$

Then $\sum a'_n$ and $\sum a''_n$ are absolutely convergent series and

$$\sum a_n = \sum a'_n + \sum a''_n$$

The proof is left for Exercise 4, Section 4.2. $\quad \blacksquare$

Theorem 7.15: Let $\sum a_n$ be an absolutely convergent series, and let $\sum a'_n$ and $\sum a''_n$ be defined as in Theorem 7.14. Suppose $\sum a'_n = P$ and $\sum a''_n = M$. (Note that $M \leq 0$.) Then any rearrangement of $\sum a_n$ converges to $P + M$.

Proof: Let $\sum b_n$ be any arrangement of $\sum a_n$ and let f be the 1-1 onto function that defines the rearrangement; that is, $b_n = a_{f(n)}$. In Exercise 2(b) Section 7.2 we show that $\sum b_n$ converges absolutely and thus $\sum b_n$ converges. Let $\epsilon > 0$ be given. W shall show that

$$\left| \sum b_n - (P + M) \right| < \epsilon.$$

Since $\sum a'_n = P$ and $\sum a''_n = M$, there is a number N_1 such that if $n > N_1$, then

$$\left| \sum_{k=1}^{n} a'_k - P \right| < \frac{\epsilon}{3} \qquad (1)$$

and a number N_2 such that if $n > N_2$, then

$$\left| \sum_{k=1}^{n} a''_k - M \right| < \frac{\epsilon}{3}. \qquad (2)$$

Since $\sum b_n$ converges absolutely, there is a number N_3 such that if $n \geq N_3$, then

$$\sum_{k=n}^{\infty} |b_k| < \frac{\epsilon}{3}. \qquad (3)$$

Choose $N > N_3$ such that $\{b_1, ..., b_N\}$ contains

$$\{a_1, ..., a_{N_1}\} \cup \{a_1, ..., a_{N_2}\} = \{a_1, ..., a_{\overline{N}}\}$$

where $\overline{N} = \max\{N_1, N_2\}$. It is left for Exercise 5, Section 7.2, to show this is always possible.) Now

$$\left| \sum_{n=1}^{N} b_n - (P+M) \right| \leq \left| \sum_{n=1}^{N} b_n - (P+M) \right| + \left| \sum_{n=N+1}^{\infty} b_n \right|$$

$$= \left| \left(\sum{}^{'} b_n - P \right) + \left(\sum{}^{''} b_n - M \right) \right| + \left| \sum_{n=N+1}^{\infty} b_n \right|$$

where $\sum' b_n$ indicates we sum over the b_n's that are nonnegative and whose subscripts are smaller than N and $\sum'' b_n$ is the sum over b_n's that are negative and whose subscripts are smaller than N. Now

$$\sum{}^{'} b_n \geq \sum_{n=1}^{N_1} a'_n \quad \text{and} \quad \sum{}^{''} b_n \leq \sum_{n=1}^{N_2} a''_n \qquad \text{(Why?)}$$

so that

$$\left| \sum{}^{'} b_n - P \right| < \frac{\epsilon}{3} \quad \text{and} \quad \left| \sum{}^{''} b_n - M \right| < \frac{\epsilon}{3}.$$

Thus

$$\left| \left(\sum{}^{'} b_n - P \right) + \left(\sum{}^{''} b_n - M \right) \right| + \left| \sum_{n=N+1}^{\infty} b_n \right|$$

$$\leq \left| \sum{}^{'} b_n - P \right| + \left| \sum{}^{''} b_n - M \right| + \sum_{n=N+1}^{\infty} |b_n|$$

$$< \frac{\epsilon}{3} + \frac{\epsilon}{3} + \frac{\epsilon}{3} = \epsilon. \quad \blacksquare$$

In summary, we can add parentheses to a convergent series without affecting the convergence of the series or the number to which the series converges. Also, we can rearrange absolutely convergent series without affecting these properties.

Definition: If $a_n > 0$ for every n, the series $\sum(-1)^n a_n$ and $\sum(-1)^{n+1} a_n$ are called *alternating series*.

That is, an alternating series is one where each term has a different sign from the preceding term.

Theorem 7.16 (Alternating Series Test): Let $\sum(-1)^n a_n$ be an alternating series such that
(i) $a_n \geq a_{n+1} > 0$ for every n.
(ii) $\lim a_n = 0$.
Then $\sum(-1)^n a_n$ and $\sum(-1)^{n+1} a_n$ converge.
   Proof: We give the proof for the series $\sum(-1)^{n+1} a_n$.
   Let $\{S_n\}$ be the sequence of partial sums associated with $\sum(-1)^{n+1} a_n$. Then

$$S_{2k} = (a_1 - a_2) + (a_3 - a_4) + \cdots + (a_{2k-1} - a_{2k}).$$

Since the numbers a_n are decreasing, each parenthetical term is nonnegative so that $S_{2k} \geq 0$. Also $\{S_{2k}\}$ is an increasing sequence since the parenthetical terms are nonnegative. But

$$S_{2k} = a_1 - (a_2 - a_3) - (a_4 - a_5) - \cdots - (a_{2k-2} - a_{2k-1}) - a_{2k} \leq a_1$$

since the parenthetical terms are all positive. Thus, $\{S_{2n}\}$ is a bounded monotone increasing sequence that must converge. Suppose the limit is L.
   Now

$$S_{2k+1} = S_{2k} + a_{2k+1}$$

so that

$$\lim S_{2k+1} = \lim S_{2k} + \lim a_{2k+1} = \lim S_{2k} + 0 = L$$

since $\lim a_n = 0$. Thus $\{S_n\}$ converges to L. (See Exercise 14, Section 2.2.) $\blacksquare$

Corollary 7.16:
(a) If $\sum(-1)^{n+1} a_n$ is an alternating series that satisfies the hypotheses of Theorem 7.16 and converges to L, then $L < a_1$.

(b) Let $\sum(-1)^{n+1}a_n$ be an alternating series that satisfies the hypotheses of Theorem 7.16 and converges to L. If $\{S_n\}$ is the sequence of partial sums associated with the series, then $|L - S_n| < |a_{n+1}|$.

The proof of part (b) is left for Exercise 7, Section 7.2. ∎

Product of Series

The last topic concerning series of numbers that we discuss is the product of series. We could define the product of the series $\sum a_n$ and $\sum b_n$ to be the series $\sum a_n b_n$. However, this would not extend the idea of multiplication of two sums of numbers. For example,

$$(a_0 + a_1)(b_0 + b_1) \neq a_0 b_0 + a_1 b_1$$

in general. We shall define the *Cauchy* product of two series. As motivation for our definition, consider the product of two polynomials

$$(a_0 + a_1 x + a_2 x^2 + \cdots + a_n x^n) \cdot (b_0 + b_1 x + b_2 x^2 + \cdots + b_n x^n).$$

The coefficient of x^k in this product is $\sum_{j=0}^{k} a_j b_{k-j}$.

Definition: Let $\sum_{n=0}^{\infty} a_n$ and $\sum_{n=0}^{\infty} b_n$ be series. The *Cauchy product* of these two series is the series $\sum c_n$ where

$$c_n = \sum_{k=0}^{n} a_k b_{n-k}.$$

(Notice that the sums begin at 0. This makes the notation simpler in this case.) Now the question is: How does the convergence of the series affect the convergence of the Cauchy product? The hope might be that the Cauchy product of two convergent series converges and converges to the product of the two limits of the series. This is not the case, as the following example shows.

Example 7.10:
Let $\sum_{n=0}^{\infty} a_n$ and $\sum_{n=0}^{\infty} b_n$ be the series whose nth terms are given by

$$a_n = b_n = \begin{cases} 0 & \text{if } n = 0 \\ \frac{(-1)^n}{\sqrt{n}} & \text{if } n > 0. \end{cases}$$

Notice that both series converge by the alternating series test. We shall show that the Cauchy product of the series diverges by showing that $\lim c_n \neq 0$. Now for $n \geq 2$,

$$c_n = \sum_{k=1}^{n-1} \frac{(-1)^k}{\sqrt{k}} \cdot \frac{(-1)^{n-k}}{\sqrt{n-k}} = (-1)^n \sum_{k=1}^{n-1} \frac{1}{\sqrt{nk - k^2}}.$$

Now

$$nk - k^2 \leq n \cdot \left(\frac{n}{2}\right) - \left(\frac{n}{2}\right)^2 = \frac{n^2}{4} \quad \text{(Why?)}$$

so that

$$\frac{1}{\sqrt{nk - k^2}} \geq \frac{2}{n}.$$

Thus

$$\sum_{k=1}^{n-1} \frac{1}{\sqrt{nk - k^2}} \geq (n-1)\left(\frac{2}{n}\right) = 2 - \frac{2}{n} \quad \text{and} \quad \lim_{n\to\infty} c_n \neq 0.$$

It is not too difficult to show that the Cauchy product of two absolutely convergent series is absolutely convergent. A more difficult theorem to prove is the following.

Theorem 7.17 (Mertens' Theorem): If $\sum_{n=0}^{\infty} a_n$ is an absolutely convergent series and $\sum_{n=0}^{\infty} b_n$ is a convergent series, then the Cauchy product of these series converges to

$$\left(\sum_{n=0}^{\infty} a_n\right)\left(\sum_{n=0}^{\infty} b_n\right).$$

Proof: The first key to the proof is that if $c_n = \sum_{k=0}^{N} a_k b_{n-k}$, then

$$\sum_{k=0}^{N} c_n = \sum_{k=0}^{N} a_k(b_0 + \cdots + b_{N-k}) = \sum_{k=0}^{N} a_k \left(\sum_{j=0}^{N-k} b_j\right)$$

which can be proved inductively. Then

$$\left| \sum_{n=0}^{N} c_n - \left(\sum_{k=0}^{\infty} a_k\right)\left(\sum_{j=0}^{\infty} b_j\right) \right|$$

$$= \left| \sum_{k=0}^{N} a_k \left(\sum_{j=0}^{N-k} b_j\right) - \left(\sum_{k=0}^{\infty} a_k\right)\left(\sum_{j=0}^{\infty} b_j\right) \right|$$

$$= \left| \sum_{k=0}^{N} a_k \left(\sum_{j=0}^{N-k} b_j\right) - \sum_{k=0}^{N} a_k \left(\sum_{j=o}^{\infty} b_j\right) + \sum_{k=0}^{N} a_k \left(\sum_{j=0}^{\infty} b_j\right) - \left(\sum_{k=0}^{\infty} a_k\right)\left(\sum_{j=0}^{\infty} b_j\right) \right|$$

$$\leq \left| \sum_{k=0}^{N} a_k \left(\sum_{j=0}^{N-k} b_j - \sum_{j=0}^{\infty} b_j\right) \right| + \left| \left(\sum_{k=0}^{N} a_k - \sum_{j=0}^{\infty} a_k\right)\left(\sum_{j=0}^{\infty} b_j\right) \right|$$

$$= \left| \sum_{k=0}^{N} a_k \left(\sum_{j=N-k+1}^{\infty} b_j \right) \right| + \left| \sum_{k=N+1}^{\infty} a_k \left(\sum_{j=0}^{\infty} b_j \right) \right|.$$

Let $\epsilon > 0$ be given. Choose N_1 such that if $n \geq N_1$, then $\sum_{k=n}^{\infty} |a_k| < \epsilon$, which is possible by the absolute convergence of $\sum a_n$.

Now choose N_2 such that if $n \geq N_2$, then $|\sum_{j=n}^{\infty} b_j| < \epsilon$, which is possible by the convergence of $\sum b_n$.

Let $N_3 = $ larger of N_1 and N_2. Then if $m \geq 2N_3$ we have

$$\left| \sum_{k=0}^{m} a_k \left(\sum_{j=m-k+1}^{\infty} b_j \right) \right| + \left| \sum_{k=m+1}^{\infty} a_k \left(\sum_{j=0}^{\infty} b_j \right) \right|$$

$$\leq \left| \sum_{k=0}^{N_3} a_k \left(\sum_{j=m-k+1}^{\infty} b_j \right) \right| + \left| \sum_{k=N_3+1}^{m} a_k \left(\sum_{j=m-k+1}^{\infty} b_j \right) \right| + \sum_{k=m+1}^{\infty} |a_k| \left| \sum_{j=0}^{\infty} b_j \right|. \quad (7.4)$$

This is the second key to the proof. We have broken the term

$$\sum_{k=0}^{m} a_k \left(\sum_{j=m-k+1}^{\infty} b_j \right)$$

into two pieces. The first part,

$$\sum_{k=0}^{N_3} a_k \left(\sum_{j=m-k+1}^{\infty} b_j \right)$$

has the property that, for each $k \in \{0, 1, \ldots, N_3\}$,

$$\left| \sum_{j=m-k+1}^{\infty} b_j \right| < \epsilon$$

because $m \geq 2N_3$, which means the smallest index where the summation can begin is $j = N_3$. Thus

$$\left| \sum_{k=0}^{N_3} a_k \left(\sum_{j=m-k+1}^{\infty} b_j \right) \right| < \epsilon \left| \sum_{k=0}^{N_3} a_k \right|.$$

The second part,

$$\sum_{k=N_3+1}^{m} a_k \left(\sum_{j=m-k+1}^{\infty} b_j \right)$$

can be made arbitrarily small since

$$\left| \sum_{k=N_3+1}^{m} a_k \left(\sum_{j=m-k+1}^{\infty} b_j \right) \right| \le \sum_{k=N_3+1}^{m} |a_k| \left| \sum_{j=m-k+1}^{\infty} b_j \right|$$

$$\le \sum_{k=N_3+1}^{\infty} |a_k| B < \epsilon B$$

where

$$B = \sup \left\{ \left| \sum_{j=k}^{\infty} b_j \right| \mid k = 0, 1, \dots \right\}$$

and B is finite since $\sum b_n$ converges. Finally, the third piece of Equation (7.4) can be made arbitrarily small, since

$$\sum_{k=m+1}^{\infty} |a_k| \left| \sum_{j=0}^{\infty} b_j \right| < \epsilon B. \quad \blacksquare$$

Exercises 7.2

1. Write a formal proof of Theorem 7.11.

2. *(a)* Show that if $\sum a_n$ is a series of nonnegative terms that diverges to ∞, then any regrouping or rearrangement also diverges to ∞.
   (b) Show that if $\sum a_n$ is an absolutely convergent series and $\sum |a_n|$ converges to L, and if $\sum b_n$ is any rearrangement of $\sum |a_n|$, then $\sum |b_n|$ converges to L.

3. Show that if $\sum a_n$ is a conditionally convergent series, then it is possible to have a rearrangement of $\sum a_n$ that diverges to ∞.

4. Prove Theorem 7.14.

5. *(a)* In the proof of Theorem 7.15, justify the step that says we can choose a number N such that

$$\{b_1, \dots, b_N\} \supset \{a_1, \dots, a_{\overline{N}}\}.$$

   (b) In the proof of Theorem 7.15, why is $\sum' b_n \ge \sum_{n=1}^{N_1} a'_n$, and why does this imply $|\sum' b_n - P| < \epsilon/3$?

6. Tell whether the series whose nth term is given below converges absolutely, converges conditionally, or diverges.

(a)
$$\frac{(-1)^n}{2n^2 + 1}.$$

(b)
$$\frac{(-1)^n}{\ln n}, \quad (n \geq 2).$$

(c)
$$(-1)^n \frac{n^n}{(n+1)^{n+1}}.$$

(d)
$$(-1)^n \frac{n}{n+4}.$$

7. Prove part (b) of Corollary 7.16.

8. Find $(-1)^n/(10n)$ accurate to 0.02.

9. Justify the equation in the proof of Theorem 7.17 that says

$$\sum_{n=0}^{N} c_n = \sum_{n=0}^{N} a_k \left(\sum_{j=0}^{N-k} b_j \right).$$

10. This exercise provides a proof that the Cauchy product of two absolutely convergent series converges to the product of the series (independent of Mertens' Theorem).

Suppose that $\sum_{n=0}^{\infty} a_n$ and $\sum_{n=0}^{\infty} b_n$ are two absolutely convergent series, and $\sum_{n=0}^{\infty} c_n$ is the Cauchy product of these series.
(a) If $\sum_{n=0}^{\infty} |a_n| = L$ and $\sum_{n=0}^{\infty} |b_n| = K$ show that for any positive integer N

$$\sum_{n=0}^{N} |c_n| \leq \left(\sum_{n=0}^{N} |a_n| \right) \left(\sum_{n=0}^{N} |b_n| \right) \leq LK.$$

Thus $\sum_{n=0}^{\infty} |c_n|$ converges, so $\sum_{n=0}^{N} c_n$ converges.
(b) Show that the terms in multiplying

$$\left(\sum_{n=0}^{N} a_n \right) \left(\sum_{n=0}^{N} b_n \right) \quad \text{all appear in the terms of} \quad \sum_{n=0}^{2N} c_n,$$

and that all the terms in

$$\sum_{n=0}^{N} c_n \quad \text{appear in the product} \quad \left(\sum_{n=0}^{N} a_n \right) \left(\sum_{n=0}^{N} b_n \right).$$

(c) Show that

$$\lim_{N\to\infty} \sum_{n=0}^{N} c_n = \left(\lim_{N\to\infty} \sum_{n=0}^{N} a_n \right) \left(\lim_{N\to\infty} \sum_{n=0}^{N} b_n \right).$$

11. Here we briefly discuss double series and give the steps necessary to prove Theorem 6.9. Suppose for each ordered pair of positive integers (m, n) that $a_{m,n}$ is a real number. We say that the double series $\sum_{m,n} a_{m,n}$ converges to L if, given any $\epsilon > 0$, there is a number N such that if $N_1 > N$ and $N_2 > N$ then

$$\left| \sum_{n=1}^{N_1} \sum_{m=1}^{N_2} a_{m,n} - L \right| < \epsilon.$$

It can be shown that if
(i) $\sum_{n=1}^{\infty} a_{m,n}$ converges absolutely to A_m,
(ii) $\sum_{m=1}^{\infty} a_{m,n}$ converges absolutely to B_n, and
(iii) either $\sum A_m$ or $\sum B_n$ converges, then $\sum |a_{m,n}|$ converges and

$$\sum |a_{m,n}| = \sum_{m=1}^{\infty} A_m = \sum_{n=1}^{\infty} B_n.$$

Use these facts to show that the union of a countable number of sets of Lebesgue measure zero has Lebesgue measure zero. [*Hint:* Let each set of $\{A_1, A_2, ...\}$ have measure zero. Let $\epsilon > 0$ be given. For each set A_j, there is a countable number of open intervals $\{I_{j1}, I_{j2}, ...\}$, with $A_j \subset \cup_{k=1}^{\infty} I_{jk}$ and $\sum_{k=1}^{\infty} m(I_{jk}) < \epsilon/2^j$.]

8

Sequences and Series of Functions

In this chapter we begin a discussion about sequences and series of functions. This topic has important applications in engineering and the physical sciences, since these fields of study often utilize models in which a function can best be described by an infinite sum of functions, say $\sum f_n(x)$. Typically one would like to differentiate or integrate this function. The logical approach would be to integrate or differentiate this series term by term. For example, one would like to say

$$\frac{d}{dx} \sum f_n(x) = \sum \frac{d}{dx} f_n(x).$$

As we know from our work with series of numbers, one must be careful in performing operations on infinite sums that are legitimate with finite sums. One question we shall address is: When can a series of functions be differentiated term by term? Our approach to the problem of series of functions will be similar to the one that was used with series of numbers. That is, we shall first consider sequences of functions and then use the results we prove about sequences to prove results about series.

Finally in this chapter we shall discuss the Taylor series of a function. The results we obtain here will give conditions governing when a function may be represented as a series of the form $\sum a_n(x - c)^n$ on some open interval containing c.

8.1 Sequences of Functions

Let E be a subset of the real numbers. Suppose that for $n = 1, 2, ..., f_n$ is a function defined on E. Then for each $x \in E$, $\{f_n(x)\}$ is a sequence of real numbers. Suppose that, for each $x \in E$, the sequence of real numbers $\{f_n(x)\}$ is convergent. Then we can define a function f on E by

$$f(x) = \lim_{n \to \infty} f_n(x) \text{ for each } x \in E. \tag{8.1}$$

We want to know what properties of the functions f_n carry over to the "limit function" f. For example, if each function f_n is continuous, must f be continuous? That is, if

$$\lim_{x \to c} f_n(x) = f_n(c) \qquad (8.2)$$

for each positive integer n and each $c \in E$, must it be true that

$$\lim_{n \to \infty} \left(\lim_{x \to c} f_n(x) \right) = \lim_{x \to c} f(x) = \lim_{x \to c} \left(\lim_{n \to \infty} f_n(x) \right)? \qquad (8.3)$$

(One takes the inner limit first.)

Another question deals with the integration properties of the functions f_n and the function f. In particular, suppose that $\int_a^b f_n(x)\, dx$ exists for every n. Then must it be true that $\int_a^b f(x)\, dx$ exists and if so, must

$$\lim_{n \to \infty} \int_a^b f_n(x)\, dx = \int_a^b \left(\lim_{n \to \infty} f_n(x) \right) dx = \int_a^b f(x)\, dx? \qquad (8.4)$$

(On the left hand side of Equation (8.4), we evaluate $\int_a^b f_n\, dx$ for each positive integer n, obtaining a sequence of numbers. We then find the limit of the sequence of numbers, if it exists. On the middle and right-hand sides of Equation (8.4), we find the limiting function first, and then evaluate the integral of the limit function if that is possible.)

Likewise, we ask about the properties of differentiability. Suppose again that Equation (8.1) holds and that each function $f_n(x)$ is differentiable. Must $f(x)$ be differentiable, and if so, must it be true that

$$\lim_{n \to \infty} f_n'(x) = f'(x)?$$

Put another way, we are asking whether it is necessarily true that

$$\lim_{n \to \infty} \left(\lim_{x \to c} \frac{f_n(x) - f_n(c)}{x - c} \right) = \lim_{x \to c} \left(\lim_{n \to \infty} \frac{f_n(x) - f_n(c)}{x - c} \right). \qquad (8.5)$$

Thus the fundamental question seems to be: When can the order of taking two different limits be interchanged? (In Equation (8.4) remember that a definite integral is the limit of Riemann sums.) The following examples show that some caution is appropriate when dealing with such questions.

Example 8.1:
Let the sequence of functions $\{f_n\}$ be defined on $[0, 1]$ by

$$f_n(x) = x^n \text{ for } x \in [0, 1].$$

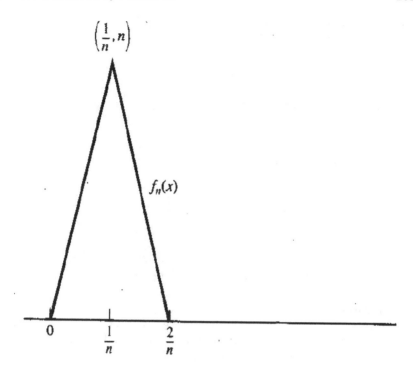

Figure 8.1
One of a sequence of functions defined in Example 8.1.

Then each function f_n is continuous, but

$$\lim f_n(x) = \lim x^n = \begin{cases} 0 & \text{if } 0 \le x < 1 \\ 1 & \text{if } x = 1 \end{cases}$$

so that the limit function is not continuous, even though each function in the sequence $\{f_n\}$ is continuous.

Example 8.2:
Let the sequence of functions $\{f_n\}$ for $n \ge 2$ be defined on $[0,1]$ by

$$f_n(x) = \begin{cases} n^2 x & \text{if } 0 \le x \le 1/n \\ 2n - n^2 x & \text{if } 1/n < x \le 2/n \\ 0 & \text{if } x > 2/n. \end{cases}$$

The graph of $f_n(x)$ is shown in Figure 8.1.
Notice that the area under the curve is 1 for each n, so that $\int_0^1 f_n(x)\, dx = 1$. However, $\lim_{n\to\infty} f_n(x) = 0$ for each $x \in [0,1]$, since, given $x \in [0,1]$,

there is a number N (which depends on x) such that $2/N < x$. Thus, for $n > N > 2/x$, $f_n(x) = 0$. Also, $f_n(0) = 0$ for every n so $f(x) = 0$. Then

$$\int_0^1 f_n(x)\, dx = 1 \text{ for every } n$$

so that

$$\lim_{n \to \infty} \int_0^1 f_n(x)\, dx = 1.$$

However, $\lim f_n(x) = 0$ on $[0, 1]$ so that

$$\int_0^1 \lim_{n \to \infty} f_n(x)\, dx = \int_0^1 0\, dx = 0.$$

Therefore, it is not true in this case that

$$\lim_{n \to \infty} \int_0^1 f_n(x)\, dx = \int_0^1 \left(\lim_{n \to \infty} f_n(x) \right) dx.$$

Example 8.3:

As far as integration properties are concerned, even worse things can happen than in Example 8.2. Enumerate the rational numbers in $[0, 1]$. (Remember, the rational numbers are countable.) Define f_n on $[0, 1]$ by

$$f_n(x) = \begin{cases} 0 & \text{if } x \neq r_1, r_2, \dots, r_n \\ 1 & \text{if } x = r_1, r_2, \dots, r_n \end{cases}$$

where r_i is the ith rational number in our enumeration. Thus

$$f(x) = \begin{cases} 0 & \text{if } x \text{ is irrational} \\ 1 & \text{if } x \text{ is rational.} \end{cases}$$

Then $\int_0^1 f_n(x)\, dx = 0$ for every n but $\int_0^1 f(x)\, dx$ does not exist.

Example 8.4:

Let

$$f_n(x) = \frac{\sin(nx)}{\sqrt{n}} \text{ for } x \in [0, 1].$$

Then $\lim_{n \to \infty} f_n(x) = 0$ for every $x \in [0, 1]$, since

$$\left| \frac{\sin(nx)}{\sqrt{n}} \right| \leq \frac{1}{\sqrt{n}}$$

but

$$f_n'(x) = \sqrt{n} \cos(nx)$$

and $\lim_{n \to \infty} \sqrt{n} \cos(nx)$ does not exist. Thus $\lim_{n \to \infty} f_n'(x) \neq f'(x)$.

By now it should be apparent that our original notion of convergence is not a strong enough condition to require that the limiting function have the same properties as the function in the sequence of functions. What is needed is a stronger idea of convergence.

Definition: For each $n = 1, 2, ...$, let f_n be a function defined on E, where E is a subset of the real numbers. Suppose that for each $x \in E$, $\{f_n(x)\}$ is a convergent sequence of real numbers. Then for each $x \in E$ we can define a function $f(x)$ on E by

$$f(x) = \lim_{n \to \infty} f_n(x).$$

That is, given $\epsilon > 0$ and $x \in E$, there is a number $N(\epsilon, x)$ such that

$$|f_n(x) - f(x)| < \epsilon \text{ if } n > N(\epsilon, x).$$

Then the function f is said to be the *pointwise limit* of the sequence of functions $\{f_n\}$.

It is a crucial point (and one that is often misunderstood initially) that in consideration of the *pointwise* limit, the number $N(\epsilon, x)$ depends on both the number ϵ and the point x. Just as in the case of convergence of sequences of numbers where no number N will necessarily work for every ϵ, no number N will necessarily work for every $x \in E$, *even with ϵ fixed*. The following example may help to clarify this.

Example 8.5:
Let $f_n(x) = x^n$ for $x \in (0, 1)$. Then $\{f_n(x)\}$ converges to 0 for every $x \in (0, 1)$.

Table 8.1 has the column headings for some values of x between 0 and 1. The row headings are for certain ϵ values. The entries in the table are the minimum values of n for the values of ϵ and x in that particular row and column to make $|x^n| < \epsilon$.

It appears from the table (and in fact this is the case) that as x becomes closer to 1 for a given value of ϵ, the number $N(\epsilon)$ increases without bound. Thus no N will work for every x, *even with ϵ fixed*.

Table 8.1
Values of n for selected values of epsilon and x in Example 8.5.

	x values				
ϵ values	.5	.8	.9	.99	.999
.1	4	11	22	230	2302
.01	7	21	44	459	4603
.001	10	31	66	688	6904
.0001	14	42	88	917	9506

The definition we gave at the beginning of the section was for the *pointwise limit* of a sequence of functions, and we have seen from some examples that this limit is not strong enough to enable us to draw many conclusions about the limit function. The next definition gives a stronger idea of convergence, which will force the limit function to inherit some of the properties in the sequence.

Definition: For each $n = 1, 2, ...$, let f_n be a function defined on E, a subset of the real numbers. Suppose that for each $x \in E$, $\{f_n(x)\}$ is a convergent sequence of real numbers. Define a function f on E by

$$f(x) = \lim_{n \to \infty} f_n(x)$$

for every $x \in E$. The function f is called the *uniform limit* of the sequence of functions $\{f_n\}$ provided that, given any $\epsilon > 0$, there is a number $N(\epsilon)$ (which depends on ϵ but *not* on x) such that if $n > N(\epsilon)$, then

$$|f_n(x) - f(x)| < \epsilon \text{ for every } x \in E.$$

If the function f is the uniform limit of the sequence of functions $\{f_n\}$, then $\{f_n\}$ is said to converge uniformly to f.

Remark: The crucial difference between the pointwise limit and the uniform limit is that in the uniform limit we are able to choose a number $N(\epsilon)$ that works for *every* $x \in E$. This is not necessarily possible in the case of pointwise convergence. Figure 8.2 illustrates this. In the uniform limit, if we draw a belt extending the amount ϵ on either side of the function f, then for $n > N(\epsilon)$, $f_n(x)$ must be in this belt for every x.

We leave the proof of the following theorem for Exercise 1, Section 8.1. It provides one of the most convenient ways to show that a sequence of functions converges uniformly.

Theorem 8.1: Suppose that $\{f_n\}$ converges pointwise to f on a set E. Let

$$M_n = \sup_{x \in E} |f_n(x) - f(x)|.$$

Then $\{f_n\}$ converges uniformly to f on E if and only if $\lim M_n = 0$. ∎

Example 8.6:
Let $f(x) = x^n$ for $x \in [0, 1/2]$. We claim that $\{f_n\}$ converges uniformly to 0. To prove this, note that

$$|x^n - 0| \le |(1/2)^n - 0| = (1/2)^n$$

for every $x \in [0, 1/2]$. So given $\epsilon > 0$, if $1/2^n < \epsilon$ (i.e., if $n > |\ln \epsilon| / \ln 2$ for $\epsilon < 1$), then

$$|x^n - 0| < \epsilon \text{ for every } x \in [0, 1/2].$$

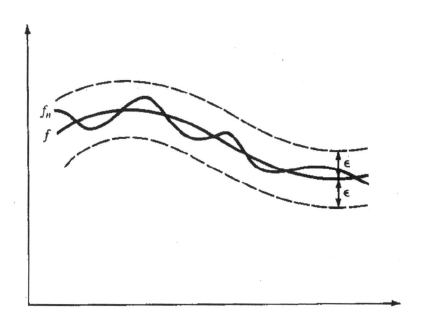

Figure 8.2
The function f^n is within ϵ of f uniformly.

This same method will work to show that $\{x^n\}$ converges uniformly to 0 on $[0, 1 - \alpha]$ for any α with $0 < \alpha < 1$. However $\{x^n\}$ does *not* converge uniformly to 0 on $[0, 1)$.

Theorem 8.2: Let f be the *uniform* limit of a sequence of continuous functions $\{f_n\}$. (That is, each function f_n is continuous for $n = 1, 2, ...$) Then f is continuous.

Remark: Thus while the *pointwise* limit of a sequence of continuous may *not* be continuous, the *uniform* limit of a sequence of continuous functions, if it exists, is continuous.

Proof: Suppose that each function f_n is defined on E, a subset of the real numbers. Choose $x_0 \in E$, and let $\epsilon > 0$ be given. We need to show that there is a $\delta > 0$ such that if $x \in E$ and $|x - x_0| < \delta$, then $|f(x) - f(x_0)| < \epsilon$.

We know there is a number N_1 such that if $n \geq N_1$, then

$$|f_n(z) - f(z)| < \frac{\epsilon}{3} \qquad (8.6)$$

for *every* $z \in E$. This is true by the *uniform* convergence of $\{f_n\}$ to f. Now each function f_n is continuous. In particular, f_{N_1} is continuous so there is a number $\delta > 0$, which depends both on the point x_0 and the function f_{N_1}, such that

$$|f_{N_1}(x) - f_{N_1}(x_0)| < \frac{\epsilon}{3} \qquad (8.7)$$

if $|x - x_0| < \delta$ and $x \in E$.

By the triangle inequality,

$$|f(x_0) - f(x)| = |f(x_0) - f_{N_1}(x_0) + f_{N_1}(x_0) - f_{N_1}(x) + f_{N_1}(x) - f(x)|$$

$$\leq |f(x_0) - f_{N_1}(x_0)| + |f_{N_1}(x_0) - f_{N_1}(x)| + |f_{N_1}(x) - f(x)|. \tag{8.8}$$

By Equation (8.6)

$$|f(x_0) - f_{N_1}(x_0)| < \frac{\epsilon}{3} \text{ and } |f(x) - f_{N_1}(x)| < \frac{\epsilon}{3}.$$

This is where *uniform convergence* is crucial – the same N_1 works for every $x \in E$.

By (8.7)

$$|f_{N_1}(x_0) - f_{N_1}(x)| < \frac{\epsilon}{3}$$

if $x \in E$ and $|x - x_0| < \delta$.

Thus if $x \in E$ and $|x - x_0| < \delta$, then

$$|f(x) - f(x_0)| < \frac{\epsilon}{3} + \frac{\epsilon}{3} + \frac{\epsilon}{3} = \epsilon$$

as required.   ∎

Theorem 8.3: Suppose $\{f_n\}$ is a sequence of Riemann integrable functions on $[a, b]$. If $\{f_n\}$ converges uniformly to f on $[a, b]$, then f is a Riemann integrable function on $[a, b]$ and

$$\lim_{n \to \infty} \int_a^b f_n(x) \, dx = \int_a^b \left(\lim_{n \to \infty} f_n(x) \right) dx = \int_a^b f(x) dx.$$

Remarks: Again the *uniform* limit does away with two possible problems that were observed in Examples 8.2 and 8.3. Thus we may move the limit inside the integral sign *if the interval of integration is finite.* In Example 8.7, which follows the proof of Theorem 8.3, we show that even uniform convergence is not enough to ensure convergence of the integrals of the sequence of functions to the integral of the limiting function if the interval of integration is infinite.

Proof: To show that f is Riemann integrable, we shall show that f satisfies the Riemann condition for integrability.

Let $\epsilon > 0$ be given. Since $\{f_n\}$ converges to f uniformly on $[a, b]$, there is a positive integer N for which $|f(x) - f_N(x)| < \epsilon/(6(b - a))$ for all $x \in [a, b]$. Since f_N is Riemann integrable, there is a partition $P(\epsilon)$ for which $\overline{S}(f_N; P(\epsilon)) - \underline{S}(f_N; P(\epsilon)) < \epsilon/3$. In Exercise 11, Section 8.1, we show that if $|f(x) - f_N(x)| < \epsilon/(6(b - a))$ on $[a, b]$, then $|M_i(f) - M_i(f_N)| < \epsilon/(3(b - a))$ (in the notation of Chapter 6) and likewise $|m_i(f) - m_i(f_N)| < \epsilon/(3(b - a))$. Thus

$$|\overline{S}(f; P(\epsilon)) - \overline{S}(f_N; P(\epsilon))| \leq \sum_i |M_i(f) - M_i(f_N)| \Delta x_i$$

$$< \frac{\epsilon}{3(b-a)} \sum \Delta x_i = \frac{\epsilon}{3},$$

and

$$|\underline{S}(f; P(\epsilon)) - \underline{S}(f_N; P(\epsilon))| \le \sum_i |m_i(f) - m_i(f_N)| \Delta x_i$$

$$< \frac{\epsilon}{3(b-a)} \sum_i \Delta x_i = \frac{\epsilon}{3}.$$

Therefore we have

$$|\overline{S}(f; P(\epsilon)) - \underline{S}(f; P(\epsilon))| \le |\overline{S}(f; P(\epsilon)) - \overline{S}(f_N; P(\epsilon))|$$

$$+ |\overline{S}(f_N; P(\epsilon)) - \underline{S}(f_N; P(\epsilon))| + |\underline{S}(f_N; P(\epsilon)) - \underline{S}(f; P(\epsilon))| < \frac{\epsilon}{3} + \frac{\epsilon}{3} + \frac{\epsilon}{3} = \epsilon.$$

Thus f is Riemann integrable on $[a, b]$.

To prove the second part of the theorem, let $\epsilon > 0$ be given. There is a number N such that, if $n > N$, then

$$|f_n(x) - f(x)| < \epsilon$$

for every $x \in [a, b]$.

This implies

$$f_n(x) - \epsilon < f(x) < f_n(x) + \epsilon$$

for every $x \in [a, b]$.

Thus, by using elementary properties of integrals, we conclude

$$\int_a^b (f_n(x) - \epsilon) \, dx < \int_a^b f(x) \, dx < \int_a^b (f_n(x) + \epsilon) \, dx$$

which implies

$$\int_a^b f_n \, dx - \epsilon(b-a) < \int_a^b f(x) \, dx < \int_a^b f_n(x) \, dx + \epsilon(b-a).$$

Therefore

$$\left| \int_a^b f_n(x) \, dx - \int_a^b f(x) \, dx \right| < \epsilon(b-a) \text{ for } n > N. \quad \blacksquare$$

Example 8.7:

Let

$$f(x) = \begin{cases} 1/n & \text{if } 0 \le x \le n \\ 0 & \text{if } x > n. \end{cases}$$

Then $\{f_n(x)\}$ converges uniformly to 0 on $[0, \infty)$, but

$$\int_0^\infty f_n(x) \, dx = \int_0^n \frac{1}{n} \, dx = 1 \ne 0 = \int_0^\infty 0 \, dx$$

Thus, as stated earlier, the interval of integration must be finite for Theorem 8.3 to apply.

Notice that Example 8.4 shows a uniformly convergent sequence of differentiable functions for which $\lim_{n\to\infty}(f_n'(x))$ does not exist. Thus, even stronger conditions than uniform convergence are needed to ensure that differentiability conditions carry through to the limiting function.

Before addressing that problem, we give a condition that ensures the uniform convergence of a sequence of functions that is very useful.

Theorem 8.4(Cauchy Criterion for Uniform Convergence): Let $\{f_n\}$ be a sequence of functions defined on E, a subset of the real numbers. Then $\{f_n\}$ converges uniformly to a function f on E if and only if, given $\epsilon > 0$ there is a number $N(\epsilon)$ (which depends on ϵ but not on x) such that

$$|f_n(x) - f_m(x)| < \epsilon$$

for every $x \in E$ whenever $m, n > N(\epsilon)$.

Remark: Such a sequence of functions is said to be uniformly Cauchy on E.

Proof: First suppose that $\{f_n\}$ converges uniformly to f on E. Let $\epsilon > 0$ be given. Then there is a number N such that $|f_n(x) - f_m(x)| < \epsilon/2$ for every $x \in E$ if $n > N$. Thus if both m and n exceed N, then

$$|f_n(x) - f_m(x)| \leq |f_n(x) - f(x)| + |f(x) - f_m(x)| < \frac{\epsilon}{2} + \frac{\epsilon}{2} = \epsilon.$$

Conversely, let $\epsilon > 0$ be given. The hypotheses say that there is a number N such that $|f_n(x) - f_m(x)| < \epsilon/2$ for every x if $n, m > N$. Thus for each $x \in E$, $\{f_n\}$ is a Cauchy sequence of real numbers, so we may define a function $f(x)$ on E by $f(x) = \lim_{n\to\infty} f_n(x)$ for $x \in E$. Clearly $\{f_n\}$ converges pointwise to f on E; the problem is to show the convergence is uniform. Now

$$|f_n(x) - f(x)| \leq |f_n(x) - f_m(x)| + |f_m(x) - f(x)|$$

for any n and m. Note that as long as $n, m > N$, $|f_n(x) - f_m(x)| < \epsilon/2$ for any $x \in E$. Now for the crucial step: For a given x there is a number N_x that depends on both x and ϵ such that if $m > N_x$, then $|f_m(x) - f(x)| < \epsilon/2$, since $\{f_n\}$ converges to f pointwise. Thus, *by letting m vary with x*, we have shown that

$$|f_n(x) - f(x)| \leq |f_n(x) - f_m(x)| + |f_m(x) - f(x)| < \frac{\epsilon}{2} + \frac{\epsilon}{2} = \epsilon$$

if $n > N$ and $x \in E$. Since N is independent of x, $\{f_n\}$ converges uniformly to f on E. ∎

As we noted earlier, in Example 8.4 with the sequence of functions

$$f_n(x) = \frac{\sin nx}{\sqrt{n}},$$

the functions converge uniformly to 0, but $\lim f'_n(x) \neq 0$. Thus to get the desired result for differentiation, we need more than uniform convergence. The result that tells us when the derivative of the limit of a sequence of functions is the limit of the derivatives of the functions is considerably more involved. To derive our result, we shall use the following theorem.

Theorem 8.5: Let $\{f_n\}$ be a sequence of functions that converge uniformly to f on $[a, b] \setminus \{x_0\}$ where $x_0 \in [a, b]$. Suppose that, for each $n = 1, 2, ...$, $\lim_{x \to x_0} f_n(x)$ exists. Then

$$\lim_{x \to x_0} \left(\lim_{n \to \infty} f_n(x) \right) = \lim_{n \to \infty} \left(\lim_{x \to x_0} f_n(x) \right).$$

Proof: Let $\alpha_n = \lim_{x \to x_0} f_n(x)$. We first show that $\{\alpha_n\}$ is a Cauchy (and thus a convergent) sequence. Let $\epsilon > 0$ be given. Since $\{f_n\}$ converges *uniformly* to f on $[a, b] \setminus \{x_0\}$, by Theorem 8.4 $\{f_n\}$ is uniformly Cauchy on $[a, b] \setminus \{x_0\}$. Thus, there is a number N such that if $n, m > N$, then

$$|f_n(x) - f_m(x)| < \epsilon/3 \text{ for every } x \in [a, b] \setminus \{x_0\}. \tag{8.9}$$

Now

$$|\alpha_n - \alpha_m| \leq |\alpha_n - f_n(x)| + |f_n(x) - f_m(x)| + |f_m(x) - \alpha_m|.$$

Choose m and n larger than N so that Equation (8.9) holds. Since

$$\lim_{x \to x_0} f_n(x) = \alpha_n \text{ for } n = 1, 2, \ldots$$

then for each positive integer k, there is a number $\delta_k > 0$ such that

$$|\alpha_k - f_k(x)| < \epsilon/3 \text{ if } 0 < |x - x_0| < \delta_k \text{ and } x \in [a, b].$$

Thus if

$$0 < |x - x_0| < \delta = \min\{\delta_m, \delta_n\} \text{ and } x \in [a, b],$$

we have

$$|\alpha_n - \alpha_m| < \epsilon/3 + \epsilon/3 + \epsilon/3 = \epsilon$$

so that $\{\alpha_n\}$ is a Cauchy sequence. Let $\alpha_0 = \lim_{n \to \infty} \alpha_n$. Next we note that

$$|f(x) - \alpha_0| \leq |f(x) - f_n(x)| + |f_n(x) - \alpha_n| + |\alpha_n - \alpha_0| \tag{8.10}$$

We shall show that each term on the right hand side of Equation (8.10) can be made arbitrarily small if x is sufficiently close to x_0.

By the *uniform* convergence of $\{f_n\}$ to f on $[a,b] \setminus \{x_0\}$, there is a number N_1 such that if $n > N_1$ then

$$|f(x) - f_n(x)| < \epsilon/3 \text{ for every } x \in [a,b] \setminus \{x_0\}.$$

Choose N_2 so that if $n > N_2$, then $|\alpha_n - \alpha_0| < \epsilon/3$. Let $N = \max\{N_1, N_2\}$. Then

$$|f(x) - \alpha_0| < \epsilon/3 + |f_N(x) - \alpha_N| + \epsilon/3.$$

Also, there is a $\delta_N > 0$ such that

$$|f_N(x) - \alpha_N| < \epsilon/3 \text{ if } 0 < |x - x_0| < \delta_N \text{ and } x \in [a,b].$$

Thus
$$|f(x) - \alpha_0| < \epsilon \text{ if } 0 < |x - x_0| < \delta_N \text{ and } x \in [a,b].$$

Therefore

$$\lim_{x \to x_0} \left(\lim_{n \to \infty} f_n(x) \right) = \lim_{x \to x_0} f(x) = \alpha_0 = \lim_{n \to \infty} \alpha_n = \lim_{n \to \infty} \left(\lim_{x \to x_0} f_n(x) \right). \quad \blacksquare$$

Theorem 8.6: Let $\{f_n\}$ be a sequence of functions that are differentiable on an open interval containing $[a,b]$. Suppose that
(i) There is a point $x_0 \in [a,b]$ where $\{f_n(x_0)\}$ converges, and
(ii) $\{f_n'\}$ converges uniformly on $[a,b]$.
Then
(a) $\{f_n\}$ converges uniformly to a function f on $[a,b]$, and
(b) $f'(x) = \lim_{n \to \infty} f_n'(x)$ on (a,b).
    That is, uniform convergence of the derivatives and convergence of the functions at one point ensure that the derivative of the limit function exists and is equal to the limit of the derivatives of the sequence of functions.

Proof: This proof is somewhat longer than the others and requires three steps.

Step 1. Here we show that the sequence of functions $\{f_n\}$ converges uniformly on $[a,b]$. This will be done by using the Mean Value Theorem to show that the sequence is uniformly Cauchy.
    Let $\epsilon > 0$ be given. Since $\{f_n(x_0)\}$ is a convergent sequence, it is a Cauchy sequence. Thus there is a number N_1 such that if $m,n > N_1$, then

$$|f_n(x_0) - f_m(x_0)| < \epsilon. \tag{8.11}$$

The *uniform* convergence of $\{f_n'(x)\}$ means there is a number N_2 such that if $m,n > N_2$, then

$$|f_n'(x) - f_m'(x)| < \epsilon \text{ for every } x \in [a,b]. \tag{8.12}$$

Let $g_{n,m}(x) = f_n(x) - f_m(x)$. Notice that $g_{n,m}(x)$ is differentiable and Equation (8.12) says that $|g_{m,n}'(x)| < \epsilon$ for every $x \in [a,b]$ if $n,m > N_2$.

The Mean Value Theorem applied to $g_{n,m}(x)$ says that if $x, y \in [a, b]$ then

$$|g_{n,m}(x) - g_{n,m}(y)| = |g'_{n,m}(z)||x - y| \text{ for some } z \in (x, y).$$

Thus

$$|g_{n,m}(x) - g_{n,m}(y)| < \epsilon(b - a) \qquad (8.13)$$

if $x, y \in [a, b]$ and $n, m > N_2$.

Now choose an arbitrary $w \in [a, b]$. We shall show that

$$|f_n(w) - f_m(w)| < \epsilon(b - a)$$

if $x, y \in [a, b]$ and $n, m > N_2$. par Now choose an arbitrary $w \in [a, b]$. We shall show that

$$|f_n(w) - f_m(w)| < \epsilon + \epsilon(b - a)$$

if $n, m > N = \max\{N_1, N_2\}$, which does not depend on w. That is, we can make $|f_n(w) - f_m(w)|$ arbitrarily small by making n and m sufficiently large, *independent of w*, so that $\{f_n\}$ is a uniformly Cauchy sequence on $[a, b]$. To do this, we observe

$$|f_n(w) - f_m(w)| = |f_n(w) - f_n(x_0) + f_n(x_0) - f_m(w) + f_m(x_0) - f_m(x_0)|$$

$$\leq |(f_n(w) - f_m(w)) - (f_n(x_0) - f_m(x_0))| + |(f_n(x_0) - f_m(x_0)|$$

$$= |g_{m,n}(w) - g_{m,n}(x_0)| + |(f_n(x_0) - f_m(x_0)|$$

$$< \epsilon|w - x_0| + \epsilon(b - a) + \epsilon$$

if $m, n > N$.

Thus $\{f_n\}$ is a uniformly Cauchy sequence on (a, b), and so there is a function $f(x) = \lim_{n \to \infty} f_n(x)$ to which $\{f_n\}$ converges uniformly on $[a, b]$. This completes the first step of the proof.

Step 2: Fix a point $z_0 \in (a, b)$ and define

$$h_n(x) = \frac{f_n(x) - f_n(z_0)}{x - z_0}$$

and

$$h(x) = \frac{f(x) - f(z_0)}{x - z_0} \text{ for } x \in [a, b], \; x \neq z_0. \qquad (8.14)$$

We want to show that $\{h_n\}$ converges uniformly to h on $[a, b] \setminus \{z_0\}$. It may appear that since $\{f_n\}$ converges uniformly to f on $[a, b]$, one could immediately conclude this. There is, however, a problem: Namely, the denominator $x - z_0$ may be made arbitrarily small. Thus, a little more work is required.

If $x \neq z_0$, then

$$|h_n(x) - h_m(x)| = \left| \frac{f_n(x) - f_n(z_0)}{x - z_0} - \frac{f_m(x) - f_m(z_0)}{x - z_0} \right|$$

$$= \frac{1}{|x - z_0|} |(f_n(x) - f_m(x)) - (f_n(z_0) - f_m(z_0))|$$

$$= \frac{1}{|x - z_0|} |g_{n,m}(x) - g_{n,m}(z_0)| \text{ (using our earlier notation)}$$

$$\leq \frac{1}{|x - z_0|} |x - z_0| \sup_{t \in [a,b]} |g'_{n,m}(t)| \leq \epsilon \text{ if } n, m > N_2.$$

Thus $\{h_n\}$ is a uniformly convergent Cauchy sequence on $[a, b] \setminus \{z_0\}$, so that $\{h_n\}$ converges uniformly to some function, say ϕ, on $[a, b] \setminus \{z_0\}$. From Equation (8.14) and the fact that $\{f_n\}$ converges to f on $[a, b]$, we can conclude that $\{h_n\}$ converges pointwise to h on $[a, b] \setminus \{z_0\}$. Thus $\{h_n\}$ converges uniformly to ϕ and pointwise to h on $[a, b] \setminus \{z_0\}$, so $h(x) = \phi(x)$ on $[a, b] \setminus \{z_0\}$. Hence $\{h_n\}$ converges uniformly to h on $[a, b] \setminus \{z_0\}$.

Step 3: We apply Theorem 8.5, which tells us that

$$\lim_{x \to z_0} \left(\lim_{n \to \infty} h_n(x) \right) = \lim_{n \to \infty} \left(\lim_{x \to z_0} h_n(x) \right). \tag{8.15}$$

Now $\lim_{x \to z_0} h_n(x) = f'_n(z_0)$, since f_n is differentiable and $\lim_{n \to \infty} h_n(x) = h(x)$ if $x \neq z_0$. So from (8.15) we conclude

$$\lim_{x \to z_0} h(x) = \lim_{n \to \infty} f'_n(z_0). \tag{8.16}$$

We know from Theorem 8.5 that $\lim_{n \to \infty} f'_n(z_0)$ exists. (This is important!) Thus

$$\lim_{n \to \infty} f'_n(z_0) = \lim_{x \to z_0} h(x) = \lim_{x \to z_0} \frac{f(x) - f(z_0)}{x - z_0} = f'(z_0).$$

Since z_0 was an arbitrary point of (a, b), this completes the proof. ∎

Example 8.8:
We show how Theorem 8.6 can be used to show that for an irrational number r,

$$\frac{d}{dx}(x^r) = rx^{r-1}.$$

The verification of the details of this argument is left as Exercise 7, Section 8.1. Choose $M > 1$. We first restrict our attention to the interval $[1, M]$ and show that the hypotheses of Theorem 8.6 can be satisfied if the functions f_n are chosen suitably. Let $\{r_n\}$ be a sequence of rational numbers converging to r. (Notice that if it is convenient to require the sequence to be monotone, we may do so.)

$$f_n(x) = x^{r_n}.$$

Then

$$f'_n(x) = r_n x^{r_n - 1}.$$

Our candidate for $h(x)$ is rx^{r-1}. The sequence $\{f_n(x_0)\}$ converges when $x_0 = 1$. The effort comes in showing that $\{f'_n\}$ converges uniformly to h on $[1, M]$. Now

$$|f'_n(x) - h(x)| = |r_n x^{r_n - 1} - rx^{r-1}|$$
$$\leq |r_n x^{r_n - 1} - rx^{r_n - 1}| + |r| \cdot |x^{r_n - 1} - x^{r-1}|$$
$$= |r_n - r||x^{r_n - 1}| + |r| \cdot |x^{r_n - 1} - x^{r-1}|$$
$$= |r_n - r||x^{r_n - 1}| + |r| \cdot |x^{r_n - 1}| \cdot |1 - x^{r - r_n}|.$$

Now one can show that $|x^{r_n - 1}|$ is bounded on $[1, M]$, say by K. Then

$$|f'_n(x) - h(x)| \leq K \cdot |r_n - r| + K \cdot |r| \cdot |1 - x^{r - r_n}|.$$

Next one can show that $|1 - x^{r - r_n}|$ may be made uniformly arbitrarily close to 0 on $[1, M]$ by taking r_n sufficiently close to r. Thus both $K \cdot |r^n - r|$ and $K \cdot |r| \cdot |1 - x^{r - r_n}|$ may be made uniformly arbitrarily close to 0 on $[1, M]$ by taking r_n sufficiently close to r. This shows that $\{r_n x^{r_n - 1}\}$ converges uniformly to rx^{r-1} for $x \in [1, M]$. Thus x^r is differentiable on $(1, M)$ with derivative rx^{r-1} for any $M > 1$. Therefore, we may conclude

$$\frac{d}{dx}(x^r) = rx^{r-1}$$

on $(1, \infty)$ for any irrational number r. The case for x between 0 and 1 is left for Exercise 7, Section 8.1.

Exercises 8.1

1. Suppose $\{f_n(x)\}$ converges to $f(x)$ pointwise on a set E. Let

$$M_n = \sup_{x \in E} |f_n(x) - f(x)|.$$

   Show that $\{f_n\}$ converges to f uniformly on E if and only if $\lim_{n \to \infty} M_n = 0$.

2. Show that if $\{f_n\}$ converges uniformly on (a, b) and if $\{f_n(x)\}$ converges at $x = a$ and $x = b$ then $\{f_n\}$ converges uniformly on $[a, b]$.

3. Show that if $f_n(x) = x^n$ on $[0, 1]$, then

$$\lim \int_0^1 f_n(x)\, dx = \int_0^1 \lim f_n(x)\, dx.$$

   Thus uniform convergence is not a *necessary* condition for convergence of the integrals.

4. Give an example of a sequence of functions that is not continuous at any point, but which converges uniformly to a continuous function.

5. Show that the following sequences of functions converge uniformly to 0 on the given sets:
   (a) $\{(\sin nx)/(nx)\}$ on $[\alpha, \infty)$ where $\alpha > 0$.
   (b) $\{xe^{-nx}\}$ on $[0, \infty)$.
   (c) $\{x/(1 + nx)\}$ on $(0,1)$.
   (d) $\{[\ln(1 + nx)]/n\}$ on $[0, M]$.

6. Let $f_n(x) = xe^{-nx} + ((n + 1)/n)\sin(x)$.
   (a) Find a function f such that $\{f_n\}$ converges to f on $[0, \infty)$.
   (b) Show that the convergence in part (a) is uniform.

7. Complete the work to show that

   $$\frac{d}{dx}x^r = rx^{r-1} \text{ on } (0, \infty)$$

   by showing
   (a) $\{x^{r_n-1}\}$ is uniformly bounded on $[1, M]$ (see Exercise 12).
   (b) $|1 - x^{r-r_n}|$ may be made uniformly small on $[1, M]$ by making $r - r_n$ sufficiently close to 0.
   (c) Show that for $x \in (0, 1)$

   $$\frac{d}{dx}x^r = rx^{r-1}.$$

   [*Hint*: For $x \in (0, 1)$, x=1/y for some $y \in (1, \infty)$.]

8. *(a)* Show that the sequence of functions $f_n(x) = x/n$ converges pointwise but not uniformly to 0 on $[0, \infty)$.
   (b) Show that the sequence of functions $f_n(x) = x/n$ converges uniformly to 0 on the interval $[0, M]$ for any number M.

9. Show that for any $\alpha \in (0, 1)$, the sequence of functions $f_n(x) = x^n$ converges uniformly to 0 on $[0, 1 - \alpha]$ but does not converge uniformly on $[0, 1)$.

10. *(a)* Prove that if $\{f_n\}$ converges uniformly to the function f on the sets $A_1, A_2, ..., A_n$, then it converges uniformly on $\cup_1^n A_i$.
    (b) If $\{f_n\}$ converges to the function f uniformly on the sets $A_1, A_2, \cdots$, must it do so on $\cup_1^\infty A_i$?

11. *(a)* Show that if $\{f_n\}$ converges uniformly to the function f, and if each function f_n is bounded, then f is bounded.
    (b) If $\{f_n\}$ converges pointwise to the function f, and if each function f_n is bounded, must f be bounded?
    (c) Let $\{f_n\}$ be a sequence of bounded functions that converges uniformly to f on an interval $[a, b]$. Suppose that for some $\epsilon' > 0$ and some N, $|f_N(x) - f(x)| < \epsilon'/2$ for all $x \in [a, b]$. Let $M(f) = \sup\{f(x) \mid x \in [a, b]\}$ and $M(f_N) = \sup\{f_N(x) \mid x \in [a, b]\}$. Show that $|M(f) - M(f_N)| < \epsilon'$.

(d) In the proof of Theorem 8.3, we claimed that since $|f(x) - f_N(x)| < \epsilon/(6(b-a))$ for all $x \in [a, b]$, then $|M_i(f) - M_i(f_N)| < \epsilon/(3(b-a))$. Use part (c) to justify this claim.

(e) (This exercise provides a stronger result than part (c).) Suppose that $\{f_n\}$ is a sequence of bounded functions that converges uniformly to the function f on $[a, b]$. Suppose that for some $\epsilon' > 0$ and some N, $|f_N(x) - f(x)| < \epsilon'/2$ for all $x \in [a, b]$. If $M(f)$ and $M(f_N)$ are as in part (c), show that $|M(f) - M(f_N)| \leq \epsilon'/2$.

12. (a) A sequence of functions $\{f_n\}$ is said to be *uniformly bounded* on a set E if there is a number M such that $|f_n(x)| \leq M$ for every $n = 1, 2, \ldots$ and every $x \in E$. Show that a uniformly convergent sequence of bounded functions is uniformly bounded.

(b) Is the result in part (a) true if "uniformly convergent" is replaced with "pointwise convergent"?

13. (a) If $\{f_n\}$ and $\{g_n\}$ are sequences of functions that converge uniformly to the functions f and g, respectively, on a set E, show that $\{f_n + g_n\}$ converges uniformly to $f + g$ on E.

(b) Find an example where $\{f_n\}$ converges uniformly to f and $\{g_n\}$ converges uniformly to g, but $\{f_n g_n\}$ does not converge uniformly to fg.

(c) Show that if $\{f_n\}$ converges to 0 uniformly on a set E and if g is bounded on E, then $\{gf_n\}$ converges uniformly to 0 on E.

(d) Suppose that $\{f_n\}$ converges uniformly to f and $\{g_n\}$ converges uniformly to g on a set E, and that each f_n and g_n is bounded on E. Show that $\{f_n g_n\}$ converges uniformly to fg on E.

14. (a) Give an example of a sequence of functions $\{f_n\}$ that satisfies all of the following conditions:

(i) Each function f_n is continuous on $[0, 1]$.

(ii) The sequence $\{f_n\}$ converges to a function f on $[0, 1]$.

(iii) There is a sequence $\{x_n\} \subset [0, 1]$ with $\{x_n\}$ converging to x, but $\{f_n(x_n)\}$ does not converge to $f(x)$.

(b) If we require the sequence of functions in part (a) to be uniformly convergent, is it necessarily true that $\{f_n(x_n)\}$ must converge to $f(x)$?

15. (a) Suppose $\{f_n\}$ is a sequence of monotone increasing functions that converges to a function f on $[a, b]$. Show that f is monotone increasing.

(b) Show that if $\{f_n\}$ is a monotone increasing sequence of functions that converges to a continuous function f on $[a, b]$, then the convergence is uniform.

16. Let $\{f_n\}$ be the sequence of functions defined by

$$f_n(x) = \begin{cases} ne^{-nx} & \text{if } x \geq 0 \\ 0 & \text{if } x < 0. \end{cases}$$

Note that $f_n(x) \geq 0$. Show that:

(a) $\int_{-\infty}^{\infty} f_n(x)\, dx = 1$.

(b) If g is a function that is continuous at $x = 0$, then

$$\lim_{n \to \infty} \int_{-\infty}^{\infty} f_n(x) g(x)\, dx = g(0).$$

(Such a sequence of functions is called an *approximate identity*.)

8.2   Series of Functions

Convergence of Series of Functions

In this section we want to form series of functions. We begin with a sequence of functions $\{f_n\}$, with each function defined on a set E. Then, for each $x \in E$ and each positive integer n, define

$$S_n(x) = f_1(x) + \cdots + f_n(x).$$

Thus, for each $x \in E$, $\{S_n(x)\}$ is the sequence of partial sums for the series of numbers $\sum f_n(x)$. Suppose that $\lim_{n \to \infty} S_n(x)$ exists for each $x \in E$. Then we can define a function f on E by

$$f(x) = \lim_{x \to \infty} S_n(x).$$

Definition: With the preceding notation, we say that the series of functions $\sum f_n$ *converges to f on E* and write

$$f = \sum f_n(x).$$

If $\{S_n\}$ converges uniformly to f on E, then we say the series $\sum f_n$ *converges uniformly to f on E.*

As with series of numbers, we shall refer to the sequence $\{S_n\}$ as the sequence of partial sums for the series $\sum f_n$. Therefore, the question of convergence or uniform convergence of a series of functions is exactly the same question as the convergence or uniform convergence associated with the sequence of partial sums. Thus it should not be surprising when we appeal to the results of Section 8.1 to prove many of our results about series, similar to our use of results of Chapter 2 on sequences of numbers to prove theorems about series in Chapter 7.

As we stated earlier, series of functions have important applications in the physical and social sciences as well as mathematics. In these applications one often wants to perform an operation to each term of the series. As we know from our work with sequences, this is not always legitimate, and usually uniform convergence (and possibly more) is needed to ensure that this procedure is valid. We now give some conditions that guarantee uniform convergence of a series of functions.

Theorem 8.7 (Cauchy Criterion for Uniform Convergence of a Series): Let $\{f_n\}$ be a sequence of functions defined on a set E. The series $\sum f_n$ converges uniformly on E if and only if, given $\epsilon > 0$, there is a number $N(\epsilon)$ such that if $m > n > N(\epsilon)$, then

$$\left| \sum_{i=n+1}^{m} f_i(x) \right| < \epsilon$$

for every $x \in E$.

Remark: Again notice that in uniform convergence $N(\epsilon)$ depends only on ϵ and not on the choice of $x \in E$, and that the inequality must hold for every $x \in E$.

Proof: We apply the Cauchy criterion for uniform convergence to the sequence of partial sums $\{S_n\}$ for $\sum f_n$. Now

$$S_n = f_1(x) + \cdots + f_n(x)$$

$$S_m(x) = f_1(x) + \cdots + f_n(x) + f_{n+1}(x) + \cdots + f_m(x)$$

since $m > n$. Thus

$$|S_m(x) - S_n(x)| = \left| \sum_{i=n+1}^{m} f_i(x) \right|$$

and the sequence of partial sums $\{S_n(x)\}$ converges uniformly on E if and only if, given $\epsilon > 0$, there is a number $N(\epsilon)$ such that if $m > n > N(\epsilon)$, then

$$|S_m(x) - S_n(x)| = \left| \sum_{i=n+1}^{m} f_i(x) \right| < \epsilon \text{ for every } x \in E. \quad \blacksquare$$

While the Cauchy criterion provides one way to tell whether we have uniform convergence, in practice it is often difficult to apply. The next theorem that we prove shows the technique most often used. Notice, however, that to prove it we use the result of Theorem 8.7.

Theorem 8.8 (Weierstrass M-test): Let $\{f_n\}$ be a sequence of functions defined on a set E. Suppose that, for each positive integer n, there is a number M_n such that

$$|f_n(x)| \leq M_n \text{ for every } x \in E.$$

If the series $\sum M_n$ converges, then the series of functions $\sum f_n$ converges uniformly on E.

Remark: Notice that the Weierstrass M-test gives a sufficient, but not necessary, condition for uniform convergence. In Exercise 2, Section 8.2, we give a series of functions that converge uniformly, but $\sum M_n$ diverges for that series.

Proof: We have, for any m and n with $m > n$,

$$\left| \sum_{i=n+1}^{m} f_i(x) \right| \leq \sum_{i=n+1}^{m} |f_i(x)| \leq \sum_{i=n+1}^{m} M_i \text{ for every } x \in E.$$

By the Cauchy criterion for convergence of real numbers, given $\epsilon > 0$, there is a number $N(\epsilon)$, such that if $m > n > N(\epsilon)$, then

$$\sum_{i=n+1}^{m} M_i < \epsilon.$$

Thus

$$\left| \sum_{i=n+1}^{m} f_i(x) \right| < \epsilon$$

for every $x \in E$, if $m > n > N(\epsilon)$. So by Theorem 8.7 the series $\sum f_n$ converges uniformly on E. ∎

Example 8.9:

Let $f_n(x) = n^2 e^{-nx}$ on $[1, \infty)$. We show that $\sum f_n$ converges uniformly. Since e^{-nx} is a decreasing function, $|f_n(x)| \leq n^2 e^{-n}$ on $[1, \infty)$. The series $\sum n^2 e^{-n}$ converges. (Why?) So by the Weierstrass M-text, the series $\sum n^2 e^{-nx}$ converges uniformly on $[1, \infty)$. In fact, using an almost identical argument, one can show that the series $\sum n^k e^{-nx}$ converges uniformly on $[\alpha, \infty)$ for any $\alpha > 0$ and any number k.

We now state some corollaries, which give conditions on series that ensure certain operations on series are legitimate.

Corollary 8.2: If $\{f_n\}$ is a sequence of continuous functions defined on a set E and $\sum f_n$ converges uniformly to a function f on E, then f is continuous on E.

Proof: Let $S_N(x) = f_1(x) + \cdots + f_N(x)$. Each function f_i is continuous, so S_N is continuous, since it is the sum of a finite number of continous functions. Thus by Theorem 8.2

$$\lim_{N \to \infty} S_N = \sum f_n = f$$

is continuous, since this is the *uniform* limit of a sequence of continuous functions. ∎

Corollary 8.3: Let $\{f_n\}$ be a sequence of Riemann integrable functions on a finite interval $[a, b]$. If $\sum f_n$ converges uniformly to f on $[a, b]$, then $\int_a^b f(x)\, dx$ exists and

$$\int_a^b f(x)\, dx = \int_a^b \left(\sum f_n(x) \right)\, dx = \sum \left(\int_a^b f_n(x)\, dx \right).$$

Proof: Again, let $S_N(x) = f_1(x) + \cdots + f_N(x)$. Since each function f_i is integrable on $[a, b]$, we have that S_N is integrable on $[a, b]$. Thus $\{S_n\}$ is a sequence of integrable functions that converges uniformly to $\sum f_n$, so by Theorem 8.3

$$\sum \int_a^b f_n(x)\, dx = \lim_{N \to \infty} \sum_1^N \int_a^b f_n(x)\, dx = \lim_{N \to \infty} \int_a^b \sum_1^N f_n(x)\, dx$$

$$= \lim_{N \to \infty} \int_a^b S_N(x)\, dx = \int_a^b \left(\lim_{N \to \infty} S_N(x) \right)\, dx = \int_a^b \sum f_n(x)\, dx.$$

The step

$$= \lim_{N \to \infty} \int_a^b S_N(x)\, dx = \int_a^b \left(\lim_{N \to \infty} S_N(x) \right)\, dx$$

is where uniform convergence is used. ■

Corollary 8.5: Let $\{f_n\}$ be a sequence of functions that are differentiable on an open interval containing $[a, b]$ such that
(i) $\sum f_n(x_0)$ converges for some $x_0 \in [a, b]$, and
ii $\sum f_n'(x)$ converges uniformly to a function $h(x)$ on $[a, b]$.
Then
(a) $\sum f_n$ converges uniformly to a function f on $[a, b]$, and
(b) $\sum f_n'(x) = h(x) = f'(x)$ for $x \in (a, b)$.
The proof of this corollary is similar to the previous two. ■

Power Series

Definition: Let $\{a_n\}_{n=0}^\infty$ be a sequence of real numbers. An expression of the form

$$a_0 + a_1(x - c) + a_2(x - c)^2 + \cdots = \sum_0^\infty a_n(x - c)^n$$

is called a *power series in x - c*. The numbers a_n are the *coefficients* for the power series.

To simplify the notation somewhat, we shall first consider power series with $c = 0$.

For a given set of coefficients, we want to answer the following questions:

(i) For what values of x does the power series converge? For values of x where the power series converges, define

$$f(x) = a_0 + a_1 x + a_2 x^2 + \cdots = \sum_{n=1}^{\infty} a_n x^n.$$

Then:
(ii) When is $f(x)$ continuous?
(iii) When is $\int \sum a_n x^n \, dx = \sum \int x^n \, dx$?
(iv) When is $f'(x) = \sum n a_n x^{n-1}$?

To answer these questions, we shall apply our earlier theorems with $f_n(x) = a_n x^n$. Notice that $f_n(x) = a_n x^n$ is infinitely differentiable and continuous and is integrable over any finite interval. Also notice that a power series converges if $x = 0$. Beyond that, there may be values of x for which the power series converges and other values of x for which it diverges. For example, the power series

$$1 + x + x^2 + x^3 + \cdots$$

is one with $a_n = 1$ for every n. It is also a geometric series for a given x with $r = x$. Thus the series converges if $|x| < 1$ and diverges if $|x| \geq 1$.

The next theorem characterizes the behavior of power series.

Theorem 8.9: Let $\sum a_n x^n$ be a power series, and let

$$\lambda = \overline{\lim}(|a_n|^{1/n}) \text{ and } R = \frac{1}{\lambda},$$

(where $R = 0$ if $\lambda = \infty$ and $R = \infty$ if $\lambda = 0$). Then the power series $\sum a_n x^n$
(a) Converges absolutely if $|x| < R$.
(b) Diverges if $|x| > R$.
(c) Assume $0 < R < \infty$. Then the power series $\sum a_n x^n$ converges *uniformly* on $[-R + \epsilon, R - \epsilon]$ for any $\epsilon > 0$. If $R = \infty$, the series converges uniformly on any compact interval.

Proof: To prove parts (a) and (b) for $0 < \lambda < \infty$, we apply the root test for real numbers to $\sum a_n x^n$. Then

$$\overline{\lim} |a_n x^n|^{1/n} = \overline{\lim} |a_n|^{1/n} |x^n|^{1/n} = \overline{\lim} |x| |a_n|^{1/n}$$

$$= |x| \overline{\lim} |a_n|^{1/n} = |x| \lambda.$$

Thus the series converges absolutely if

$$|x| \lambda < 1 \text{ or } |x| < \frac{1}{\lambda} = R$$

and diverges if

$$|x| \lambda > 1 \text{ or } |x| > \frac{1}{\lambda} = R.$$

Figure 8.3
Points in the proof of Theorem 8.9.

If $\lambda = 0$, we note that $|x|\lambda < 1$ for all values of x, and if $\lambda = \infty$, the series diverges if $|x| > 0$. This proves parts (a) and (b).

To prove part (c) in the case $R < \infty$, let $\epsilon > 0$ be given, and let $\rho = R - \epsilon/2$. Consider Figure 8.3.

Now

$$\overline{\lim}|a_n|^{1/n} = \frac{1}{R}$$

and

$$\frac{1}{R} < \frac{1}{R - \epsilon/2}.$$

Thus, there is a number N such that if $n > N$, then

$$|a_n|^{1/n} < \frac{1}{R - \epsilon/2} = \frac{1}{\rho}.$$

Thus for $n > N$,

$$|a_n| < \left(\frac{1}{\rho}\right)^n.$$

So

$$\sum_{n>N} |a_n x^n| = \sum_{n>N} |a_n||x^n| < \sum_{n>N} \frac{|x^n|}{\rho^n} < \sum_{n>N} \left(\frac{R - \epsilon}{\rho}\right)^n \quad \text{if } |x| < R - \epsilon.$$

Since $(R - \epsilon)/\rho < 1$, the right-hand side is a convergent geometric series. We can now apply the Weierstrass M-test with $M_n = ((R-\epsilon)/\rho)^n$ for $n > N$ and $M_n = |a_n|R^n$ for $n \leq N$. We know

$$\sum_{n>N} \left(\frac{R - \epsilon}{\rho}\right)^n$$

converges and adding a finite number of terms to a series does not affect its convergence, so $\sum M_n$ converges. Thus, by the Weierstrass M-test, the series $\sum a_n x^n$ converges uniformly on $[-R+\epsilon, R-\epsilon]$. We prove part (c) in the case $R = \infty$ in Exercise 16(a), Section 8.2. ∎

A repetition of a previous warning: Even though the power series converges uniformly on $[-R + \epsilon, R - \epsilon]$ for every $\epsilon > 0$, we *cannot* conclude that the power series converges uniformly on $(-R, R)$ as the example $\sum x^n$ shows.

Definition: The number R in Theorem 8.9 is called the *radius of convergence* of a power series.

Theorem 8.10: Let $\sum a_n x^n$ be a power series with $a_n \neq 0$. Suppose

$$\lambda = \lim_{n \to \infty} \frac{|a_{n+1}|}{|a_n|}$$

exits and $R = 1/\lambda$ (where again $R = 0$ if $\lambda = \infty$ and $R = \infty$ if $\lambda = 0$). Then the power series $\sum a_n x^n$
(a) Converges absolutely if $|x| < R$.
(b) Diverges if $|x| > R$.
(c) Converges uniformly on $[-R + \epsilon, R - \epsilon]$ for any $\epsilon > 0$ if $0 < R < \infty$. If $R = \infty$, the series converges uniformly on any compact interval.

Remark: Notice that this theorem has the same conclusions as Theorem 8.9 and, in fact, is often easier to apply. The difference is that in Theorem 8.10 we need a *limit* to exist and this may not happen, whereas in Theorem 8.9 we use the limit superior, which always exists.

Proof: The proof follows the pattern of Theorem 8.9. We do the proof for $0 < \lambda < \infty$. For a given value of x,

$$\frac{|a_{n+1} x^{n+1}|}{|a_n x^n|} = \frac{|a_{n+1}|}{|a_n|} |x|.$$

Using the ratio test, we take the limit as $n \to \infty$ (if it exists) and note that the series converges if

$$\lambda |x| = \lim \frac{|a_{n+1}|}{|a_n|} |x|$$

is less than 1 and diverges if $\lambda |x| > 1$.

Thus the power series converges absolutely for $|x| < 1/\lambda = R$ and diverges if $|x| > R$. This proves the first two parts of the theorem, for $0 < \lambda < \infty$.

To prove part (c) in the case $R < \infty$, let $\epsilon > 0$ be given. Since

$$\lim \frac{|a_{n+1}|}{|a_n|} = \frac{1}{R} \text{ and } \frac{1}{R} < \frac{1}{R - \epsilon/2}$$

there is a number N such that if $n \geq N$, then

$$\frac{|a_{n+1}|}{|a_n|} < \frac{1}{R - \epsilon/2}.$$

Thus

$$\frac{|a_{n+k}|}{|a_N|} < \frac{1}{(R - \epsilon/2)^k}$$

for any positive integer k. (Why?) Then for any x with $|x| < R - \epsilon$,

$$|a_{N+k}||x|^{N+k} < (R-\epsilon)^N|a_N|\frac{(R-\epsilon)^k}{(R-\epsilon/2)^k} = \rho^k|a_N|(R-\epsilon)^N$$

where $\rho = (R-\epsilon)/(R-\epsilon/2) < 1$. Thus, beyond the Nth term, the power series is bounded by a convergent geometric series if $|x| < R - \epsilon$, so by the Weierstrass M-test, the power series converges uniformly on $[R+\epsilon, R-\epsilon]$. We prove part (c) in the case $R = \infty$ in Exercise 16(b), Section 8.2. ∎

Example 8.10:
Find the radius of convergence of the power series $\sum n!x^n$. We use Theorem 8.10. Then

$$\frac{a_{n+1}}{a_n} = \frac{(n+1)!}{n!} = n+1.$$

Thus $\lim_{n\to\infty} a_{n+1}/a_n = \infty$, so the radius of convergence is 0.

Example 8.11
Find the radius of convergence of the series

$$1 + 2x + 3^2x^2 + 2^3x^3 + 3^4x^4 + 2^5x^5 + \cdots.$$

Here $\lim_{n\to\infty} a_{n+1}/a_n$ does not exist. (Why?) However, $\overline{\lim}|a_n|^{1/n} = 3$ so the radius of convergence is $\frac{1}{3}$.

Corollary 8.10(a): Let $\sum a_n x^n$ be a power series with radius of convergence R. If $x \in (-R, R)$, then the power series is continuous at x.
Proof: We do the proof in the case $R < \infty$. Choose $x_0 \in (-R, R)$. Then $x_0 \in (-R+\epsilon, R-\epsilon)$ for some $\epsilon > 0$. But

$$f_n(x) = a_n x^n$$

is continuous, and $\sum a_n x^n$ converges uniformly on $[-R+\epsilon, R-\epsilon]$ by Theorem 8.10. By Corollary 8.2 a uniformly convergent series of continuous functions is continuous, so that $\sum a_n x^n$ is continuous on $[-R+\epsilon, R-\epsilon]$, in particular, at x_0. ∎

Corollary 8.10(b): Let $\sum a_n x^n$ be a power series with radius of convergence R, and suppose $[a,b] \subseteq (-R, R)$. Then

$$\int_a^b \sum a_n x^n \, dx = \sum a_n \int_a^b x^n \, dx.$$

Proof: We again do the proof in the case $R < \infty$. Since $[a,b] \subseteq (-R, R)$, let

$$\epsilon = \min\{a + R, R - b\}.$$

Figure 8.4
$[a, b]$ is contained in $(-R, R)$.

(See Figure 8.4.)
   Then
$$[a, b] \subseteq \left[-R + \frac{\epsilon}{2}, R - \frac{\epsilon}{2}\right]$$
and the power series converges uniformly on $[-R + \epsilon/2, R - \epsilon/2]$ and thus on $[a, b]$. Therefore we can apply Corollary 8.3. ■.

   The next question is when can power series be differentiated term by term; that is, when is
$$\frac{d}{dx}\left(\sum a_n x^n\right) = \sum a_n n x^{n-1} \ ?$$
We know from Theorem 8.4 that more is involved than uniform convergence. To answer this question, we need a preliminary result.

   Theorem 8.11: If $\sum a_n x^n$ has radius of convergence R, then so does $\sum a_n n x^{n-1}$.
   Proof: Notice that
$$\overline{\lim}|a_n n|^{1/n} = \overline{\lim}|a_n|^{1/n}|n|^{1/n}.$$

Using L'Hôpital's Rule, one can show
$$\lim_{n \to \infty} n^{1/n} = 1.$$

We showed in Exercise 11, Section 2.3, if $\lim_{n \to \infty} a_n = a \geq 0$ and $\overline{\lim}_{n \to \infty} b_n = b \geq 0$ and $a \cdot b$ is not of the form $0 \cdot \infty$, then
$$\overline{\lim}_{n \to \infty} a_n b_n = ab$$

so
$$\overline{\lim}_{n \to \infty} |a_n n|^{1/n} = \overline{\lim}_{n \to \infty} |a_n|^{1/n}.$$

   Also
$$\lim_{n \to \infty} |x^{n-1}|^{1/n} = \lim_{n \to \infty} \frac{|x|}{|x|^{1/n}} = |x| \text{ for any finite } x.$$

Thus the same proof as given in Theorem 8.9 works to prove this theorem. ■

Corollary 8.11: Let $\sum a_n x^n$ be a power series with radius of convergence $R > 0$. Then if $x \in (-R, R)$

$$\frac{d}{dx}\left(\sum a_n x^n\right) = \sum a_n n x^{n-1}.$$

Proof: We do the proof in the case $R < \infty$. We verify the hypotheses of Corollary 8.5. To verify (i), choose $x_0 = 0$. For $x \in (-R, R)$, there is an $\epsilon > 0$ such that $x \in (-R + \epsilon, R - \epsilon)$. By Theorem 8.11

$$\sum a_n n x^{n-1} \text{ converges uniformly on } [-R + \epsilon, R - \epsilon].$$

Thus the hypotheses of the corollary are satisfied, so

$$\frac{d}{dx}\left(\sum a_n x^n\right) = \sum n a_n x^{n-1}. \quad \blacksquare$$

The next theorem is the basis for many applications involving series. For example, to solve differential equations by power series, one solves an equation using the fact that two power series represent the same function if and only if the coefficients are equal term by term. The next theorem proves this fact.

Theorem 8.12: Suppose $\sum a_n x^n$ and $\sum b_n x^n$ are two power series that converge on an interval $(-R, R)$ for some $R > 0$. If

$$\sum a_n x^n = \sum b_n x^n$$

on $(-R, R)$, then $a_n = b_n$, $n = 1, 2, \ldots$. Thus two power series are equal if and only if the coefficients are equal term by term.

Proof: Let

$$f(x) = \sum a_n x^n \text{ and } g(x) = \sum b_n x^n$$

on $(-R, R)$, and suppose $f(x) = g(x)$. Then

$$a_0 = f(0) = g(0) = b_0$$

so that $a_0 = b_0$. Also

$$a_1 = f'(0) = g'(0) = b_1$$

so $a_1 = b_1$. In fact, for any positive integer n,

$$n! a_n = f^{(n)}(0) = g^{(n)}(0) = n! b_n$$

so that

$$a_n = \frac{f^{(n)}(0)}{n!} = \frac{g^{(n)}(0)}{n!} = b_n. \quad \blacksquare$$

The relation $a_n = f^{(n)}(0)/n!$ for a power series $f(x) = a_n x^n$ is extremely important and will be used in Section 8.3.

The results that we have obtained for power series of the form $\sum a_n x^n$ translate to the more general power series $\sum a_n (x-c)^n$ simply by making the change of variable $z = x - c$. The only difference will be in calculating the interval of convergence–that is, the values of x for which the series converges.

Example 8.12:
We calculate the interval of convergence for the series $\sum (x-1)^n/2^n$.
Let $z = x - 1$. Then the series becomes $\sum z^n/2^n$. Then $a_n = (\frac{1}{2})^n$ so that $\overline{\lim} |a_N|^{1/n} = \frac{1}{2} = \lambda$. Thus the radius of convergence is $R = 1/\lambda = 2$. So the series converges absolutely when $|z| < R = 2$, and diverges when $|z| > 2$. When $|z| = 2$, we must check further. If $z = 2$,

$$\sum \frac{z^n}{2^n} = \sum \frac{2^n}{2^n} = \sum 1$$

which diverges. If $x = -2$,

$$\sum \frac{z^n}{2^n} = \sum \frac{(-2)^n}{2^n} = \sum (-1)^n$$

which also diverges. Thus $\sum z^n/2^n$ converges if $|z| < 2$ and diverges if $|z| \geq 2$. Converting back to x, we have the series converging if and only if

$$-2 < z < 2 \text{ or } -2 < x - 1 < 2,$$

that is, when

$$-1 < x < 3.$$

Exercises 8.2

1. Show that the following series converge uniformly on the given interval:
   (a)
   $$\sum \frac{\sin nx}{n^2} \text{ on } (-\infty, \infty).$$

   (b)
   $$\sum \frac{x}{n^{1+\alpha}} \text{ for } \alpha > 0 \text{ on } [-M, M].$$

   (c)
   $$\sum \frac{n^k}{x^n} \text{ on } [2, \infty).$$

   (d)
   $$\sum \left(\frac{1}{n}\right)^x \text{ on } [1 + \epsilon, \infty).$$

   (e)
   $$\sum \left(\frac{\ln x}{x}\right)^n \text{ on } [1, \infty).$$

2. Define a series of functions $\sum f_n$ on $[0,1]$ by

$$f_n(x) = \begin{cases} 1/n & \text{if } 1/2^n < x \le 1/2^{n-1} \\ 0 & \text{otherwise.} \end{cases}$$

Show that $\sum f_n$ converges uniformly, but that the Weierstrass M-test fails.

3. Show that the series $x + x(1-x) + x(1-x)^2 + x(1-x)^3 + \cdots$ converges on $[0,1]$, but the convergence is not uniform.

4. Prove Corollary 8.5.

5. Use L'Hôpital's Rule to show $\lim n^{1/n} = 1$.

6. Show that if $\sum f_n$ converges uniformly on a set E, then $\{f_n\}$ converges uniformly to 0 on E.

7. Suppose $f(x) = \sum a_n(x-c)^n$ is a power series with radius of convergence $R > 0$. Find $f^{(k)}(c)$ in terms of the coeffieicnts a_k for each nonnegative integer k.

8. For $|x| < 1$, $\sum x^n = 1/(1-x)$.
   (a) Find $\sum nx^n$ for $|x| < 1$.
   (b) Find $\sum n^2 x^n$ for $|x| < 1$.
   (c) Find $\sum(n^2/2^n)$.

9. Use the fact that $\int (1+x)^{-1}\, dx = \tan^{-1} x$ to show:
   (a) $\tan^{-1} x = x - (x^3/3) + (x^5/5) - (x^7/7) + \cdots$ for $x \in (-1,1)$.
   (b) Find a series expression for π.
   (c) Use this series to find π accurate to 0.2.
   (d) Is it true that $\tan^{-1}(1) = 1 - \frac{1}{3} + \frac{1}{5} - \frac{1}{7} + \cdots$? Justify your answer.

10. Find the values of x for which the following power series converge:
    (a) $\sum n^k x^n$ for k an integer.
    (b) $\sum(x+2)^n/(\ln n)$.
    (c) $\sum x^n/(n!)$.
    (d) $\sum[1 \cdot 2 \cdot 3 \cdots n \cdot x^{2n}]/[1 \cdot 3 \cdot 5 \cdots (2n-1)]$ (see Exercise 11).

11. Show that if the power series $\sum a_n x^n$ has radius of convergence R, then, for any positive integer k, the series $\sum a_n x^{kn}$ has radius of convergence $\sqrt[k]{R}$.

12. If $\sum a_n x^n$ converges to $f(x)$ for $|x| < R_1$ and $\sum b_n x^n$ converges to $g(x)$ for $|x| < R_2$, show that

$$f(x)g(x) = \sum c_n x^n \text{ for } |x| < R = \min\{R_1, R_2\}$$

where $c_n = \sum_{k=0}^{n} a_k b_{n-k}$.

13. Show that if $|x| < 1$, then $\ln(1+x) = x - (x^2/2) + (x^3/3) - (x^4/4) + \cdots$.

14. Show that if $\sum a_n x^n$ is a power series such that $0 < m \le |a_n| \le M < \infty$ for every positive integer n, then the radius of convergence of the power series is 1.

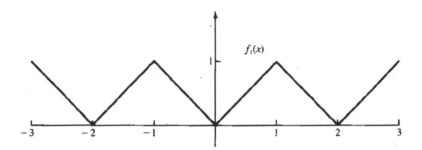

Figure 8.5
Graphs for Exercise 15.

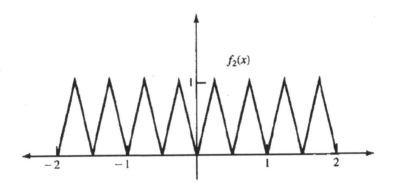

Figure 8.6
Graphs for Exercise 15.

15. Here we construct a continuous function that is not differentiable at any point. Define $f_1(x) = |x|$ for $-1 \leq x \leq 1$ and $f_1(x + 2) = f_1(x)$ for all other x. The graph of $f_1(x)$ is given in Figure 8.5.

Define $f_n(x) = f_1(4^{n-1}x)$ for $n \geq 2$. The graph of $f_2(x)$ is given in Figure 8.6. Let $f(x) = \sum_{n=1}^{\infty}(3/4)^n f_n(x)$.

(a) Show that $f(x)$ is continuous.
(b) Fix a real number x_0. For each positive integer m, there is an integer k_m with
$$k_m \leq 4^m x_0 < k_m + 1.$$
Set $\alpha_m = k_m/4^m$ and $\beta_m = (k_m + 1)/4^m$. Note that $\alpha_m \leq x_0 < \beta_m$. Show that if $n \leq m$, then
$$|f_1(4^n \alpha_m) - f_1(4^n \beta_m)| = \left(\frac{1}{4}\right)^{m-n}.$$

(c) Show that

$$\lim_{m \to \infty} \frac{f(\beta_m) - f(\alpha_m)}{\beta_m - \alpha_m}$$

does not exist. Thus by Exercise 15, Section 5.1, $f'(x_0)$ does not exist.

16. *(a)* Suppose that $\sum a_n x^n$ is a power series for which $\overline{\lim}(|a_n|^{1/n}) = 0$. Show that the power series converges uniformly on any compact interval $[a, b]$.

(b) Suppose that $\sum a_n x^n$ is a power series with $a_n \neq 0$, and

$$\lim_{n \to \infty} \frac{|a_{n+1}|}{|a_n|} = 0.$$

Show that the power series converges uniformly on any compact interval $[a, b]$.

8.3   Taylor Series

In this section and in Chapter 9 on Fourier series, we shall apply some of our previous results to particular types of series. In both sections we shall be asking similar questions, which are, in a sense, an extension of ideas from linear algebra. The problem is, given a set of functions $\{f_0, f_1, f_2, ...\}$, what functions can be expressed as a series of these functions $f(x) = \sum a_n f_n$ where a_n is a real number for each nonnegative integer n. If f and g are two such functions, then any linear combination of f and g, $\alpha f + \beta g$ where α and β are real numbers, can also be expressed as a series of these functions. Thus, in the language of linear algebra, we have a linear space.

From Theorem 8.12 we know if our set of functions is $\{1, x, x^2, ...\}$ and $f(x) = \sum a_n x^n$, then the numbers a_n are unique, so one can think of $\{1, x, x^2, ...\}$ as a basis for this linear space of functions. Thus we have an example of an infinite dimensional vector space. We leave this abstract point of view for a while and concentrate on a particular situation.

In Section 8.2 we showed that if $f(x) = \sum a_n x^n$ is a power series of radius $R > 0$, then $f(x)$ has derivatives of all orders and $a_n = f^{(n)}(0)/n!$. We want to consider the following problem: Given a function f, when can f be represented as a power series? That is, when can we find numbers $a_0, a_1, a_2, ...$ such that $f(x) = \sum a_n(x - c)^n$ for all x on some open interval containing c?

Definition: A function $f(x)$ is said to be of *class* C^n on an open interval I if $f^{(n)}(x)$ exists and is continuous for each $x \in I$. In this case we write $f \in C^n(I)$. If $f \in C^n(I)$ for every $n = 1, 2, ...$, then we say that f is of *class* C^∞ on I and write $f \in C^\infty(I)$.

Notice that if $f \in C^n(I)$ and $m < n$, then $f \in C^m(I)$.

Example 8.13:

Let

$$f(x) = \begin{cases} x^5 \sin(1/x) & \text{if } x \neq 0 \\ 0 & \text{if } x = 0. \end{cases}$$

Then $f(x)$ is of class $C^2(I)$ but not of class $C^n(I)$ if $n > 2$ and if I is any interval containing 0. (Why?) The function $g(x) = e^x$ is of class $C^\infty(I)$ for any interval I.

If $f \in C^\infty(I)$, $c \in I$ and $f(x) = \sum_{n=0}^{\infty} a_n (x - c)^n$, then, using our results from Section 8.2 as a guide, we can find that $a_n = f^{(n)}(c)/n!$ and write the series

$$\sum_{n=0}^{\infty} \frac{f^{(n)}(c)}{n!} (x - c)^n.$$

This series is called the *Taylor series for f(x) about x=c*. If $c = 0$, the series is called the *Maclaurin series for f(x)*.

We want to find conditions on $f(x)$ under which the Taylor series of a function is actually equal to the function on some nondegenerate interval. A function f is said to be *analytic* at the point c if there is an open interval I containing c such that

$$f(x) = \sum_{n=0}^{\infty} \frac{f^{(n)}(c)}{n!} (x - c)^n \text{ for every } x \in I.$$

Example 8.14:

Find the Maclaurin series for $f(x) = \cos x$.

$$\begin{array}{lll} f(x) = \cos x & f(0) = 1 & a_0 = 1 \\ f'(x) = -\sin x & f'(0) = 0 & a_1 = 0 \\ f''(x) = -\cos x & f''(0) = -1 & a_2 = 1/2! \\ f'''(x) = \sin x & f'''(0) = 0 & a_3 = 0 \\ f^{(4)}(x) = \cos x & f^{(4)}(0) = 1 & a_4 = 1/4! \end{array}$$

One can can check that

$$f^{(n)}(0) = \begin{cases} 0 & \text{if } n \text{ is odd} \\ (-1)^{n/2} & \text{if } n \text{ is even.} \end{cases}$$

So the Maclaurin series for $\cos x$ is

$$\sum_{n=0}^{\infty} \frac{f^{(n)}(0)}{n!} x^n = 1 - \frac{x^2}{2!} + \frac{x^4}{4!} - \frac{x^6}{6!} + \cdots$$

$$= \sum_{n=0}^{\infty} \frac{(-1)^n x^{2n}}{(2n)!}.$$

We are *not* saying, although it will turn out to be the case, that

$$\cos x = \sum_{n=0}^{\infty} \frac{(-1)^n x^{2n}}{(2n)!}$$

but only that the series on the right is the Maclaurin series for $\cos x$.

Example 8.15:
Find the Taylor series for $f(x) = \ln x$ about $c = 1$.

$f(x) = \ln x$ $f(1) = 0$ $a_0 = 0$

$f'(x) = \frac{1}{x}$ $f'(1) = 1$ $a_1 = 1$

$f''(x) = \frac{-1}{x^2}$ $f''(1) = -1$ $a_2 = -1/2!$

$f'''(x) = \frac{2}{x^3}$ $f'''(1) = 2$ $a_3 = \frac{2}{3!} = \frac{1}{3}$

$f^{(4)}(x) = \frac{2(-3)}{x^4}$ $f^{(4)}(1) = -6$ $a_4 = \frac{-(2)(3)}{4!} = \frac{-1}{4}$

$\vdots$ $\vdots$ $\vdots$

$f^{(n)}(x) = \frac{(-1)^{n-1}(n-1)!}{x^n}$ $f^{(n)}(1) = (-1)^{n-1}(n-1)!$ $a_n = \frac{(-1)^{n-1}}{n}$

So the Taylor series for $f(x) = \ln x$ about $c = 1$ is

$$(x-1) - \frac{(x-1)^2}{2} + \frac{(x-1)^3}{3} - \cdots = \sum_{n=1}^{\infty} \frac{(-1)^{n-1}(x-1)^n}{n}.$$

We again emphasize that we are *not* saying (although it will turn out to be the case if $|x - 1| < 1$) that

$$\ln x = \sum_{n=1}^{\infty} \frac{(-1)^{n-1}(x-1)^n}{n}.$$

One may wonder if we are not being overly cautious by not immediately writing

$$f(x) = \sum_{n=1}^{\infty} \frac{f^{(n)}(c)}{n!}(x-c)^n$$

for functions with infinitely many derivatives at $x = c$. However, the function

$$f(x) = \begin{cases} e^{-1/x^2} & \text{if } x \neq 0 \\ 0 & \text{if } x = 0 \end{cases}$$

has the property $f^{(n)}(0) = 0$ for every n as we verify in Exercise 2, Section 8.3. This is an example of a function that is infinitely differentiable at 0, and yet the Maclaurin series does not equal the function on any open interval containing 0.

We now address the problem of finding when a function is equal to its Taylor series. We observe that if $f \in C^n(I)$ where I is an open interval containing the point a, we can form the *Taylor polynomial of degree n* generated by f at the point $x = a$, denoted $T_n(f; a)$:

$$T_n(f; a) = \sum_{k=0}^{n} \frac{f^{(k)}(a)}{k!} (x - a)^k.$$

As we have seen, this polynomial has the property that the polynomial and its first n derivatives agree with the function and its first n derivatives at $x = a$.

To find when a function is actually equal to its Taylor series on an interval about $x = a$, we form the difference between the function $f(x)$ and the nth Taylor polynomial

$$R_n(x) = f(x) - \sum_{k=0}^{n} \frac{f^{(k)}(a)}{k!} (x - a)^k.$$

The function $R_n(x)$ is called the *remainder* or *error* function. A function will be equal to its Taylor series on an interval when $\lim_{n \to \infty} R_n(x) = 0$ on that interval. To find conditions that ensure this, we shall find it convenient to express $R_n(x)$ as an integral.

Theorem 8.13 (Integral Form of the Remainder): Assume that $f \in C^{n+1}(I)$ where I is some interval containing the point a. Then for $x \in I$,

$$R_n(x) = \frac{1}{n!} \int_a^x (x - t)^n f^{(n+1)}(t) \, dt. \qquad (8.17)$$

Proof: The proof is by induction and uses integration by parts and the Fundamental Theorem of Calculus. For the case $n = 1$ by definition of $R_1(x)$, we have

$$f(x) = f(a) + f'(a)(x - a) + R_1(x). \qquad (8.18)$$

We must show that

$$R_1(x) = \int_a^x (x - t) f''(t) \, dt.$$

From Equation (8.18)
$R_1(x) = f(x) - f(a) - f'(a)(x - a)$ (by the Fundamental Theorem of Calculus)

$$= \int_a^x f'(t) \, dt - f'(a) \int_a^x dt = \int_a^x (f'(t) - f'(a)) \, dt.$$

We now use the integration by parts formula $\int u\, dv = uv - \int v\, du$ with

$$u = f'(t) - f''(a), \quad du = f''(t),$$
$$v = t - x, \quad dv = dt$$

to get

$$\int_a^x (f'(t) - f'(a))\, dt = (f'(t) = f'(a))(t - x)|_{t=a}^x - \int_a^x (t - x) f''(t)\, dt$$

$$= 0 + \int_a^x (x - t) f''(t)\, dt$$

which establishes the theorem for the case $n = 1$.

Now we assume the theorem is true when $n = k$ and show that this implies the theorem is true for the case $n = k + 1$. That is, we assume

$$R_k(x) = \frac{1}{k!} \int_a^x (x - t)^k f^{(k+1)}(t)\, dt \tag{8.19}$$

and must show that

$$R_{k+1}(x) = \frac{1}{(k+1)!} \int_a^x (x - t)^{k+1} f^{(k+2)}(t)\, dt.$$

Notice that

$$R_{k+1}(x) = f(x) - \sum_{n=0}^{k+1} \frac{f^{(n)}(a)}{n!}(x - a)^n$$

$$= f(x) - \sum_{n=0}^{k} \frac{f^{(n)}(a)}{n!}(x - a)^n - \frac{f^{(k+1)}(a)}{(k+1)!}(x - a)^{k+1}$$

$$= R_k(x) - \frac{f^{(k+1)}(a)}{(k+1)!}(x - a)^{k+1}.$$

Now use the induction hypothesis to substitute the integral in (8.19) for $R_k(x)$ and the fact that

$$f^{(k+1)}(a)\frac{(x - a)^{k+1}}{(k+1)!} = \frac{f^{(k+1)}(a)}{k!} \int_a^x (x - t)^k\, dt$$

to obtain

$$R_{k+1}(x) = \frac{1}{k!} \int_a^x (f^{(k+1)}(t) - f^{(k+1)}(a))(x - t)^k\, dt.$$

Again use integration by parts with

$$u = f^{(k+1)}(t) - f^{(k+1)}(a)$$

$$dv = (x - t)^k \, dt$$

$$du = f^{(k+2)}(t) \, dt$$

$$v = \frac{-(x - t)^{k+1}}{k + 1}$$

to get

$$R_{k+1}(x) = \frac{1}{k!}(f^{(k+1)}(t) - f^{(k+1)}(a))\left(\frac{-(x-t)^{k+1}}{k+1}\right)\Big|_{t=a}^{x}$$

$$-\frac{1}{k!}\int_a^x f^{(k+2)}(t)\left(\frac{-(x-t)^{k+1}}{k+1}\right)dt$$

$$= \frac{1}{(k+1)!}\int_a^x (x-t)^{k+1} f^{(k+2)}(t) \, dt.$$

This establishes the result for $n = k + 1$ and proves the theorem. ∎

Corollary 8.13(a): Assume the hypotheses of Theorem 8.13 hold and that $m \leq f^{(n+1)}(t) \leq M$ on the interval I. Then

$$m\frac{(x-a)^{n+1}}{(n+1)!} \leq R_n(x) \leq M\frac{(x-a)^{n+1}}{(n+1)!} \text{ if } x > a. \qquad (8.20a)$$

and

$$m\frac{(x-a)^{n+1}}{(n+1)!}(-1)^{n+1} \leq R_n(x) \leq M\frac{(x-a)^{n+1}}{(n+1)!}(-1)^{n+1} \text{ if } x < a. \qquad (8.20b)$$

The proof is left for Exercise 10, Section 8.3. ∎

Corollary 8.13(b) (Lagrange Form of the Remainder): Suppose that $f \in C^n(I)$. Fix $a \in I$, and choose $x \in I$ with $x \neq a$. Then there is a point ξ_n between x and a (where ξ_n depends on a, x, and n) such that

$$f(x) = f(a) + f'(a)(x - a) + \frac{f''(a)}{2}(x - a)^2$$

$$+ \cdots + \frac{f^{(n-1)}(a)}{(n-1)!}(x - a)^{n-1} + \frac{f^{(n)}(\xi_n)}{n!}(x - a)^n.$$

Proof: We shall prove the result in the case $x > a$. We have

$$f(x) = \sum_{k=0}^{n-1} \frac{f^{(k)}(a)}{k!}(x - a)^k + R_{n-1}(x).$$

Let $m = \min\{f^{(n)}(t) \mid t \in [a, x]\}$ and $M = \max\{f^{(n)}(t) \mid t \in [a, x]\}$. Then

$$m\int_a^x (x - t)^{n-1} \, dt \leq \int_a^x (x - t)^{n-1} f^{(n)}(t) \, dt \leq M\int_a^x (x - t)^{n-1} \, dt.$$

Integrating the left and right sides of the inequality gives

$$\frac{m(x-a)^n}{n} \le \int_a^x (x-t)^{n-1} f^{(n)}(t)\, dt \le \frac{M(x-a)^n}{n}.$$

Therefore by Theorem 8.13

$$\frac{m(x-a)^n}{n!} \le R_{n-1}(x) \le \frac{M(x-a)^n}{n!}.$$

Thus, by the Intermediate Value Theorem for continuous functions, there is a point ξ_n between a and x where

$$R_{n-1}(x) = \frac{f^{(n)}(\xi_n)}{n!}(x-a)^n. \quad \blacksquare$$

This gives the next result.

Corollary 8.13(c) (Taylor's Law of the Mean): Suppose $f \in C^n(I)$ where I is an open interval containing $[a,b]$. Then there is a point ξ_n with $a < \xi_n < b$ where

$$f(b) = f(a) + f'(a)(b-a) + \cdots + \frac{f^{(n-1)}(a)}{(n-1)!}(b-a)^{n-1} + \frac{f^{(n)}(\xi_n)}{n!}(b-a)^n. \quad \blacksquare$$

Corollary 8.13(d) (Cauchy Form of the Remainder): Let $f \in C^n(I)$ with $a \in I$. Then if $x \in I$, there is a point ξ_n between a and x where

$$R_{n-1}(x) = \frac{f^{(n)}(\xi_n)}{(n-1)!}(x-a)(x-\xi_n)^{n-1}.$$

Remark: Again the notation ξ_n is used to emphasize that the point between a and x may change as n changes. We also note that this point depends on a and x.

Proof: By Theorem 8.13

$$R_{n-1}(x) = \frac{1}{(n-1)!} \int_a^x (x-t)^{n-1} f^{(n)}(t)\, dt.$$

The Mean Value Theorem for Integrals states that if F(x) is continuous on $[a,b]$, then there is a point $c \in (a,b)$ for which

$$\int_a^b F(t)\, dt = F(c)(b-a).$$

We apply this to the previous integral where $F(t) = (x-t)^{n-1} f^{(n)}(t)$ and conclude that there is a point ξ_n where

$$\int_a^x (x-t)^{n-1} f^{(n)}(t)\, dt = (x-\xi_n)^{n-1} f^{(n)}(\xi_n)(x-a). \quad \blacksquare$$

Because of its importance, we restate an earlier observation as a theorem.

Theorem 8.14: Let $f \in C^\infty(I)$, and let a be an interior point of I. Then

$$f(x) = \sum_{n=0}^{\infty} \frac{f^{(n)}(a)}{n!}(x-a)^n \text{ if and only if } \lim_{n\to\infty} R_n(x) = 0. \quad \blacksquare$$

Thus, for the Taylor series of the function f to be equal to the function on an interval $[a, b]$, by Corollary 8.13(b) it is enough that $f^{(n)}(x)(x-a)^n/n!$ converges to 0 uniformly on $[a, b]$ as $n \to \infty$. We show in Exercise 1, Section 8.3, that for any number A

$$\lim_{n\to\infty} \frac{|A|^n}{n!} = 0.$$

Thus, if there is a number A such that $f^{(n)}(x)| < A^n$ for every $x \in [a, b]$ and every positive integer n, then the Taylor series for the function is equal to the function. We state this formally.

Theorem 8.15: Let $f \in C^\infty((a, b))$, and let $c \in (a, b)$. Suppose there is an open interval I with $c \in I$ and some number A for which $|f^{(n)}(x)| < A^n$ for every $x \in I$ and every positive integer n. Then for any $x \in I \cap (a, b)$,

$$f(x) = \sum_{n=0}^{\infty} \frac{f^{(n)}(c)}{n!}(x-c)^n. \quad \blacksquare$$

One of the uses of Taylor series is to calculate the values of many functions to any desired degree of accuracy. This is how tables of trigonometric, logarithmic, and exponential functions are constructed. Also it is one way computers arrive at the values of these functions. We give two examples.

Example 8.16:
We calculate the value of e accurate to 10^{-6}, assuming that we know $e < 3$. Let $f(x) = e^x$. We use Corollary 8.13(c) with $a = 0$ and $b = 1$ to write

$$f(1) = f(0) + f'(0)(1-0) + \frac{f''(0)}{2!}(1-0)^2$$

$$+ \cdots + \frac{f^{(n-1)}(0)}{(n-1)!}(1-0)^{n-1} + \frac{f^{(n)}(c)}{n!}(1-0)^n$$

where c is between 0 and 1. Now $f^{(n)}(x) = e^x$ for all n, so $f^{(n)}(0) = 1$ for all n. Thus

$$f(1) = e^1 = 1 + 1 + \frac{1}{2!} + \frac{1}{3!} + \cdots + \frac{1}{(n-1)!} + \frac{e^c}{n!}$$

and since $c \le 1$, $e^c \le e < 3$. Thus the error after n terms is less than $3/n!$ For $n = 10$ this number is $3/10! = .0000008$; so if we want e accurate to $.0000008$, we compute $1 + 1 + 1/2! + 1/3! + \cdots + 1/9!$

Example 8.17:

We calculate $\sin 5°$ accurate to .00001.

First we must convert $5°$ to $\pi/36$ radians. Then we use Corollary 8.13(c) with $a = 0$, $b = \pi/36$, and $f(x) = \sin x$. It can be shown that

$$f^{(n)}(0) = \begin{cases} 0 & \text{if } n \text{ is even} \\ (-1)^{(n-1)/2} & \text{if } n \text{ is odd.} \end{cases}$$

So

$$f\left(\frac{\pi}{36}\right) = f(0) + f'(0)\left(\frac{\pi}{36} - 0\right)$$

$$+ \cdots + \frac{f^{(n-1)}(0)}{(n-1)!}\left(\frac{\pi}{36} - 0\right)^{n-1} + \frac{f^{(n)}(c)}{n!}\left(\frac{\pi}{36} - 0\right)^n$$

$$= \frac{\pi}{36} - \frac{(\pi/36)^3}{3!} + \frac{(\pi/36)^5}{5!} - \cdots + \frac{f^{(n)}(c)}{n!}\left(\frac{\pi}{36}\right)^n$$

and since $|f^{(n)}(x)|$ is $|\sin x|$ or $|\cos x|$, $|f^{(n)}(x)| \leq 1$. Thus the error after n terms is bounded by $(\pi/36)^n/n!$. For $n = 3$, this is 0.00011, and for $n = 4$, it is 0.0000024. Thus

$$\sin 5° = \frac{\pi}{36} - \frac{(\pi/36)^3}{3!} = .0872665 - .0001108 = .0871557$$

accurate to .0000024.

Exercises 8.3

1. Show that, for any number A, the series $\sum A^n/n!$ converges absolutely. Conclude that $\lim_{n\to\infty} A^n/n! = 0$.

2. Let

$$f(x) = \begin{cases} e^{-1/x^2} & \text{if } x \neq 0 \\ 0 & \text{if } x = 0. \end{cases}$$

(a) Show that, for every positive integer n,

$$f^{(n)}(x) = \begin{cases} e^{-1/x^2}(2^n x^{-3n} + g(x)), & \text{if } x \neq 0 \\ 0 & \text{if } x = 0 \end{cases}$$

where $g(x)$ is a function consisting of terms of smaller absolute value near $x = 0$ than $|x|^{-3n}$.

(b) Show that $\lim_{x\to 0} f^{(n)}(x) = 0$. Thus $f^{(n)}(x)$ is continuous.

3. Calculate the Maclaurin series for the following functions:
   (a) $\sin x$.
   (b) $\ln[(1+x)/(1-x)]$.

4. **(a)** For m a positive integer, show that the Maclaurin series for $(1+x)^m$ is

$$\sum_{n=0}^{m} \binom{m}{n} x^n \text{ where } \binom{m}{n} = \frac{m!}{n!(m-n)!}.$$

(b) For some number p, which is not a nonnegative integer, show that the Maclaurin expansion for $(1+x)^p$ is given by

$$1 + \sum \frac{p(p-1)(p-2)\cdots(p-n+1)}{n!} x^n.$$

Show that this series converges if $|x| < 1$ and diverges if $|x| > 1$. Show that it diverges if $x = 1$ and $p \le -1$, and diverges if $x = -1$ and $p < 0$. [*Hint:* For the case $x = -1$ and $p < 0$, use *Raabe's Test*. It says that if $\sum a_n$ is a series of positive terms with

$$\lim n[(a_n/a_{n+1}) - 1] = L,$$

then the series $\sum a_n$ converges if $L > 1$ and diverges if $L < 1$.]

5. Find the Taylor series for the following functions about the given point:
   (a) $\cos x$ about $x = \pi/3$.
   (b) $x^{-1/2}$ about $x = 4$.
   (c) $\ln x$ about $x = 2$.
   (d) x^2 about $x = 2$.

6. Let $f(x)$ be a function that can be represented by a Taylor series with error function $R_n(x)$.
   (a) Show that for $c \in [a, b]$

$$\left| \int_a^b f(x)\, dx - \sum_{k=0}^{n} \int_a^b \frac{f^{(k)}(c)}{k!}(x-c)^k\, dx \right| \le R_n(\theta)(b-a)$$

   where θ is the point between a and b where $R_n(x)$ is maximized.
   (b) Find $\int_0^{.5} (\sin x)/x\, dx$ accurate to .001.
   (c) Calculate the Maclaurin expansion for e^{-x^2}.
   (d) Calculate $\int_0^1 e^{-x^2}\, dx$ accurate to .0001.

7. Use the Maclaurin expansion for the necessary functions to show

$$\frac{d}{dx} e^x = e^x$$

$$\frac{d}{dx} \sin x = \cos x.$$

8. Use the remainder function of Taylor's Theorem to find $\cos 10°$ accurate to .0001.

9. Show that e is irrational. [*Hint:* Write $e = 1 + 1/1! + 1/2! + \cdots$. If $e = p/q$ where p and q are integers, then

$$eq! = \text{integer} + \frac{1}{(q+1)} + \frac{1}{(q+1)(q+2)} + \cdots$$

if $q > 1$.]

10. Prove Corollary 8.13(a).

11. Use the Maclaurin series expansion for the exponential function to show that $e^x e^y = e^{x+y}$.

8.4   The Cantor Set and Cantor Function

In this section we study two topics that were fundamental to the development of set theory in the latter part of the nineteenth century, the Cantor set and the Cantor function. Among other things, these caused mathematicians to reexamine some ideas about the size of sets and about properties that functions could possess.

The Cantor Set

To construct the Cantor set, we begin with the closed interval $[0, 1]$ and remove the open interval $(1/3, 2/3)$. This leaves the closed set consisting of $[0, 1/3] \cup [2/3, 1]$. We name this set A_1.

$$A_1 = [0, 1/3] \cup [2/3, 1]$$

The length of the interval we removed is $1/3$. This is the first step in the process. In subsequent steps, we apply this construction to each of the surviving closed intervals of the previous step. So in step 2, we remove the open intervals $(1/9, 2/9)$ and $(7/9, 8/9)$. This leaves

$$A_2 = [0, 1/9] \cup [2/9, 3/9] \cup [6/9, 7/9] \cup [8/9, 9/9]$$

In step 2 we have removed two intervals of length $1/9$.
Step 3 creates

$$A_3 = [0, 1/27] \cup [2/27, 3/27] \cup [6/27, 7/27] \cup [8/27, 9/27]$$

$$\cup [18/27, 19/27] \cup [24/27, 25/27] \cup [26/27, 27/27]$$

Let us analyze what happens on each step. We double the number of closed intervals from the previous step, but the length of each of the new closed intervals is $1/3$ the length of each of the closed intervals of the previous step. For the open intervals being removed, we have twice as many as in the previous step, and each is $1/3$ as long as each individual open interval from the previous step. So the total length removed after two steps is $1/3$ from step one plus $2(1/9)$ from the second step, for a total length of $1/3 + 2/9$. With similar reasoning, we can determine how much is removed in the third step. We will be working with twice the number of closed intervals as in step 1; i.e. 2×2 and each open interval that is removed will have length $1/3$ the length of the a single interval in the previous step; i.e. $(1/3) \times (1/3) \times (1/3) = 1/27$. The total length removed in the third step is thus

$$2^2 \left(\frac{1}{3}\right)^3 .$$

The total removed up to and including the third step is

$$\frac{1}{3} + 2\frac{1}{9} + 4\frac{1}{27} = 2^0 \left(\frac{1}{3}\right)^1 + 2^1 \left(\frac{1}{3}\right)^2 + 2^2 \left(\frac{1}{3}\right)^3 .$$

It appears (and is, in fact, true) that the total removed up to and including the fourth step is

$$\frac{1}{3} + 2\frac{1}{9} + 4\frac{1}{27} + \frac{1}{81} = 2^0 \left(\frac{1}{3}\right)^1 + 2^1 \left(\frac{1}{3}\right)^2 + 2^2 \left(\frac{1}{3}\right)^3 + 2^3 \left(\frac{1}{3}\right)^4 .$$

Following the pattern, step k creates 2^{k-1} intervals, each of length $(1/3)^k$. The total length deducted at step k is $2^{k-1}/3^k$.
The length removed up to and including step k is

$$2^0(1/3)^1 + 2^1(1/3)^2 + \dots + 2^{k-1}(1/3)^k = \sum_{n=1}^{k} 2^{n-1}(1/3)^n .$$

The Cantor set is $\cap_{k=1}^{\infty} A_k$. In other words, the Cantor set is what is left after we remove all the middle-third intervals. The total length removed is

$$\sum_{j=1}^{\infty} 2^{j-1} \left(\frac{1}{3}\right)^j = \sum_{j=0}^{\infty} 2^j \left(\frac{1}{3}\right)^{j+1} = \frac{1}{3} \sum_{j=0}^{\infty} 2^j \left(\frac{1}{3}\right)^j = \frac{1}{3(1 - \frac{2}{3})} = 1 .$$

Thus, the Cantor set has measure zero. It is non-empty because the endpoints of the excised intervals remain in the set.

Now we show that the Cantor set is uncountable. This is noteworthy because it means we have constructed an uncountable set of measure zero.

Theorem 8.16: The Cantor set is uncountable.
Proof: To show that the Cantor set is uncountable, we first note how to write a number between 0 and 1 in the base three system. That is, if x is a number between 0 and 1, we can write x as

$$x = \frac{a_1}{3} + \frac{a_2}{3^2} + \frac{a_3}{3^3} + \frac{a_4}{3^4} + \cdots$$

where $a_i = 0$, 1, or 2. For example,

$$.875 = \frac{875}{1000} = 2 \times \frac{1}{3} + 1 \times \frac{1}{9} + 2 \times \frac{1}{27} + 1 \times \frac{1}{81} + \cdots.$$

So in base 3, $\frac{875}{1000}$ would be expressed as .2121...

Now we note that in the first step of constructing the Cantor set, we removed all numbers between 1/3 and 2/3 and kept all the rest. This means we removed all the numbers, and only those numbers, whose first ternary digit was 1. In step 2 we removed numbers between 1/9 and 2/9 and those between 7/9 and 8/9; i.e., those that had a 1 in the second ternary digit. At this point, we have removed any ternary number that had a 1 in the first or second digit. Continuing this process, we remove all numbers that have a 1 in their ternary expansion. What is left is the Cantor set.

In the binary system, any number between 0 and 1 can be written as

$$\frac{a_1}{2} + \frac{a_2}{2^2} + \frac{a_3}{2^3} + \cdots$$

where $a_1 = 0$ or 1. Let $x = a_1 a_2 a_3 \ldots$ be a number in the binary system. Define $b_i = 2a_i$. So $b_1 b_2 b_3 \ldots$ is a number in the ternary system consisting of 0's and 2's. That is precisely the Cantor set. Thus there is a 1-1 onto mapping from the numbers in [0,1] in the binary system to the Cantor set. Since the real numbers in [0,1] are uncountable, the Cantor set is uncountable. ∎

Thus we have demonstrated that the Cantor set is an uncountable set of measure 0.

The Cantor set has a number of interesting properties that you are asked to prove in the exercises, including

(a) The Cantor set contains no intervals.

(b) All points in the Cantor set are limit points of the Cantor Set. (This says the Cantor set is a perfect set.)

The Cantor Function

The Cantor function is related to the Cantor set and likewise has some interesting properties. We will define the Cantor function to be the limit of a sequence of functions $\{f_n\}$. This will give us a chance to apply some of the ideas of Section 8.1 to obtain a difficult and important result.

We define $\{f_n\}$ most conveniently using graphs.

Define f_1 as follows:

$f_1(x) = 1/2$ if $1/3 < x < 2/3$. On the remainder of the values of x in $[0, 1]$ define f_1 so that it is a linear continuous function from $(0, 0)$ to $(1/3, 1/2)$ and from $(2/3, 1/2)$ to $(1, 1)$. (See Figure 8.7.)

Let f_2 be the function that is

1/4 if $1/9 \leq x \leq 2/9$
1/2 if $1/3 \leq x \leq 2/3$
3/4 if $7/9 \leq x \leq 8/9$

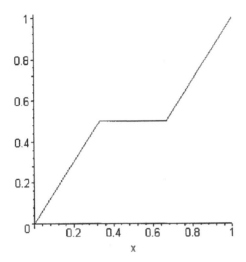

Figure 8.7
One in a sequence of functions that converges to the Cantor function.

On the remainder of the values of x in $[0,1]$ define f_2 so that it is a continuous function from $(0,0)$ to $(1,1)$ and linear where it is not constant. (See Figure 8.8.) Continue as suggested in Figure 8.9.

Facts about the sequence we are constructing:

(1) If $m > n$, then a flat area for $f_n(x)$ is also a flat area for $f_m(x)$.

(2) The vertical jump between two consecutive flat intervals of $f_n(x)$ is $1/2^n$.

Theorem 8.17: The sequence $f_n(x)$ is a uniformly Cauchy sequence of continuous functions.

Proof: Let $\epsilon > 0$ be given. There is a positive integer N such that $\frac{1}{2^{N-1}} < \epsilon$. Claim: If $n, m > 4N$ then $|f_n(x) - f_m(x)| < \epsilon$. Choose $x \in (0,1)$. If x is on a flat area of f_N, then $f_n(x) = f_m(x)$. If x is not on a flat area of f_N then it lies between two adjoining flat areas of f_N. Now

$$|f_n(x) - f_N(x)| < 1/2^N \text{ and } |f_m(x) - f_N(x)| < 1/2^N$$

so

$$|f_n(x) - f_m(x)| < \frac{2}{2^N} < \epsilon.$$

Thus $f_n(x)$ is a uniformly Cauchy sequence of continuous functions, so the limit function, which we define as the Cantor function, is continuous. ∎

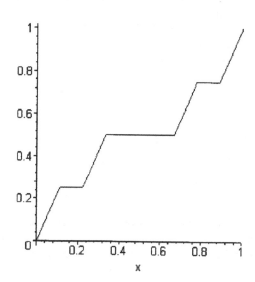

Figure 8.8
One in a sequence of functions that converges to the Cantor function.

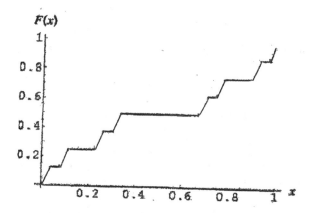

Figure 8.9
One in a sequence of functions that converges to the Cantor function.

Corollary 8.17: The Cantor function is uniformly continuous.

Proof: The Cantor function is continous on the compact interval $[0, 1]$, and thus is uniformly continuous. ∎

Definition: A function $f(x)$ is absolutely continuous on $[a, b]$ if, given $\epsilon > 0$, there is a $\delta > 0$ such that

$$\sum_{i=1}^{n} |f(x_i') - f(x_i)| < \epsilon$$

for every finite collection $\{(x_i, x_i')\}$ of non-overlapping intervals with

$$\sum_{i=1}^{n} |x_i' - x_i| < \delta.$$

Theorem 8.18: The Cantor function is not absolutely continuous.

Proof: Let $\epsilon = 1/2$ and choose any $\delta > 0$. There is a function $f_n(x)$ in the construction of the Cantor function for which the total length of the intervals where $f_n(x)$ is constant exceeds $1 - \delta$. By the construction of $f(x)$, $f(x)$ is constant on intervals where $f_n(x)$ is constant. The complement of the set where $f_n(x)$ is constant is a finite set of intervals, the sum of whose lengths is less than δ. Denote these intervals as $\{(x_i, x_i')\}$. Thus we have

$$\sum_{i=1}^{n} |x_i' - x_i| < \delta \text{ and } \sum_{i-1}^{n} |f(x_i') - f(x_i)| = 1$$

so $f(x)$ is not absolutely continuous. ∎

Theorem 8.19: The derivative of the Cantor function is zero, except on a set of measure zero. The proof is left to Exercise 4, Section 8.4. ∎

Exercises 8.4

1. Show that the Cantor set is compact.

2. Show that the Cantor set contains no intervals.

3. Show that every point in the Cantor set is a limit point of the Cantor set. Such a set is called a perfect set.

4. Prove Theorem 8.19.

9

Fourier Series

9.1 Fourier Coefficients

In Section 8.3 we showed that a function $f(x)$ that was infinitely differentiable on an open interval containing the number c, and satisfied certain conditions, could be represented by the Taylor series

$$f(x) = \sum_{n=0}^{\infty} \frac{f^{(n)}(c)}{n!}(x-c)^n$$

on some interval containing c.

In this chapter we want to do something similar. That is, we want to determine what functions can be expressed as a series of the form $\sum_{n=0}^{\infty} c_n \phi_n(x)$ where $\{\phi_n\}_{n=1}^{\infty}$ is a collection of functions that we shall specify later. Some of the techniques of linear algebra will give us a clue on how to begin. We give a short review of these methods now.

Suppose that we have a vector space with basis $X_1, \ldots X_n$ and that we have defined a dot (inner) product on this vector space so that the vectors in our basis are orthogonal; that is,

$$X_i \cdot X_j = 0 \text{ if } i \neq j.$$

Now let V be a vector in the vector space, say

$$V = a_1 X_1 + \cdots + a_n X_n.$$

Suppose one can find $V \cdot X_i$ for any vector X_i in the basis, and we want to find the scalars a_i. This can be done by observing

$$V \cdot X_i = (a_1 X_1 + \cdots + a_n X_n) \cdot X_i = a_1 X_1 \cdot X_i + \cdots + a_i X_i \cdot X_i + \cdots + a_n X_n \cdot X_i = a_i X_i \cdot X_i$$

since $X_i \cdot X_j = 0$ if $i \neq j$. Thus

$$a_i = \frac{V \cdot X_i}{X_i \cdot X_i}.$$

If $\{X_1, \ldots, X_n\}$ is an *orthonormal* basis (that is, if the set is orthogonal and $X_i \cdot X_i = 1$ for $1, \ldots, n$), then $a_i = V \cdot X_i$.

We want to construct a vector space where the vectors are functions with a common domain. If we define vector addition and scalar multiplication as the usual addition of functions and multiplication of a function by a real number, then we have a vector space. Next we would like to define an inner product on this vector space. A convenient way to do this is with an integral. We shall use the Riemann integral. If f and g are Riemann integrable functions on $[a,b]$, we can define the inner product (or dot product) of f and g by

$$f \cdot g = (f,g) = \int_a^b fg.$$

In our work we shall take $a = -\pi$ and $b = \pi$. This class of functions forms a vector space, so we are in a workable setting. One can verify that this "inner product" we have defined is almost an inner product, but not quite. In particular

$$(f,g) = (g,f)$$

$$(\alpha f, g) = \alpha(f,g) = (f, \alpha g)$$

$$(f, g_1 + g_2) = (f, g_1) + (f, g_2)$$

$$(f,f) \geq 0.$$

The only problem is that it is possible that $(f,f) = 0$ without f being identically 0. For example, if

$$f(x) = \begin{cases} 0 & \text{if } x \neq 1 \\ 1 & \text{if } x = 1, \end{cases}$$

then $\int_0^2 f \cdot f = 0$, but f is not identically zero. It turns out that this is not a great problem. One can merely group into a single class all the functions f_i that have the property $\int_a^b (f_i - f_j)^2 = 0$ and take the class as one vector. Two Riemann integrable functions are in the same class if and only if they are equal, except on a set of measure zero. This is a rather minor point and will have very little bearing in our development.

Now comes the problem of the selection of a basis. There are several ways to do this. The basis that we shall use gives rise to what is known as Fourier series. These ideas were introduced in the early nineteenth century when Joseph Fourier announced that an arbitrary function could be expressed as a series using sine and cosine functions. This turned out to be not entirely correct, although it is the case for a large and important class of functions. Our "basis" will be these sine and cosine functions. There are several other bases one could use. Some of these go by the names Bessel functions, Legendre polynomials, and Hermite polynomials. Fourier analysis has proved to be an extremely useful tool in the analysis of problems where wave phenomena are observed (e.g., heat flow problems, acoustics, and electromagnetic radiation) and has developed into a central topic of mathematical research.

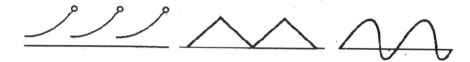

Figure 9.1
Three periodic functions

Definition: A function $f(x)$ is said to be *periodic* of period $T > 0$ if $f(x + T) = f(x)$ for every x in the domain of f.

Periodic functions are those that have a repetitious form. Some examples of periodic functions are shown in Figure 9.1.

The functions $\sin x$ and $\cos x$ are examples of functions that are periodic of period 2π. Notice that in the definition of a periodic function, it was not required that T be the smallest positive value for which $f(x+T) = f(x)$. Thus $\sin(nx)$ and $\cos(nx)$ are also periodic of period 2π for any positive integer n.

Definition: A *trigonometric polynomial* is a function $f(x)$ of the form

$$f(x) = \bar{a}_0 + \sum_{n=1}^{N} [a_n \cos(nx) + b_n \sin(nx)]$$

where $\bar{a}_0, a_1, \ldots, a_N$ and $b_1, \ldots, b_N$ are constants. A *trigonometric series* is a series of the form

$$\bar{a}_0 + \sum_{n=1}^{\infty} [a_n \cos(nx) + b_n \sin(nx)].$$

The reason for the notation $\bar{a}_0$ is that we shall want to change the constant soon.

The "basis" to which we alluded earlier will consist of

$$\{1, \sin x, \cos x, \sin 2x, \cos 2x, \ldots \}.$$

This is not an orthonormal basis, but it is orthogonal as can be seen from the formulas

$$(\cos nx, \cos mx) = \int_{-\pi}^{\pi} \cos(nx) \cos(mx) \, dx = \begin{cases} 0 & \text{if } n \neq m \\ \pi & \text{if } n = m \neq 0 \\ 2\pi & \text{if } n = m = 0 \end{cases} \quad (9.1)$$

$$(\cos nx, \sin mx) = \int_{-\pi}^{\pi} \cos(nx) \sin(mx) \, dx = 0 \quad (9.2)$$

$$(\sin nx, \sin mx) = \int_{-\pi}^{\pi} \sin(nx) \sin(mx) \, dx = \begin{cases} 0 & \text{if } n \neq m \text{ or } n = m = 0 \\ \pi & \text{if } n = m \geq 1 \end{cases} \quad (9.3)$$

$$(1, \sin nx) = \int_{-\pi}^{\pi} \sin(nx)\, dx = 0 \tag{9.4}$$

$$(1, \cos nx) = \int_{-\pi}^{\pi} \cos(nx)\, dx = 0 \quad (n > 0) \tag{9.5}$$

where m and n are nonnegative integers.

Now suppose

$$f(x) = \bar{a}_0 + \sum_{n=1}^{N} (a_n \cos nx + b_n \sin nx).$$

Then one can solve for a_n and b_n as we described earlier. In particular

$$(f, \cos kx) = \int_{-\pi}^{\pi} f(x) \cos kx\, dx$$

$$= \int_{-\pi}^{\pi} \bar{a}_0 \cos kx\, dx + \int_{-\pi}^{\pi} a_1 \cos x \cos kx\, dx$$

$$+ \int_{-\pi}^{\pi} a_2 \cos 2x \cos kx\, dx + \cdots + \int_{-\pi}^{\pi} a_N \cos Nx \cos kx\, dx$$

$$+ \int_{-\pi}^{\pi} b_1 \sin x \cos kx\, dx + \cdots + \int_{-\pi}^{\pi} b_N \sin Nx \cos kx\, dx.$$

Using formulas (9.1)–(9.5), we see that all integrals are 0 except

$$\int_{-\pi}^{\pi} a_k \cos kx \cos kx\, dx = \begin{cases} a_k \pi & \text{if } k \neq 0 \\ \bar{a}_0 2\pi & \text{if } k = 0. \end{cases}$$

Similarly,

$$(f, \sin kx) = \int_{-\pi}^{\pi} f(x) \sin kx\, dx = \pi b_k \quad k = 1, 2, \ldots$$

Thus if we let $a_0/2 = \bar{a}_0$, we have

$$f(x) = \frac{a_0}{2} + \sum_{n=1}^{N} (a_n \cos nx + b_n \sin nx)$$

where

$$a_n = \frac{1}{\pi} \int_{-\pi}^{\pi} f(x) \cos nx\, dx \text{ for } n \geq 0 \tag{9.6a}$$

$$b_n = \frac{1}{\pi} \int_{-\pi}^{\pi} f(x) \sin nx\, dx \text{ for } n \geq 1. \tag{9.6b}$$

Definition: The numbers a_n and b_n of Equations (9.6a) and (9.6b) are called the *Fourier coefficients* of the function $f(x)$. A series of the form

$$\frac{a_0}{2} + \sum_{n=1}^{\infty}(a_n \cos nx + b_n \sin nx)$$

where the a_n's and b_n's are determined by Equations (9.6a) and (9.6b) is called the *Fourier series* for f.

Notice that we can define the Fourier series for any function that is Riemann integrable on $[-\pi, \pi]$. If f is Riemann integrable and

$$\frac{a_0}{2} + \sum_{n=1}^{\infty}(a_n \cos nx + b_n \sin nx)$$

is the Fourier series for f, then we shall write

$$f(x) \sim \frac{a_0}{2} + \sum_{n=1}^{\infty}(a_n \cos nx + b_n \sin nx).$$

We shall be interested in determining what functions can be expressed as a Fourier series. These functions will have period 2π, since $\sin nx$ and $\cos nx$ have period 2π. This not a necessary restriction. If the period of the function were $2L$, we would modify the formulas to

$$f(x) = \frac{a_0}{2} + \sum_{n=1}^{\infty}\left(a_n \cos\left(\frac{n\pi x}{L}\right) + b_n \sin\left(\frac{n\pi x}{L}\right)\right),$$

where

$$a_n = \frac{1}{L}\int_{-\pi}^{\pi} f(x)\cos\left(\frac{n\pi x}{L}\right)\,dx, \quad b_n = \frac{1}{L}\int_{-\pi}^{\pi} f(x)\sin\left(\frac{n\pi x}{L}\right)\,dx.$$

To simplify the notation we shall restrict our attention to functions of period 2π.

Definition: A function f is said to be *even* if $f(x) = f(-x)$ and f is said to be *odd* if $f(-x) = -f(x)$ for every x in the domain of f.

Some examples of even functions are $\cos x$ and x^{2n} for n an integer. Some odd functions are $\sin x$ and x^{2n+1} for n an integer. In Exercise 5, Section 9.1, we show that for an even function the Fourier coefficients b_n will be 0, and for an odd function the Fourier coefficients a_n will be 0. There are functions that are neither even nor odd.

We calculate the Fourier coefficients for two examples.

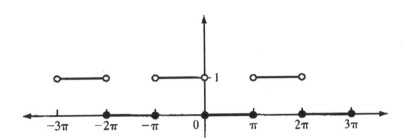

Figure 9.2
The periodic function defined in Example 9.1.

Example 9.1:
Let
$$f(x) = \begin{cases} 0 & \text{if } 0 \le x \le \pi \\ 1 & \text{if } -\pi < x < 0 \end{cases}$$
and $f(x + 2\pi) = f(x)$. The graph of f is given in Figure 9.2.
Now
$$a_n = \frac{1}{\pi} \int_{-\pi}^{\pi} f(x) \cos nx \, dx = \frac{1}{\pi} \int_{-\pi}^{0} 1 \, \cos nx \, dx$$
$$= \frac{1}{n\pi} \sin nx \Big|_{-\pi}^{0} = 0, \quad n = 1, 2, \dots$$
$$a_0 = \frac{1}{\pi} \int_{-\pi}^{0} 1 \, dx = 1$$
$$b_n = \frac{1}{\pi} \int_{-\pi}^{\pi} f(x) \sin nx \, dx = \frac{1}{\pi} \int_{-\pi}^{0} 1 \, \sin nx \, dx$$
$$= \frac{-1}{n\pi} \cos nx \Big|_{-\pi}^{0} = \begin{cases} 0 & \text{if } n \text{ is even} \\ -2/(n\pi) & \text{if } n \text{ is odd.} \end{cases}$$
Thus
$$f(x) \sim \frac{a_0}{1} + \sum_{n=1}^{\infty} (a_n \cos nx + b_n \sin nx)$$
$$= \frac{1}{2} - \left(\frac{2}{\pi}\right) \sin x - \left(\frac{2}{3\pi}\right) \sin 3x - \left(\frac{2}{5\pi}\right) \sin 5x - \cdots .$$

Example 9.2:
Let $f(x) = x$. Notice that this is an odd function, so it should be the case that $a_n = 0$ for every n. Now
$$a_0 = \frac{1}{\pi} \int_{-\pi}^{\pi} x \, dx = \frac{1}{\pi} \left(\frac{x^2}{2}\right) \Big|_{-\pi}^{\pi} = 0$$

$$a_n = \frac{1}{\pi} \int_{-\pi}^{\pi} x \cos nx \, dx \qquad \text{(integrate by parts)}$$

$$= \frac{1}{\pi} \left(\frac{x}{n} \sin nx + \frac{1}{n^2} \cos nx \right) \Big|_{-\pi}^{\pi} = 0$$

and

$$b_n = \frac{1}{\pi} \int_{-\pi}^{\pi} x \sin nx \, dx$$

$$= \frac{1}{\pi} \left(\frac{-x}{n} \cos nx + \frac{1}{n^2} \sin nx \right) \Big|_{-\pi}^{\pi}$$

$$= \frac{-2}{n} \cos n\pi = \begin{cases} -2/n & \text{if } n \text{ is even} \\ 2/n & \text{if } n \text{ is odd.} \end{cases}$$

Thus for $f(x) = x$,

$$f(x) \sim 2 \sin x - \sin 2x + \frac{2}{3} \sin 3x - \frac{1}{2} \sin 4x + \cdots .$$

Exercises 9.1

1. Verify that for n and m nonnegative integers,
   (a)

$$\int_{-\alpha}^{\pi} \cos nx \cos mx = \begin{cases} 0 & \text{if } n \neq m \\ \pi & \text{if } n = m \neq 0 \\ 2\pi & \text{if } n = m = 0. \end{cases}$$

   (b) $\int_{-\pi}^{\pi} \cos nx \sin mx \, dx = 0.$
   (c)

$$\int_{-\pi}^{\pi} \sin nx \sin mx \, dx = \begin{cases} 0 & \text{if } n \neq m \text{ or } n = m = 0 \\ \pi & \text{if } n = m \geq 1. \end{cases}$$

2. Show that if $f(x) = \bar{a}_0 + \sum_{n=1}^{N} (a_n \cos nx + b_n \sin nx)$, then $(f(x), \sin kx) = \pi b_k$, and thus $b_k = (1/\pi) \int_{-\pi}^{\pi} f(x) \sin kx \, dx$ for $k = 1, \ldots, N$.

3. (a) Show that a linear combination of even functions is even.
   (b) Show that a linear combination of odd functions is odd.

4. Find the Fourier series for the following functions that are defined on the interval $(-\pi, \pi]$ and are periodic of period 2π:
   (a)

$$f(x) = \begin{cases} 1 & \text{if } 0 < x \leq \pi \\ -1 & \text{if } -\pi < x \leq 0. \end{cases}$$

   (b) $f(x) = \cos x.$
   (c) $f(x) = x^2.$
   (d) $f(x) = |x|.$
   (e) $f(x) = \sin^2 x.$

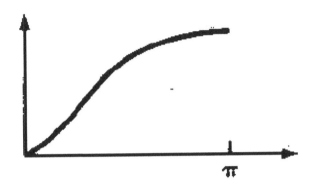

Figure 9.3
Graph of function used in Exercise 6.

5. *(a)* Prove that if $f(x)$ is odd, then the Fourier coefficient $a_n = 0$.
   (b) Prove that if $f(x)$ is even, then the Fourier coefficient $b_n = 0$.
   [*Hint:* In part (a) write

   $$\int_{-\pi}^{\pi} f(x) \cos nx \, dx = \int_{-\pi}^{0} f(x) \cos nx \, dx + \int_{0}^{\pi} f(x) \cos nx \, dx,$$

   then use the fact that $f(-x) = -f(x)$, and use the change of variables $y = -x$ in the first integral.]

6. Suppose that $f(x)$ is defined on $[0, \pi]$ and has the graph shown in Figure 9.3.
   (a) Draw the rest of the graph of $f(x)$ if $f(x)$ is to be an odd function of period 2π.
   (b) Repeat part (a), except make $f(x)$ an even function.

7. Construct a function that is neither even nor odd.

8. If you have access to a computer graphics program, graph $f(x)$ and the trigonometric polynomial of $f(x)$ for $n = 1, 5, 10$ for the function of Exercise 4.

9.2   Representation by Fourier Series

Convergence of Fourier Series

We now want to determine a class of functions that can be expressed as Fourier series. The situation is similar to that of Taylor series. That is, Equations (9.6a) and (9.6b) of Section 9.1 tell us how to calculate the Fourier

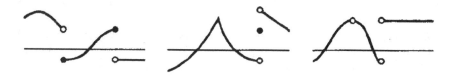

Figure 9.4
Some piecewise continuous functions

coefficients of a function so that we can write the Fourier series for a Riemann integrable function, but we do not know when the Fourier series is actually equal to the function.

Definition: A function f is said to be *piecewise continuous* on $[a, b]$ if there are finitely many points $x_0, x_1, \ldots, x_n \in [a, b]$ such that
(i) f is continuous on $[a, b]$ except at $x_0, x_1, \ldots, x_n$, and
(ii) f has jump or removable discontinuities at $x_0, x_1, \ldots, x_n$.

Recall that if f has a jump discontinuity at $x_i \in (a, b)$, then

$$\lim_{x \downarrow x_i} f(x) \text{ and } \lim_{x \uparrow x_i} f(x)$$

exist and are finite. It is not required that f be defined at $x_0, x_1, \ldots, x_n$.

Some examples of the graphs of piecewise continuous functions are given in Figure 9.4.

Notation: Let $\mathcal{F}$ denote the class of functions of period 2π that are piecewise continuous.

Notice that $\sin nx$ and $\cos nx$, $n = 0, 1, 2, \ldots$, are in $\mathcal{F}$ as are finite linear combinations of these functions.

Let's return to our experience with vector spaces. If f is a vector in an inner product space, then $(f, f) = ||f||^2$ is often described as the square of the length of f. If f and g are vectors in this vector space, then $(f - g, f - g)$ is the square of the "distance" between the vectors f and g.

Theorem 9.1: Let $\{\phi_1, \phi_2, \ldots, \phi_n\}$ be an orthonormal set of functions, and let $\{c_1, \ldots, c_n\}$ be the Fourier coefficients for a function $f \in \mathcal{F}$ with respect to this set; that is,

$$c_i = \int_{-\pi}^{\pi} f(x)\phi_i(x) \, dx = (f, \phi_i) \quad i = 1, \ldots n.$$

If $\{d_1, \ldots, d_n\}$ is any set of constants and

$$S_n(x) = \sum_{i=1}^{n} c_i \phi_i(x) \text{ and } T_n(x) = \sum_{i=1}^{n} d_i \phi_i(x),$$

then $(f - S_n, f - S_n) \le (f - T_n, f - T_n)$ with equality holding if and only if $c_i = d_i$ for each $i = 1, \dots, n$.

Remark: It is important to realize what this theorem says because it tells why Fourier coefficients are important. Since $(f - S_n, f - S_n)$ is the square of the distance between the functions f and S_n, the smaller this number is, the better the approximation S_n is to the function f. The theorem says that if one wants to choose the constants $c_1, \dots, c_n$ so that $c_1\phi_1 + \cdots + c_n\phi_n$ gives the best approximation to f, then one should choose the c_i to be the Fourier coefficients.

Proof: By definition

$$(f - T_n, f - T_n) = \int_{-\pi}^{\pi} [f(x) - T_n(x)][f(x) - T_n(x)]\, dx$$

$$= \int_{-\pi}^{\pi} [f(x)]^2 - 2f(x)T_n(x) + [T_n(x)]^2\, dx. \tag{9.7}$$

Now

$$T_n(x) = \sum_{i=1}^{n} d_i\phi_i(x).$$

It will be shown in Exercise 2, Section 9.2, that if $\{\phi_i\}$ *is an orthonormal set,* then

$$\int_{-\pi}^{\pi} \left(\sum_{i=1}^{n} d_i\phi_i(x) \right) \left(\sum_{i=1}^{n} d_i\phi_i(x) \right)\, dx = \sum_{i=1}^{n} d_i^2.$$

Thus the right-hand side of Equation (9.7) is

$$\int_{-\pi}^{\pi} [f(x)]^2\, dx - 2\sum_{i=1}^{n} d_i \int_{-\pi}^{\pi} f(x)\phi_i(x)\, dx + \sum_{i=1}^{n} d_i^2. \tag{9.8}$$

But $c_i = \int_{-\pi}^{\pi} f(x)\phi_i(x)\, dx$, so Equation (9.8) becomes

$$\int_{-\pi}^{\pi} [f(x)]^2\, dx - 2\sum_{i=1}^{n} d_ic_i + \sum_{i=1}^{n} d_i^2$$

$$= \int_{-\pi}^{\pi} [f(x)]^2\, dx - \sum_{i=1}^{n} c_i^2 + \sum_{i=1}^{n} c_i^2 - 2\sum_{i=1}^{n} c_id_i + \sum_{i=1}^{n} d_i^2 \tag{9.9}$$

$$\int_{-\pi}^{\pi} [f(x)]^2\, dx - \sum_{i=1}^{n} c_i^2 + \sum_{i=1}^{n} (c_i - d_i)^2.$$

Now

$$\int_{-\pi}^{\pi} [f(x)]^2\, dx - \sum_{i=1}^{n} c_i^2 = (f - S_n, f - S_n) \tag{9.10}$$

so that

$$(f - T_n, f - T_n) = (f - S_n, f - S_n) + \sum_{i=1}^{n}(c_i - d_i)^2$$

$$\geq (f - S_n, f - S_n)$$

with equality if and only if $c_i = d_i$ for every $i - 1, \ldots, n$. ∎

Corollary 9.1(a) (Bessel's Inequality): Let $\{\phi_n\}$ be an orthonormal set on $[-\pi, \pi]$, and let $\sum_{i=1}^{\infty} c_i \phi_i(x)$ be the Fourier series for $f \in \mathcal{F}$; that is,

$$c_i = \int_{-\pi}^{\pi} f(x)\phi_i(x) \, dx.$$

Then

$$\sum_{i=1}^{\infty} c_i^2 \leq \int_{-\pi}^{\pi} [f(x)]^2 \, dx.$$

Proof: From Equation (9.10) we have

$$\int_{-\pi}^{\pi} [f(x)]^2 \, dx - \sum_{i=1}^{n} c_i^2 = (f - S_n, f - S_n) \geq 0 \text{ for all } n.$$

Thus

$$\lim_{n \to \infty} \sum_{i=1}^{n} c_i^2 = \sum_{i=1}^{\infty} c_i^2 \leq \int_{-\pi}^{\pi} [f(x)]^2 \, dx. \quad ∎$$

We now determine the relationship between the c_i's of the orthonormal basis

$$\left\{ \frac{1}{\sqrt{2\pi}}, \frac{\cos x}{\sqrt{\pi}}, \frac{\sin x}{\sqrt{\pi}}, \frac{\cos 2x}{\sqrt{\pi}}, \frac{\sin 2x}{\sqrt{\pi}}, \cdots \right\}$$

and the numbers a_n and b_n associated with the basis

$$\{1, \cos x, \sin x, \cos 2x, \sin 2x, \ldots\}.$$

We have (renumbering to begin with c_0)

$$c_0 = \frac{1}{\sqrt{2\pi}} \int_{-\pi}^{\pi} f(x) \, dx \text{ and } a_0 = \frac{1}{\pi} \int_{-\pi}^{\pi} f(x) \, dx$$

$$c_{2n-1} = \frac{1}{\sqrt{\pi}} \int_{-\pi}^{\pi} f(x) \cos nx \, dx \text{ and } a_n = \frac{1}{\pi} \int_{-\pi}^{\pi} f(x) \cos nx \, dx$$

$$c_{2n} = \frac{1}{\sqrt{\pi}} \int_{-\pi}^{\pi} f(x) \sin nx \, dx \text{ and } b_n = \frac{1}{\pi} \int_{-\pi}^{\pi} f(x) \sin nx \, dx$$

so

$$c_0 = \frac{\sqrt{\pi}}{\sqrt{2}} a_0, \quad c_{2n} = \sqrt{\pi} b_n, \quad c_{2n-1} = \sqrt{\pi} a_n \quad n = 1, 2, \ldots$$

and

$$\sum_{n=0}^{\infty} c_n^2 = \frac{\pi}{2} a_0^2 + \pi \sum_{n=1}^{\infty} (a_n^2 + b_n^2).$$

Corollary 9.1(b): Let a_n and b_n be the Fourier coefficients for a function $f(x)$ as defined by Equations (9.6a) and (9.6b) in Section 9.1. Then

$$\frac{a_0^2}{2} + \sum_{n=1}^{\infty} (a_n^2 + b_n^2)) \le \frac{1}{\pi} \int_{-\pi}^{\pi} [f(x)]^2 \, dx.$$

Proof: We have

$$\frac{a_0^2}{2} + \sum_{n=1}^{\infty} (a_n^2 + b_n^2) = \frac{1}{\pi} \sum_{n=0}^{\infty} c_n^2 \le \frac{1}{\pi} \int_{-\pi}^{\pi} [f(x)]^2 \, dx. \quad \blacksquare$$

Corollary 9.1(c) (Riemann-Lebesgue Lemma): If $\int_{-\pi}^{\pi} [f(x)]^2 \, dx$ exists and is finite, and f is Riemann integrable, then

(a)

$$\lim_{n \to \infty} \int_{-\pi}^{\pi} f(x) \cos nx \, dx = 0$$

and

(b)

$$\lim_{n \to \infty} \int_{-\pi}^{\pi} f(x) \sin nx \, dx = 0.$$

Proof: By Corollary 9.2(a)

$$\sum_{n=0}^{\infty} c_n^2 \le \frac{1}{\pi} \int_{-\pi}^{\pi} [f(x)]^2 \, dx$$

and by hypothesis, the integral $\int_{-\pi}^{\pi} [f(x)]^2 \, dx$ is finite. Thus $\sum_{n=0}^{\infty} c_n^2$ is a convergent series, so $\lim_{n \to \infty} c_n^2 = 0$ and thus $\lim_{n \to \infty} c_n = 0$. But

$$c_{2n-1} = \frac{1}{\sqrt{\pi}} \int_{-\pi}^{\pi} f(x) \cos nx \, dx \text{ and } c_{2n} = \frac{1}{\sqrt{\pi}} \int_{-\pi}^{\pi} f(x) \sin nx \, dx. \quad \blacksquare$$

In Exercise 10, Section 9.2, we show that under the hypotheses of the Riemann-Lebesgue Lemma,

$$\lim_{n \to \infty} \int_a^b f(x) \cos nx \, dx = 0 \text{ and } \lim_{n \to \infty} \int_a^b f(x) \sin nx \, dx = 0$$

for any interval $[a, b] \subset [-\pi, \pi]$.

We now begin to answer the question that was posed earlier. When can a function be represented by its Fourier series? Let $f(x)$ be a function such that

$$a_n = \int_{-\pi}^{\pi} f(x) \cos nx \, dx \text{ and } b_n = \int_{-\pi}^{\pi} f(x) \sin nx \, dx$$

are defined and $f(x)$ is periodic of period 2π. If $f \in \mathcal{F}$, then these conditions are satisfied.

Before we begin this task, it may be worthwhile to point out the existence of some examples where the Fourier series does not behave as one might hope. The specific examples are somewhat complicated, and we shall not list them here. An excellent (but advanced) reference in this area is *Trigonometric Series* by Antoni Zygmund (1967).

(1) There exists a continuous function whose Fourier series diverges at a point.

(2) There exists a continuous function whose Fourier series converges everywhere, but not unformly.

(3) Given any countable set E in $(-\pi, \pi)$, there is a continuous function whose Fourier series diverges in E and converges on $(-\pi, \pi) \setminus E$.

To begin our investigation, we let

$$S_N(x) = \frac{a_0}{2} + \sum_{n=1}^{N}(a_n \cos nx + b_n \sin nx)$$

and let

$$R_N(x) = f(x) - S_N(x)$$

so that $R_N(x)$ is a sort of "remainder" term that measures how much the Nth trigonometric polynomial of f differs from f. We would like $\lim_{N \to \infty} S_N(x) = f(x)$ or, equivalently, $\lim_{N \to \infty} R_N(x) = 0$. Notice that we have three possible ways to interpret $\lim_{N \to \infty} R_N(x) = 0$:

(i) $R_N(x)$ converges to 0 pointwise.

(ii) $R_N(x)$ converges to 0 uniformly.

(iii) $\lim(R_N, R_N) = 0$ where (,) is the "inner product" that we defined earlier.

The main results that we shall prove for $f \in \mathcal{F}$ are:

(1) If f is differentiable at x_0, then the Fourier series for f at x_0 converges to $f(x_0)$. Thus if f is differentiable, the Fourier series of f converges pointwise to f.

(2) If f is continuous and $f' \in \mathcal{F}$, then the Fourier series of f converges uniformly to f.

(3) If f is continuous and has Fourier series

$$\frac{a_0}{2} + \sum_{n=1}^{\infty}(a_n \cos nx + b_n \sin nx),$$

then

$$\frac{a_0^2}{2} + \sum_{n=1}^{\infty}(a_n^2 + b_n^2) = \frac{1}{\pi}\int_{-\pi}^{\pi}[f(x)]^2\,dx.$$

This will mean that $\lim(R_N, R_N) = 0$. One way to begin to answer the question of representation is to obtain an explicit formula for $R_N(x)$.

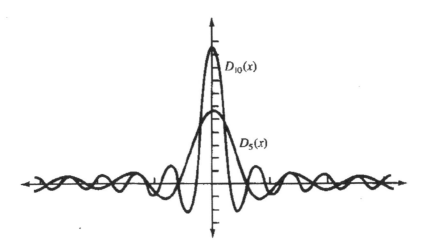

Figure 9.5
Two Dirichlet kernels

Definition: The *Dirichlet kernel* of index N is the function $D_N(x)$ defined by

$$D_N(x) = 1/2 + \cos x + \cos 2x + \cdots + \cos Nx.$$

The graphs of $D_5(x)$ and $D_{10}(x)$ are shown in Figure 9.5. As n becomes larger, the peak of $D_n(x)$ at $x = 0$ becomes more pronounced.

In Exercise 1, Section 9.2, we show that

$$D_N(x) = \begin{cases} \frac{\sin(N+1/2)x}{2\sin(x/2)} & \text{if } x \neq 2\pi m, \; m \text{ an integer} \\ N + 1/2 & \text{if } x = 2\pi m. \end{cases}$$

Also, it should be clear that

$$\frac{1}{\pi} \int_{-\pi}^{\pi} D_N(t) \, dt = 1$$

for every N.

Theorem 9.2: Let $f \in \mathcal{F}$, and let S_N, R_N, and D_N be as previously defined. Then

$$S_N(x) = \frac{1}{\pi} \int_{-\pi}^{\pi} f(u) D_N(x - u) \, du$$

$$= \frac{1}{\pi} \int_{-\pi}^{\pi} f(x - t) D_N(t) \, dt = \frac{1}{\pi} \int_{-\pi}^{\pi} f(x + t) D_N(t) \, dt$$

and

$$R_N(x) = \frac{1}{\pi} \int_{-\pi}^{\pi} [f(x) - f(x-t)] D_N(t)\, dt. \tag{9.11}$$

Remark: Equation (9.11) will be very useful in showing when $\lim_{N \to \infty} R_N(x) = 0$, and it is one reason why the Dirichlet kernel is an important tool in Fourier analysis.

Proof: We first note that

$$\frac{1}{\pi} \int_{-\pi}^{\pi} f(u) D_N(x-u)\, du = \frac{1}{\pi} \int_{-\pi}^{\pi} f(u) \left[\frac{1}{2} + \sum_{k=1}^{N} \cos k(x-u) \right] du$$

$$= \frac{1}{\pi} \int_{-\pi}^{\pi} f(u) \left[\frac{1}{2} + \sum_{k=1}^{N} (\cos kx \cos ku + \sin kx \sin ku) \right] du$$

(using $\cos(\alpha - \beta) = \cos\alpha\cos\beta + \sin\alpha\sin\beta$)

$$= \frac{1}{2\pi} \int_{-\pi}^{\pi} f(u)\, du + \sum_{k=1}^{N} \cos kx \left(\frac{1}{\pi} \int_{-\pi}^{\pi} f(u)\cos ku\, du \right)$$

$$+ \sum_{k=1}^{N} \sin kx \left(\frac{1}{\pi} \int_{-\pi}^{\pi} f(u)\sin ku\, du \right)$$

$$= \frac{a_0}{2} + \sum_{k=1}^{N} (a_k \cos kx + b_k \sin kx) = S_N(x).$$

To show

$$\int_{-\pi}^{\pi} f(u) D_N(x-u)\, du = \int_{-\pi}^{\pi} f(x-t) D_N(t)\, dt$$

we make the change of variables $u = x - t$. Then $du = -dt$. (Remember that x is fixed.) If $u = \pi$, then $t = x - \pi$ and if $u = -\pi$, then $t = x + \pi$. Thus

$$\int_{-\pi}^{\pi} f(u) D_N(x-u)\, du = - \int_{x+\pi}^{x-\pi} f(x-t) D_N(t)\, dt = \int_{x-\pi}^{x+\pi} f(x-t) D_N(t)\, dt.$$

Now f and D_N are periodic of period 2π. Thus any integral of the product of these functions over an interval of length 2π is the same, so

$$\int_{x-\pi}^{x+\pi} f(x-t) D_N(t)\, dt = \int_{-\pi}^{\pi} f(x-t) D_N(t)\, dt.$$

Showing that

$$\int_{-\pi}^{\pi} f(u) D_N(x-u)\, du = \int_{-\pi}^{\pi} f(x+t) D_N(t)\, dt$$

is done in a similar manner and is left for Exercise 3, Section 9.2.

Now

$$R_N(x) = f(x) - S_N(x) \text{ and } \frac{1}{\pi}\int_{-\pi}^{\pi} D_N(t)\, dt = 1$$

so

$$f(x) = \frac{1}{\pi}\int_{-\pi}^{\pi} f(x)D_N(t)\, dt. \quad \blacksquare$$

We can now give a sufficient condition for $\lim_{N\to\infty} R_N(x) = 0$.

Theorem 9.3: Let $f \in \mathcal{F}$, and suppose that f is differentiable at the point x_0. Then

$$f(x_0) = \lim_{N\to\infty}\left[\frac{a_0}{2} + \sum_{n=1}^{N}(a_n \cos nx_0 + b_n \sin nx_0)\right]$$

where a_n and b_n are the Fourier coefficients defined by Equations (9.6a) and (9.6b) of Section 9.1.

Proof: By Theorem 9.2 we have

$$R_N(x) = \frac{1}{\pi}\int_{-\pi}^{\pi}(f(x) - f(x-t))D_N(t)\, dt$$

$$= \frac{1}{\pi}\int_{-\pi}^{\pi}(f(x) - f(x-t))\frac{\sin(N+1/2)t}{2\sin(t/2)}\, dt$$

$$= \frac{1}{\pi}\int_{-\pi}^{\pi}\frac{f(x) - f(x-t)}{2\sin(t/2)}(\sin Nt\cos(t/2) + \cos Nt\sin(t/2))\, dt$$

$$= \frac{1}{\pi}\int_{-\pi}^{\pi}\frac{f(x) - f(x-t)}{2\sin(t/2)}\cos(t/2)\sin Nt\, dt$$

$$+ \frac{1}{\pi}\int_{-\pi}^{\pi}\frac{f(x) - f(x-t)}{2}\cos Nt\, dt. \quad (9.12)$$

By the Riemann-Lebesgue Lemma

$$\lim_{N\to\infty}\frac{1}{\pi}\int_{-\pi}^{\pi}(f(x) - f(x-t))\cos Nt\, dt = 0 \text{ if } f \in \mathcal{F}.$$

Next we deal with

$$\int_{-\pi}^{\pi}\frac{f(x) - f(x-t)}{2\sin(t/2)}\cos(t/2)\sin Nt\, dt.$$

Since $f \in \mathcal{F}$ and $\sin(t/2)$ is continuous, the function

$$\frac{f(x) - f(x-t)}{2\sin(t/2)}$$

is piecewise continuous (in t) except possibly at $t = 0$. When $t = 0$, the function is of the form $0/0$, so *if f is differentiable at x,* we have

$$\lim_{t \to 0} \frac{f(x) - f(x - t)}{2 \sin(t/2)} = \lim_{t \to 0} \frac{f(x) - f(x - t)}{t} \frac{t/2}{\sin(t/2)} = f'(x)$$

since $\lim_{\theta \to 0}(\sin \theta)/\theta = 1$. So if f is differentiable at x, as a function of t, the function

$$\frac{f(x) - f(x - t)}{2 \sin(t/2)} \cos(t/2)$$

is in $\mathcal{F}$, and by the Riemann-Lebesgue Lemma,

$$\lim_{N \to \infty} \int_{-\pi}^{\pi} \frac{f(x) - f(x - t)}{2 \sin(t/2)} \cos(t/2) \sin Nt \, dt = 0. \quad \blacksquare$$

Thus we see that the Fourier series of a piecewise continuous function converges to the function at points where the function is differentiable. In particular, if f is differentiable and periodic of period 2π, the Fourier series of f converges pointwise to f.

Next we shall show how the Fourier series behaves at points of discontinuity. It does the best possible thing if $f' \in \mathcal{F}$; it averages the left- and right-hand limits.

Theorem 9.4: Suppose f and f' are piecewise continuous and of period 2π. Suppose $x_0 \in (-\pi, \pi)$. Then

$$\lim_{N \to \infty} S_N(x_0) = \frac{1}{2} \left[\lim_{x \downarrow x_0} f(x) + \lim_{x \uparrow x_0} f(x) \right]$$

where

$$S_N(x) = \frac{a_0}{2} + \sum_{n=1}^{N} (a_n \cos nx + b_n \sin nx).$$

Proof: By Theorem 9.2

$$S_N(x_0) = \frac{1}{\pi} \int_{-\pi}^{\pi} f(x_0 + t) \frac{\sin(N + 1/2)t}{2 \sin(t/2)} \, dt$$

$$= \frac{1}{\pi} \int_{-\pi}^{0} f(x_0 + t) \frac{\sin(N + 1/2)t}{2 \sin(t/2)} \, dt$$

$$+ \frac{1}{\pi} \int_{0}^{\pi} f(x_0 + t) \frac{\sin(N + 1/2)t}{2 \sin(t/2)} \, dt.$$

Let

$$f(x_0^+) = \lim_{x \downarrow x_0} f(x), \quad f(x_0^-) = \lim_{x \uparrow x_0} f(x).$$

Now by symmetry

$$\frac{1}{\pi}\int_{-\pi}^{0}\frac{\sin(N+1/2)t}{2\sin(t/2)}\,dt = \frac{1}{\pi}\int_{0}^{\pi}\frac{\sin(N+1/2)t}{2\sin(t/2)}\,dt = \frac{1}{2}$$

so that

$$\frac{f(x_0^-)+f(x_0^+)}{2} = \frac{1}{\pi}\int_{-\pi}^{0}\frac{f(x_0^-)\sin(N+1/2)t}{2\sin(t/2)}\,dt + \frac{1}{\pi}\int_{0}^{\pi}\frac{f(x_0^+)\sin(N+1/2)t}{2\sin(t/2)}\,dt.$$

Thus

$$S_N(x_0) - \left(\frac{f(x_0^-)+f(x_0^+)}{2}\right) = \frac{1}{\pi}\int_{-\pi}^{0}\frac{f(x_0+t)-f(x_0^-)}{2\sin(t/2)}\sin\left(N+\frac{1}{2}\right)t\,dt$$

$$+\frac{1}{\pi}\int_{0}^{\pi}\frac{f(x_0+t)-f(x_0^+)}{2\sin(t/2)}\sin\left(N+\frac{1}{2}\right)t\,dt.$$

Now if f is piecewise continuous, then

$$\frac{f(x_0+t)-f(x_0^+)}{2\sin(t/2)}$$

is piecewise continuous on $(0,\pi]$ and

$$\lim_{t\downarrow 0}\frac{f(x_0+t)-f(x_0^+)}{2\sin(t/2)} = \lim_{t\downarrow 0}\frac{f(x_0+t)-f(x_0^+)}{t}\cdot\frac{t/2}{\sin(t/2)}$$

$$= \lim_{t\downarrow 0}\frac{f(x_0+t)-f(x_0^+)}{t}.$$

This final limit exists because f' is piecewise continuous. Thus

$$\frac{f(x_0+t)-f(x_0^+)}{2\sin(t/2)}$$

is piecewise continuous on $[0,\pi]$. Similarly,

$$\frac{f(x_0+t)-f(x_0^-)}{2\sin(t/2)}$$

is piecewise continuous on $[-\pi,0]$. Then

$$\lim_{N\to\infty} S_N(x_0) - \left(\frac{f(x_0^+)+f(x_0^-)}{2}\right)$$

$$= \lim_{N\to\infty}\left[\frac{1}{\pi}\int_{-\pi}^{0}\left(\frac{f(x_0+t)-f(x_0^-)}{2\sin(t/2)}\right)\sin\left(N+\frac{1}{2}\right)t\,dt\right.$$

$$+\frac{1}{\pi}\int_0^\pi \left(\frac{f(x_0+t)-f(x_0^+)}{2\sin(t/2)}\right)\sin\left(N+\frac{1}{2}\right)t\ dt\Bigg]$$

$$=\lim_{N\to\infty}\left[\frac{1}{\pi}\int_{-\pi}^0 \left(\frac{f(x_0+t)-f(x_0^-)}{2\sin(t/2)}\right)(\sin Nt\cos(t/2)+\cos Nt\sin(t/2))\ dt\right.$$

$$+\frac{1}{\pi}\int_0^\pi \left(\frac{f(x_0+t)-f(x_0^+)}{2\sin(t/2)}\right)(\sin Nt\cos(t/2)+\cos Nt\sin(t/2))\ dt\Bigg]=0$$

by the Riemann-Lebesgue Lemma. ∎

Corollary 9.4: If f and f' are piecewise continuous and $f\in\mathcal{F}$, and if f is continuous at x_0, then the Fourier series of f converges to $f(x_0)$ at x_0. ∎

It begins to appear that the behavior of f' is the determining factor in the convergence of the Fourier series to the desired value. Theorem 9.5 will reinforce this idea.

We now give conditions under which the Fourier series converges to f uniformly.[1]

In the proof of this theorem we shall use the Schwarz inequality, which was an earlier exercise (Section 1.2, Exercise 18). Before giving the proof we recall that result.

Schwarz Inequality: Let $a_1,\ldots,a_n$ and $b_1,\ldots,b_n$ be real numbers. Then

$$\sum_{i=1}^n a_i b_i \le \left(\sum_{i=1}^n a_i^2\right)^{1/2}\left(\sum_{i=1}^n b_i^2\right)^{1/2}.\quad ∎$$

Theorem 9.5: Suppose that f is a continuous function of period 2π, and suppose that $f'(x)\in\mathcal{F}$. Then the Fourier series of f converges uniformly to f.

Proof: Suppose that $\sum(a_n\cos nx+b_n\sin nx)$ is the Fourier series for $f(x)$, and $\sum(c_n\cos nx+d_n\sin nx)$ is the Fourier series for $f'(x)$.

From Corollary 9.1(a), we know that $\sum a_n^2$, $\sum b_n^2$, $\sum c_n^2$, and $\sum d_n^2$ are finite. Now

$$c_n=\frac{1}{\pi}\int_{-\pi}^\pi f'(x)\cos nx\ dx \quad\text{and}\quad d_n=\frac{1}{\pi}\int_{-\pi}^\pi f'(x)\sin nx\ dx.$$

Integrating the expression for c_n by parts gives

$$c_n=\frac{1}{\pi}\left(f(x)\cos nx\Big|_{-\pi}^\pi+n\int_{-\pi}^\pi f(x)\sin nx\ dx\right)$$

---

[1]The proof given here is an adaptation of a proof given by Barry Simon ("Uniform Convergence of Fourier Series." *Amer. Math. Monthly*, Vol. 76: 1, p. 55, January 1967).

$$= \frac{n}{\pi} \int_{-\pi}^{\pi} f(x) \sin nx \, dx = na_n$$

since $f(\pi) \cos n\pi = f(-\pi) \cos(-n\pi)$.

Thus $\sum_{n=1}^{\infty} n^2 a_n^2 < \infty$, and likewise $\sum_{n=1}^{\infty} n^2 b_n^2 < \infty$. By the Schwarz inequality,

$$\sum_{n=1}^{N} |x_n y_n| \leq \left(\sum_{n=1}^{N} x_n^2 \right)^{1/2} \left(\sum_{n=1}^{N} y_n^2 \right)^{1/2} \quad \text{for every } N$$

so that convergence of the series $\sum x_n^2$ and $\sum y_n^2$ implies convergence of the series $\sum_{n=1}^{\infty} |x_n y_n|$. Hence

$$\sum_{n=1}^{\infty} |a_n| = \sum_{n=1}^{\infty} \left| \left(\frac{1}{n} \right)(na_n) \right| \leq \left[\sum_{n=1}^{\infty} \frac{1}{n^2} \right]^{1/2} \left[\sum_{n=1}^{\infty} (n^2 a_n^2) \right]^{1/2}$$

and both $\sum_{n=1}^{\infty} 1/n^2$ and $\sum_{n=1}^{\infty} n^2 a_n^2$ are convergent. Thus $\sum_{n=1}^{\infty} |a_n|$ converges. For every x, $|a_n \cos nx| \leq |a_n|$, so the series $\sum_{n=1}^{\infty} |a_n \cos nx|$ converges uniformly by the Weierstrass M-test. Similarly $\sum_{n=1}^{\infty} |b_n \sin nx|$ converges uniformly. Thus the Fourier series converges absolutely and uniformly. Since f is differentiable at all but a finite number of points, the Fourier series converges to f except possibly at those points. However, the uniform limit of a series of continuous functions is continuous, so it follows that the Fourier series converges uniformly to f.   ∎

Thus we have been able to determine conditions under which $R_N(x)$ converges to 0, pointwise and uniformly. Finally, we would like to know when it is true that $\lim_{N \to \infty}(R_N(x), R_N(x)) = 0$. This is known as L^2 convergence of the Fourier series, and in applications we often need to answer questions regarding convergence in terms of this L^2 convergence. Deriving the answer turns out to be somewhat involved.

In the course of proving Theorem 9.1 and Bessel's inequality, we found

$$0 \leq (f - S_N, f - S_N) = \int_{-\pi}^{\pi} [f(x)]^2 \, dx - \sum_{n=1}^{N} c_i^2$$

and that

$$\sum_{i=0}^{\infty} c_i^2 = \int_{-\pi}^{\pi} [f(x)]^2 \, dx.$$

Thus a necessary and sufficient condition for $\lim_{N \to \infty}(R_N, R_N) = 0$ is that

$$\sum_{i=0}^{\infty} c_i^2 = \int_{-\pi}^{\pi} [f(x)]^2 \, dx.$$

Our aim is to determine conditions on f that will ensure this.

First, let us establish an outline of what will be done. Let $f \in \mathcal{F}$ have Fourier series

$$\frac{a_0}{2} + \sum_{n=1}^{\infty} (a_n \cos nx + b_n \sin nx)$$

and let

$$S_m(x) = \frac{a_0}{2} + \sum_{n=1}^{m} (a_n \cos nx + b_n \sin nx)$$

and

$$\sigma_m(x) = \frac{S_0(x) + S_1(x) + \cdots + S_m(x)}{m+1}.$$

A sum of the type $\sigma_m(x)$, where we average the partial sums of a series, is called a "Cesàro sum" and will be very useful.

Cesàro sums take a little getting used to, but they will be an important tool for us. One can show (see Randolph, p. 147) that if $\sum c_n$ is a series with sequence of partial sums $\{S_n\}$, then the sequence of Cesàro sums $\{\sigma_n\}$ has the property that

$$\underline{\lim} \, S_n \leq \underline{\lim} \, \sigma_n \leq \overline{\lim} \, \sigma_n \leq \overline{\lim} \, S_n.$$

Thus if the sequence of partial sums of a series converges, the sequence of Cesàro sums converges to the same number. However, it may be that the sequence of partial sums diverges and the sequence of Cesàro sums converges.

Example 9.3:
Consider a series whose sequence of partial sums is $\{C, 0, C, 0, C, \dots\}$ for some number C. Then

$$\sigma_1 = C, \quad \sigma_2 = \frac{C+0}{2}, \quad \sigma_3 = \frac{C+0+C}{3}$$

and one can check that

$$\sigma_{2n} = \frac{nC}{2n} = \frac{C}{2}, \quad \sigma_{2n-1} = \frac{nC}{2n-1}$$

so that $\lim_{n \to \infty} \sigma_n = C/2$, even though the sequence of partial sums diverges. Notice that the series $\sum_{n=1}^{\infty} (-1)^{n+1} C$ gives rise to the sequence of partial sums $\{C, 0, C, 0, \dots\}$. Thus, even though this series diverges, the sequence of Cesàro sums converges.

We noted earlier that it is possible to have a continuous function whose Fourier series diverges. One of our important results says that the sequence of Cesàro sums of the Fourier series of a continuous periodic function converges uniformly to the function.

Once we have shown that $\{\sigma_n\}$ converges to f uniformly, it will be fairly simple to show

$$\int_{-\pi}^{\pi} f^2 \, dx = \sum_{m=0}^{\infty} c_m^2$$

or equivalently

$$\frac{1}{\pi} \int_{-\pi}^{\pi} f^2 \, dx = \frac{a_0^2}{2} + \sum_{n=1}^{\infty} (a_n^2 + b_n^2).$$

Either of these equations is known as "Parseval's Equality." The key to the whole procedure is to show that $\{\sigma_n\} \to f$ converges uniformly if f is continuous and periodic of period 2π. To prove this, it will be convenient to use a function that is the Cesàro sum of Dirichlet kernels.

Definition: Let $D_N(t)$ be the Dirichlet kernel of index N;

$$D_N(t) = 1/2 + \cos t + \cos 2t + \cdots + \cos Nt.$$

The *Fejer kernel* of index N is the function $K_N(t)$ defined by

$$K_N(t) = \frac{D_0(t) + D_1(t) + \cdots + D_N(t)}{N+1}.$$

We make some observations about the Fejer kernel, the proofs of which are left for Exercise 5, Section 9.2.

(1)

$$\frac{1}{\pi} \int_{-\pi}^{\pi} K_N(t) \, dt = 1$$

(2)

$$K_N(x) = \frac{1}{2(N+1)} \frac{\sin^2[(N+1)x/2]}{\sin^2(x/2)} \geq 0$$

(3)

$$K_N(x) \leq \frac{1}{(N+1)(1-\cos\delta)} \quad \text{for } 0 < \delta \leq |x| \leq \pi$$

(4)

$$\lim_{x \to 0} K_N(x) = \frac{N+1}{2}.$$

The graphs of $K_5(x)$ and $K_{10}(x)$ are shown in Figure 9.6. Notice that, as n becomes large, virtually all the area under the graph of $K_n(x)$ is concentrated near $x = 0$.

Then if we form the integral $\int_{-\pi}^{\pi} g(t) K_n(t) \, dt$ for n very large and g a continuous function, the value of this integral will be primarily determined by

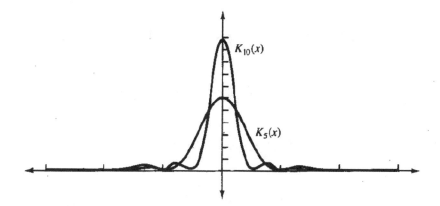

Figure 9.6
Two Fejer kernels

the value of $g(t)$ close to zero, since $K_n(t)$ becomes very small once t moves away from 0. This is the intuition behind the proof of the next theorem.

Theorem 9.6 (Fejer's Theorem): Let f be a continuous function of period 2π and let

$$S_n(x) = \frac{a_0}{2} + \sum_{m=1}^{n} (a_m \cos mx + b_m \sin mx)$$

be the nth partial sum of the Fourier series generated by f. Let

$$\sigma_n(x) = \frac{S_0(x) + S_1(x) + \cdots + S_n(x)}{n+1}.$$

Then $\{\sigma_n\}$ converges uniformly to f.
Proof: By Theorem 9.2, we have

$$S_n(x) = \frac{1}{\pi} \int_{-\pi}^{\pi} f(x-t) D_n(t)\, dt$$

so that

$$\sigma_n(x) = \frac{1}{\pi} \int_{-\pi}^{\pi} f(x-t) \frac{D_0(t) + D_1(t) + \cdots + D_n(t)}{n+1}\, dt$$

$$= \frac{1}{\pi} \int_{-\pi}^{\pi} f(x-t) K_n(t)\, dt.$$

Since $(1/\pi) \int_{-\pi}^{\pi} K_n(t)\, dt = 1$, we can write

$$f(x) = \frac{1}{\pi} \int_{-\pi}^{\pi} f(x) K_n(t)\, dt$$

so that
$$\sigma_n(x) - f(x) = \frac{1}{\pi} \int_{-\pi}^{\pi} (f(x-t) - f(x)) K_n(t) \, dt.$$

Thus we can write
$$\sigma_n(x) - f(x) = \frac{1}{\pi} \int_{-\delta}^{\delta} (f(x-t) - f(x)) K_n(t) \, dt$$

$$+ \frac{1}{\pi} \int_{\delta}^{\pi} (f(x-t) - f(x)) K_n(t) \, dt$$

$$+ \frac{1}{\pi} \int_{-\pi}^{-\delta} (f(x-t) - f(x)) K_n(t) \, dt$$

for $0 < \delta < \pi$. We want $\{\sigma_n\}$ to converge uniformly to f. We need to show
that each of the integrals may be made arbitrarily small, *independent of x*, by
making n large. Thus let $\epsilon > 0$ be given. By observation 3 about the Fejer
kernel,
$$\frac{1}{\pi} \int_{\delta}^{\pi} |f(x-t) - f(x)| K_n(t) \, dt$$

$$\leq \frac{1}{\pi} \frac{1}{n+1} \frac{1}{1 - \cos \delta} \int_{\delta}^{\pi} |f(x-t) - f(x)| \, dt.$$

Since $f(x)$ is continuous on $[-\pi, \pi]$, it is bounded; so $|f(x-t) - f(x)| \leq 2M$
for some number M. Thus
$$\int_{\delta}^{\pi} |f(x-t) - f(x)| \, dt \leq 2\pi M$$

so
$$\frac{1}{\pi} \int_{\delta}^{\pi} |f(x-t) - f(x)| K_n(t) \, dt \leq \frac{2M\pi}{\pi(n+1)(1 - \cos \delta)}$$

which may be made arbitrarily small by making n sufficiently large. The identical calculation shows that
$$\frac{1}{\pi} \int_{-\pi}^{-\delta} |f(x-t) - f(x)| K_n(t) \, dt \leq \frac{2M\pi}{\pi(n+1)(1 - \cos \delta)}.$$

We now use the continuity of f to bound
$$\int_{-\delta}^{\delta} |f(x-t) - f(x)| K_n(t) \, dt.$$

Since f is continuous on $[-\pi, \pi]$, it is *uniformly* continuous there. Thus, given
$\epsilon > 0$, there is a $\delta > 0$ such that $|f(x) - f(y)| < \epsilon/2$ for every $x, y \in [-\pi, \pi]$ if
$|x - y| < \delta$. Thus if $|t| < \delta$, $|f(x-t) - f(x)| < \epsilon/2$, so
$$\int_{-\delta}^{\delta} |f(x-t) - f(x)| K_n(t) \, dt < \frac{\epsilon}{2} \int_{-\delta}^{\delta} K_n(t) \, dt$$

$$\leq \frac{\epsilon}{2} \int_{-\pi}^{\pi} K_n(t)\, dt = \frac{\epsilon}{2} \text{ for every } n.$$

Now choose N (which will depend on both ϵ and δ) large enough so that $2M/[(n+1)(1-\cos\delta)] < \epsilon/4$ when $n > N$. Note that the order of doing our procedure is crucial! *First* we find δ to make

$$\int_{-\delta}^{\delta} |f(x-t) - f(x)| K_n(t)\, dt < \frac{\epsilon}{2}$$

and *then* choose N, which depends on both δ and ϵ because of the δ in $2M/[(n+1)(1-\cos\delta)]$, to make

$$\int_{-\pi}^{-\delta} |f(x-t) - f(x)| K_n(t)\, dt < \frac{\epsilon}{4} \text{ and } \int_{\delta}^{\pi} |f(x-t) - f(x)| K_n(t)\, dt < \frac{\epsilon}{4}$$

if $n > N$.

Thus for $n > N$,

$$|\sigma_n(x) - f(x)| \leq \frac{1}{\pi} \int_{-\pi}^{\pi} |f(x-t) - f(x)| K_n(t)\, dt$$

$$= \frac{1}{\pi} \int_{-\pi}^{-\delta} |f(x-t) - f(x)| K_n(t)\, dt + \frac{1}{\pi} \int_{-\delta}^{\delta} |f(x-t) - f(x)| K_n(t)\, dt$$

$$+ \frac{1}{\pi} \int_{\delta}^{\pi} |f(x-t) - f(x)| K_n(t)\, dt$$

$$< \frac{\epsilon}{4} + \frac{\epsilon}{2} + \frac{\epsilon}{4} = \epsilon$$

independent of $x \in [-\pi, \pi]$. Thus $\{\sigma_n\}$ converges to f uniformly on $[-\pi, \pi]$. ∎

Corollary 9.6(a): Two continuous functions of period 2π that have the same Fourier series are identical. ∎

Corollary 9.6(b) (Parseval's Theorem): If $f(x)$ is continuous and periodic of period 2π and has Fourier series $a_0/2 + \sum_{n=1}^{\infty}(a_n \cos nx + b_n \sin nx)$, then

$$\frac{a_0^2}{2} + \sum_{n=1}^{\infty}(a_n^2 + b_n^2) = \frac{1}{\pi} \int_{-\pi}^{\pi} [f(x)]^2\, dx.$$

Proof: If $S_n = a_0/2 + \sum_{m=1}^{n}(a_m \cos mx + b_m \sin mx)$, then by Theorem 9.1

$$\int_{-\pi}^{\pi} [f(x) - S_n(x)]^2\, dx \leq \int_{-\pi}^{\pi} [f(x) - \sigma_n(x)]^2\, dx.$$

But since $\{\sigma_n\}$ converges to f uniformly by Fejer's Theorem, then

$$\lim_{n\to\infty} \int_{-\pi}^{\pi} [f(x) - \sigma_n(x)]^2 \, dx = 0$$

which proves the result. ∎

A word of caution is appropriate. We have *not* said that if f is continuous, then the Fourier series of f converges uniformly to f. In fact, as we noted earlier, there are examples of continuous functions whose Fourier series fail to converge on a countable set.

One use of Fourier series is to find the explicit sums of series. We give two examples.

Example 9.4: If f is the function

$$f(x) = \begin{cases} 1 & \text{if } 0 \le x < \pi \\ -1 & \text{if } -\pi \le x < 0, \end{cases}$$

then f has Fourier series

$$\frac{4}{\pi} \sum_{k=0}^{\infty} (2k+1)^{-1} \sin(2k+1)x.$$

We can use this fact to evaluate the series

$$\sum_{k=0}^{\infty} \frac{(-1)^{k+1}}{2k+1} = 1 - \frac{1}{3} + \frac{1}{5} - \frac{1}{7} + \cdots.$$

Since f is differentiable at $x_0 = \pi/2$, we know by Theorem 9.3 that

$$f(\pi/2) = \frac{4}{\pi} \sum \frac{\sin[(2k+1)\pi/2]}{2k+1} = \frac{4}{\pi} \sum \frac{(-1)^{k+1}}{2k+1}.$$

But $f(\pi/2) = 1$, so $\pi/4 = \sum (-1)^{k+1}/(2k+1)$.

Example 9.5:

Let g be the function $g(x) = \pi - |x|$ on $[-\pi, \pi]$ and $g(x+2\pi) = g(x)$. The Fourier series for $g(x)$ is

$$\frac{\pi}{2} + \frac{4}{\pi} \sum_{k=0}^{\infty} (2k+1)^{-2} \cos(2k+1)x.$$

That is, $a_0 = \pi$, and

$$a_n = \begin{cases} 4/(\pi n^2) & \text{if } n \text{ is odd} \\ 0 & \text{if } n \text{ is even} \end{cases}$$

and $b_n = 0$ for all n.

Parseval's Theorem is applicable, since g is continuous. That is

$$\frac{a_0^2}{2} + \sum(a_n^2 + b_n^2) = \frac{1}{\pi} \int_{-\pi}^{\pi} f^2 \, dx.$$

Now

$$\frac{1}{\pi} \int_{-\pi}^{\pi} f^2 \, dx = \frac{1}{\pi} \left[\int_{-\pi}^{0} (\pi + x)^2 \, dx + \int_{0}^{\pi} (\pi - x)^2 \, dx \right] = \frac{2\pi^2}{3}.$$

Thus

$$\frac{\pi^2}{2} + \left(\frac{4}{\pi}\right)^2 \sum_{k=0}^{\infty} \frac{1}{(2k+1)^4} = \frac{2\pi^2}{3}$$

so

$$\sum_{k=0}^{\infty} \frac{1}{(2k+1)^4} = \left(\frac{\pi}{4}\right)^2 \left(\frac{2\pi^2}{3} - \frac{\pi^2}{2}\right) = \frac{\pi^4}{96}.$$

Sine and Cosine Series

We have seen that certain functions of period 2π can be represented as trigonometric or Fourier series. Also we have seen how Fourier series can be used sometimes to evaluate series of numbers. Now we consider functions defined on the interval $(0, \pi)$ and show that a certain class of functions can be represented as a cosine series or a sine series, that is, a series of the form

$$\sum a_n \cos nx \text{ or } \sum b_n \sin nx.$$

This will enable us to explicitly evaluate a number of additional series.

The idea is fairly simple. If f is a function defined on $(0, \pi)$, we can make f an even function of period 2π by defining $f(x) = f(-x)$ for $-\pi < x < 0$ and $f(x + 2\pi) = f(x)$. We call the extended even function f_1. Then, using formulas (9.6a) and (9.6b) of Section 9.1, we can write the Fourier series for $f_1(x)$

$$\frac{a_0}{2} + \sum(a_n \cos nx + b_n \sin nx).$$

Since f_1 is even, we know that $b_n = 0$ for every n. Also

$$a_n = \frac{1}{\pi} \int_{-\pi}^{\pi} f_1(x) \cos nx \, dx.$$

Since $f_1(x)$ and $\cos nx$ are even functions, their product, $f_1(x) \cos nx$, is also an even function; so

$$\frac{1}{\pi} \int_{-\pi}^{\pi} f_1(x) \cos nx \, dx = \frac{1}{\pi} \left(\int_{-\pi}^{0} f_1(x) \cos nx \, dx + \int_{0}^{\pi} f_1(x) \cos nx \, dx \right)$$

$$= \frac{2}{\pi} \int_{0}^{\pi} f_1(x) \cos nx \, dx.$$

The crucial observation is that since $f_1(x)\cos nx$ is even, then

$$\int_{-\pi}^{0} f_1(x)\cos nx\, dx = \int_{0}^{\pi} f_1(x)\cos nx\, dx.$$

Also $f_1(x) = f(x)$ on $(0,\pi)$, so

$$\frac{2}{\pi}\int_{0}^{\pi} f_1(x)\cos nx\, dx = \frac{2}{\pi}\int_{0}^{\pi} f(x)\cos nx\, dx.$$

Thus

$$a_n = \frac{2}{\pi}\int_{0}^{\pi} f(x)\cos nx\, dx \quad n = 1,2,\dots \tag{9.13}$$

and we have determined a cosine series for $f(x)$.

Likewise, we can extend f to be an odd function of period 2π by defining $f(-x) = -f(x)$ for $-\pi < x < 0$ and $f(x+2\pi) = f(x)$. It is also necessary to require $f(0) = 0$. Call this new function f_2. We can write the Fourier series for $f_2(x)$ by

$$f_2(x) = \frac{a_0}{2} + \sum(a_n\cos nx + b_n\sin nx).$$

Since f_2 is odd, $a_n = 0$ for every n, and using the same techniques as before, we can show

$$b_n = \frac{2}{\pi}\int_{0}^{\pi} f(x)\sin nx\, dx \quad n = 1,2,\dots \tag{9.14}$$

We know from our earlier work that if f is piecewise continuous and is differentiable at the point x_0, then

$$f(x_0) = \frac{a_0}{2} + \sum_{n=1}^{\infty} a_n\cos nx_0 + \sum_{n=1}^{\infty} b_n\sin nx_0$$

where a_n and b_n are calculated by formulas (9.13) and (9.14).

Example 9.6:
Find the cosine and sine series for $f(x) = x(\pi - x)$, $0 \le x \le \pi$.
If $n = 0$,

$$a_0 = \frac{2}{\pi}\int_{0}^{\pi} x(\pi - x)\, dx = \frac{2}{\pi}\left(\frac{\pi x^2}{2} - \frac{x^3}{3}\right)\Big|_{0}^{\pi} = \frac{\pi^2}{3}.$$

If $n = 1,2,\dots$,

$$a_n = \frac{2}{\pi}\int_{0}^{\pi} x(\pi - x)\cos nx\, dx = \frac{2}{\pi}\left(\pi\int_{0}^{\pi} x\cos nx\, dx - \int_{0}^{\pi} x^2\cos nx\, dx\right)$$

$$= \frac{2}{\pi} \left\{ \pi \left(\frac{\cos nx}{n^2} + \frac{x \sin nx}{n^2} \right) - \left(\frac{2x \cos nx}{n} + \frac{(n^2 x^2 - 2) \sin nx}{n^3} \right) \right\} \Big|_0^{\pi}$$

$$= \frac{2}{\pi} \left\{ \frac{\pi(\cos n\pi - 1)}{n^2} + \frac{2\pi \cos n\pi}{n} \right\} = \begin{cases} 0 & \text{if } n \text{ is odd} \\ -4/n^2 & \text{if } n \text{ is even.} \end{cases}$$

Thus

$$\frac{a_0}{2} + \sum_{n=1}^{\infty} a_n \cos nx = \frac{\pi^2}{6} - 4 \left(\frac{\cos 2x}{2^2} + \frac{\cos 4x}{4^2} + \frac{\cos 6x}{6^2} + \cdots \right).$$

For $n = 1, 2, \ldots,$

$$b_n = \frac{2}{\pi} \int_0^{\pi} x(\pi - x) \sin nx \, dx$$

$$= \frac{2}{\pi} \left[x \left(\frac{\sin nx}{n^2} = \frac{x \cos nx}{n} \right) - \left(\frac{2x \sin nx}{n^2} - \frac{(n^2 x^2 - 2) \cos nx}{n^3} \right) \right] \Big|_0^{\pi}$$

$$= \begin{cases} 8/(\pi n^3) & \text{if } n \text{ is odd} \\ 0 & \text{if } n \text{ is even.} \end{cases}$$

Thus

$$\sum_{n=1}^{\infty} b_n \sin nx = \frac{8}{\pi} \left(\sin x + \frac{\sin 3x}{3^3} + \frac{\sin 5x}{5^3} + \cdots \right).$$

Exercises 9.2

1. *(a)* Show that

$$1/2 + \cos t + \cos 2t + \cdots + \cos Nt$$
$$= \begin{cases} [\sin(N+1/2)t]/[2\sin(1/2)t] & \text{if } t \neq 2\pi m, \ m \text{ an integer} \\ N + 1/2 & \text{if } t = 2\pi m. \end{cases}$$

(b) Show that

$$K_N(t) = \frac{[D_0(t) + D_1(t) + \cdots + D_N(t)]}{N+1} = \frac{1}{2(N+1)} \frac{\sin^2((N+1)t/2)}{\sin^2(t/2)}$$

by showing that

$$\sum_{j=0}^{N} \sin(j + 1/2)t = \frac{\sin^2((N+1)t/2)}{\sin(t/2)}.$$

2. *(a)* Show that if $\{\phi_1, \phi_2, \ldots, \phi_n\}$ forms an orthonormal set, then

$$\int_{-\pi}^{\pi} \left(\sum_{m=1}^{n} a_m \phi_m \right) \left(\sum_{m=1}^{n} a_m \phi_m \right) dx = \sum_{m=1}^{n} a_m^2.$$

(b) Give an example where this is not the case if $\{\phi_1, \phi_2, \ldots, \phi_n\}$ is not an orthonormal set.

(c) Show that if $\{\phi_1, \phi_2, \ldots, \phi_n\}$ is an orthonormal set and $c_m = \int f(x)\phi_m(x)$, then

$$\left(f - \sum_{m=1}^{n} c_m \phi_m, f - \sum_{m=1}^{n} c_m \phi_m \right) = \int_{-\pi}^{\pi} f(x)^2 \, dx - \sum_{m=1}^{n} c_m^2.$$

3. *(a)* Show that if f is periodic of period 2π and Riemann integrable, then

$$\int_{-\pi}^{\pi} f(u) D_N(x - u) \, du = \int_{-\pi}^{\pi} f(x + t) D_N(t) \, dt.$$

(b) Show that if $f' \in \mathcal{F}$ and f is periodic, then $f \in \mathcal{F}$. [*Hint:* The only problem is to show that $\lim_{x \uparrow x_0} f(x)$ and $\lim_{x \downarrow x_0} f(x)$ exist at points where $f'(x_0)$ is not defined. Use the fact that f' is bounded at points where it exists to show that if $\{x_n\}$ is an increasing Cauchy sequence whose limit is x_0, then $\{f(x_n)\}$ is a Cauchy sequence.]

4. *(a)* Let the sequence of partial sums of a series be given by $\{0, 1, 5, 0, 1, 5, \ldots\}$. Find $\lim_{n \to \infty} \sigma_n$.

(b) Show that if the sequence of partial sums of a series is $\{x_1, x_2, \ldots, x_k, x_1, x_2, \ldots, x_k, \ldots\}$, then

$$\lim_{n \to \infty} \sigma_n = \frac{x_1 + \cdots + x_k}{k}.$$

5. Prove the following facts about the Fejer kernel for N, a positive integer:

(a)

$$\frac{1}{\pi} \int_{-\pi}^{\pi} K_N(t) \, dt = 1.$$

(b)

$$K_N(x) = \frac{1}{2(N+1)} \frac{\sin^2[(N+1)x/2]}{\sin^2(x/2)} \geq 0.$$

(c)

$$K_N(x) \leq \frac{1}{(N+1)(1 - \cos \delta)} \quad \text{for } 0 < \delta \leq |x| \leq \pi.$$

(d)

$$\lim_{x \to 0} K_N(x) = \frac{N+1}{2}.$$

6. It was shown in Example 9.2 that if $f(x) = x$, $x \in [-\pi, \pi)$, and $f(x+2\pi) = f(x)$, then the Fourier series for $f(x)$ is given by

$$2 \left(\sin x - \frac{\sin 2x}{2} + \frac{\sin 3x}{3} - \cdots \right).$$

(*a*) Use this fact to find $1 - 1/3 + 1/5 - 1/7 + \cdots$.
(*b*) Integrate the Fourier series for $f(x) = x$ to show that the Fourier series for $g(x) = x^2$ is given by

$$\frac{\pi^2}{3} - 4 \left(\cos x - \frac{\cos 2x}{4} + \frac{\cos 3x}{9} - \cdots \right).$$

(*c*) Use the result of part (b) to find $\sum_{n=1}^{\infty} (-1)^n / n^2$.

7. Find the sine and cosine series for the following functions defined on $(0, \pi)$:
(*a*) $f(x) = 1$.
(*b*) $f(x) = x$.
(*c*) $f(x) = x^2$.
(*d*) $f(x) = \sin x$.

8. Find the sine series for $x(\pi - x)$ and use it to show that

$$1 + \frac{1}{3^3} - \frac{1}{5^3} - \frac{1}{7^3} + \frac{1}{9^3} + \frac{1}{11^3} - \frac{1}{13^3} - \cdots = \frac{3\pi^3 \sqrt{2}}{128}.$$

If you have access to a computer graphics program, print a graph of $D_N(x)$ and $K_N(x)$ for $N = 1, 2, 5, 10$, and 20.

9. Prove the following corollary to the Riemann-Lebesgue Lemma:
If $\int_{-\pi}^{\pi} [f(x)]^2 \, dx$ exists and is finite, and if f is Riemann integrable, then
(*a*) $\lim_{n \to \infty} \int_a^b f(x) \cos nx \, dx = 0$ and
(*b*) $\lim_{n \to \infty} \int_a^b f(x) \sin nx \, dx = 0$
for any interval $[a, b] \subset [-\pi, \pi]$.

Bibliography

Apostol, T.M. *Mathematical Analysis*, 2nd ed. Reading, Massachusetts: Addison-Wesley, 1974.

Bartle, R. G. *The Elements of Real Analysis*. New York: Wiley, 1964.

Buck, R. C. *Advanced Calculus*, 2nd ed. New York: McGraw-Hill, 1965.

Fraleigh, J. *A First Course in Abstract Algebra*. Reading, Massachusetts: Addison-Wesley, 1972.

Fridy, J. *Introductory Analysis*. San Diego: Harcourt Brace Javanovich, 1987.

Fulks, W. *Advanced Calculus*. New York: Wiley, 1964.

Halmos, P. R. *Naive Set Theory*. New York: Van Nostrand Reinhold, 1961.

McCoy, N., and Janusz, G. *Introduction to Modern Algebra*, 4th ed. Boston: Allyn and Bacon, 1987.

Meschkowski, H. *Evolution of Mathematical Thought*. Oakland, California: Holden Day, 1965.

Olmstead, J. M. H. *Advanced Calculus*. New York: Appleton-Century-Crofts, 1961.

Olmstead, J. M. H., and B. R. Gelbaum. *Counterexamples in Analysis*. Oakland, California: Holden Day, 1966.

Randolph, J. A. F. *Basic Real and Abstract Analysis*. Boston: Academic Press, 1968.

Royden, H. *Real Analysis*. New York: Macmillan, 1966.

Rudin, W. *Principles of Mathematical Analysis*, 2nd ed. New York: McGraw-Hill, 1964.

Sagan, H. *Advanced Calculus*. Boston: Houghton Mifflin, 1974.

Simon, B. "Uniform Convergence of Fourier Series," *The American Mathematical Monthly*, Vol. 76, No. 1, p. 55, January 1967.

Stabler, E. R. *Introduction to Mathematical Thought*. Reading, Massachusetts: Addison-Wesley, 1953.

Willard, S. *General Topology*. Reading, Massachusetts: Addison-Wesley, 1968.

Zygmund, A. *Trigonometric Series*, 2nd ed. New York: Cambridge University Press, 1967.

Hints and Answers for Selected Exercises

Most of the problems can be solved in many ways. The hints only suggest one possible method.

Section 1.1

2.(a) $\mathcal{D}(f) = \{x | x \neq 0\}$, $\mathcal{R}(f) = \{-1, 1\}$

3.(a) $f^{-1}(A) = \{\pm\sqrt{2/3}, \pm 1, \pm 2/\sqrt{3}\}$

4.(b) $f \circ g(1) = 2$ $g \circ f(1) = 4$

6. $\mathcal{D}(f \circ g) = \{x | x > 6\} \cup \{x | x \leq 5\}$

10.(b) Let $A_1 = B_1 = [0, 1]$ and $A_2 = B_2 = [1, 2]$.

13.(f) Let $f(x) = x^2$.

14. Suppose f is not 1-1. Then there are x and y, with $f(x) = f(y)$. Let $A = \{x\}$, $B = \{y\}$.

15. From Exercise 13(b), we know $f(A \cap B) \subset f(A) \cap f(B)$. Suppose there are sets A and B for which $y \in f(A) \cap f(B)$, but $y \notin f(A \cap B)$. Show that there are $x_1 \in A \setminus B$ and $x_2 \in B \setminus A$ with $f(x_1) = f(x_2) = y$.

16.(c) If f is not 1-1, find a set A for which $A \neq f^{-1}(f(A))$. Suppose $f(x_1) = f(x_2)$. Let $A = \{x_1\}$. Conversely, suppose there is a set A for which $f^{-1}(f(A)) \neq A$. Then there is an $x_1 \in f^{-1}(f(A))$ with $x_1 \notin A$ by Exercise 16(b). Show that there is an $x_2 \in A$ with $f(x_1) = f(x_2)$.

19. If f is not onto, consider $A = X$. If f is not 1-1, there are x and y with $f(x) = f(y)$. Consider $A = \{x\}$. If f is 1-1, show that for any $A \subset X$, $f(X \setminus A) = f(X) \setminus f(A)$. If f is onto, then $f(X) = Y$.

Section 1.2

5. Show that $b - a \in P$ and $-(b - a) \in P$ if $a \neq b$.

7.(a) $1 = 1^2$

7.(c) Use induction. Notice that $a^2 < ab < b^2$.

7.(d) Show that if $b > a$, then $b^n > a^n$.

7.(e) $|a| = \sqrt{a^2}$.

8.(a) Use the fact that if $a < b$ and $x > 0$, then $ax < bx$. Write $x^m = x^{m-n} \cdot x^n$.

8.(c) If $x > 1$, then $x < x^n < (x+1)^n$. If $x < 1$, then $x < 1 < (x+1)^n$.

9.(a) $x = (x^{1/n})^n$, so if $x^{1/n} \geq 1$, then $(x^{1/n})^n \geq 1$ by Exercise 8(b). If $x^{1/m} < x^{1/n}$, then $x = (x^{1/m})^m < (x^{1/n})^m$. But $x^{1/n} < (x^{1/n})^n$ by Exercise

8(a).

11. Show that

$$ab \le \left(\frac{a+b}{2}\right)^2.$$

14.(a) $|a| = |a - b + b|$

17. The difficult part is to show that $d(x, z) \le d(x, y) + d(y, z)$. Show that if a, b, and c are positive numbers with $a + b \ge c$, then

$$\frac{a}{1+a} + \frac{b}{1+b} \ge \frac{c}{1+c}.$$

18.(a) Take

$$\alpha = \left(\sum_{i=1}^{n} b_i^2\right)^{1/2}, \quad \beta = \left(\sum_{i=1}^{n} a_i^2\right)^{1/2}.$$

18.(b) When $a_i = (\beta/\alpha)b_i$ for every $i = 1, \dots, n$

19. Square both sides of the inequality.

22. Suppose $a > b$. Take $\epsilon = (a - b)/2$.

Section 1.3

1.(a) l.u.b. $A = 1$, g.l.b. $A = 0$

1.(b)(i) 1   **(ii)** $1/2^{50}$

4.(a) If $a < b$, there is a positive integer N with $a/N < b - a$. Consider the set $\{k/N \mid k \in \mathbb{Z}\}$.

8.(a) $A + B = \{-3, -2, 1, 3, 11, 12, 15, 17, 19, 20, 23, 25\}$

8.(b) Show that if $x \in A + B$, then $x \le$ l.u.b. $A +$ l.u.b. B. Show that for any $\epsilon > 0$, there is an $x \in A + B$ such that $x >$ l.u.b. $A +$ l.u.b. $B - \epsilon$.

10. $A \subset (A \setminus B) \cup B$. What about A if $A \setminus B$ and B are countable?

11.(a) For n a positive integer, let $E_n = \{m/n \mid m \in \mathbb{Z}\}$. Then E_n is countable.

13.(a) Let $B = \{b_1, b_2, \dots\}$. Then $A \times B = \cup_{n=1}^{\infty} A \times \{b_n\}$.

14. Let P_n be the set of polynomials of degree $n - 1$ with rational coefficients. Show there is a 1-1 onto correspondence from P_n to $\mathbb{Q} \times \cdots \times \mathbb{Q}$ (n copies of the rational numbers), so P_n is countable.

Section 2.1

3. Let $a_1 = 3$, $a_2 = 3.1$, $a_3 = 3.14$.

4.(a) Let $a_n = r - 1/n$.

8.(b) Let $b_n = K - a_n$. Then $b_n \ge 0$.

8.(c) Let $c_n = b_n - a_n$.

10. There is a number N such that $b_n > b/2$ if $n > N$. Let $K = \min\{b_1, \dots, b_n, b/2\}$.

11.(a) $|\,|a_n| - |a|\,| \le |a_n - a|$

12.(a) Show that $\{a_n\}$ and $\{b_n\}$ are bounded monotone sequences.

12.(d) Let $p = \lim a_n = \lim b_n$. Show that p is in every interval. Choose $x > p$, and show that there is an interval not containing x. Likewise, if $x < p$ show that there is an interval not containing x.

12.(e) Let $I_n = (0, 1/n]$.

18. Use two cases: if $a > 0$ or $a = 0$. If $a > 0$, write

$$|\sqrt{a_n} - \sqrt{a}| = \left| \frac{(\sqrt{a_n} - \sqrt{a})(\sqrt{a_n} + \sqrt{a})}{(\sqrt{a_n} + \sqrt{a})} \right|.$$

19.(a) Show that $\{a_n\}$ is a decreasing sequence of positive numbers.

20.(a) Show there is a number M for which $|a_n b_n - 0| \le M|b_n - 0|$.

22.(a) Choose N such that if $n > N$, then $|a_n - a| < \epsilon/3$. Choose K such that if $k > K$, then

$$\left| \frac{a_1 + \cdots + a_N}{N + k} \right| < \frac{\epsilon}{3} \text{ and } \left| \frac{Na}{N + k} \right| < \frac{\epsilon}{3}.$$

Consider

$$\left| \frac{a_1 + \cdots + a_{N+k}}{N + k} - a \right|.$$

23.(b) Let $a_n = n^{1/n} - 1$. Now

$$n = [1 + a_n]^n = 1 + na_n + \frac{n(n-1)}{2}a_n^2 + \cdots .$$

So

$$n = (1 + a_n)^n \ge \frac{n(n-1)}{2}a_n^2 \text{ and thus } \frac{2}{n-1} \ge a_n^2.$$

24.(a) If $L > 1$, choose $\epsilon > 0$ such that $L - \epsilon > 1$. Choose N such that if $n > N$, then $x_{n+1}/x_n > L - \epsilon$. Then $x_{N+k} > x_N(L - \epsilon)^k$.

24.(c) Use part (a) where $L < 1$.

Section 2.2

3.(b) Suppose $\lim a_n = L$. Then $L = \lim a^{n+1} = a \lim a^n = aL$ so if a and L are real numbers, then $(a - 1)L = 0$. But $a > 1$, so $L > 1$.

4.(a) If $a^{1/n} < 1$ for some n, then $(a^{1/n})^n < 1$ but $(a^{1/n})^n = a$.

4.(b) $a^{1/n}/a^{1/(n+1)} = a^{1/[n(n+1)]} > 1$

5.(a) Let $b = 1/a$. Then $b > 1$. Now apply Exercise 4.

7.(a) Show that $\{a_n\}$ is an increasing sequence that is bounded above by 2.

8.(b) $a_1 = 1, a_2 = 1, a_3 = 2, a_4 = 1, a_5 = 2, a_6 = 3, a_7 = 1, \ldots$

8.(c) Enumerate the rational numbers.

9.(a) $(1 + 1/2n)^{3n} = [(1 + 1/2n)^{2n}]^{3/2}$. Now $(1 + 1/2n)^{2n}$ is a subsequence of $(1 + 1/n)^n$.

11. Show that $|a_{n+1} - a_n| = |a_0 - a_1|/2^{n-1}$. In fact, the series converges to $a_0 + (2/3)|a_1 - a_0|$ if $a_1 > a_0$ and $a_1 + (2/3)|a_1 - a_0|$ if $a_1 < a_0$.

Section 2.3

5.(b) Let a_{n_i} be the ith term in the sequence $\{a_n\}$ that exceeds $L + \epsilon$.

12. Suppose some interval $(L - \epsilon, L + \epsilon)$ contains only finitely many points, say $x_1, \ldots, x_N$. Let $\alpha = (1/2) \min\{|L - x_i| \,|i = 1, \ldots, N\}$, and consider $(L - \alpha, L + \alpha)$.

13.(a) Show there is an N such that if $m > N$, then $|a_m - a_N| < 1$, so that $|a_m| < |a_N| + 1$. The set $\{|a_1|, \ldots, |a_N|, |a_N| + l\}$ is a finite set and is bounded.

13.(c) Suppose that L and M are subsequential limit points of $\{a_n\}$. Let $\epsilon = (1/3)|L - M|$. Show that if $a_k \in (L - \epsilon, L + \epsilon)$ and $a_m \in (M - \epsilon, M + \epsilon)$, then $|a_k - a_m| > \epsilon$.

15.(a) Let $A_n = [n, n+1] \cap A$ for n an integer. If each A_n is countable, then $\cup_{n\in\mathbb{Z}} A_n = A$ is countable. Thus some interval $[n, n+1]$ must contain an uncountable number of points.

16. Let $A = \sup\{\alpha | \alpha$ is a limit point of $\{a_n\}\}$. Suppose $\{b_n\}$ is a sequence of limit points of $\{a_n\}$ converging to A (so each b_k is a limit point of $\{a_n\}$). For every k, choose a number from $\{a_n\}$, say a_{n_k}, such that $a_{n_k} \in (b_k - 1/k, b_k + 1/k)$ and $n_j < n_k$ if $j < k$. Show that $\{a_{n_k}\}$ converges to A.

Section 3.1

3.(a) Any interval of the form $[0, a)$ or (a, b) or (a, ∞) where $a > 0$.

4. If $x + \alpha \in A$, then there is an open interval about $x + \alpha$ that is contained in A. This will contradict the definition of α.

5. If $x \in I_x \cap I_y$, then $I_x \cup I_y$ is an open interval containing x. Since I_x is the maximal open interval, $I_x = I_x \cup I_y$.

7.(b) No.

11.(a) Neither open nor closed. Limit points $= \mathbb{R}$, $\text{int}(A) = \emptyset$. $\overline{A} = \mathbb{R}$. Boundary of $A = \mathbb{R}$.

14. Enumerate the rational numbers $\{r_1, r_2, \ldots\}$. Let $E_n = \{(r_n - r_i, r_n + r_i)|i = 1, 2, \ldots\}$.

Section 4.1

3. Let x_0 be a nonzero rational number, and let $\{x_n\}$ be a sequence of irrational numbers converging to x_0. Consider $\{f(x_n)\}$ and $f(x_0)$.

6. One only need show that if $x_0 \in (0, 1]$, then for any y_1 between 0 and $f(x_0)$, there is an $x_1 \in (0, x_0)$ with $f(x_1) = y_1$. This is sufficient because f is continuous on $(0,1]$.

9. For any $\epsilon > 0$, take $\delta = 1/4$.

11.(a) Use $||f(x)| - |f(y)|| \le |f(x) - f(y)|$.

12. Take $\epsilon = |f(a)/2|$. There is a $\delta > 0$ such that

$$|f(a) - f(x)| < \epsilon \text{ if } x \in (a - \delta, a + \delta).$$

15.(a) Consider $g(x) = f(x) - x$. What about $g(0)$ and $g(1)$?

15.(c) Consider $g(x) = f(x) - f(x+1)$.

25.(b) $|x^2 - y^2| = |x - y||x + y|$. Suppose for $\epsilon = 1/4$, there is a $\delta > 0$ such that if $|x - y| < \delta$, then $|x^2 - y^2| < \epsilon$. Take $x = 1/(2\delta)$.

30. $f \vee g(x) = (1/2)[f(x) + g(x) + |f(x) - g(x)|]$.

Section 4.2

4. Consider the sequence $\{x_n\} = \{1/n\}$. What about the value of the functions when x is between $1/n$ and $1/n + 1$?

8. Let $\epsilon = (1/2)[\theta - \lim_{x \uparrow c} f(x)]$. Choose $\delta > 0$ such that if $x \in [c - \delta, c)$, then

$$|f(x) - \lim_{x \uparrow c} f(x)| < \epsilon.$$

9.(a) Suppose $x < y$ and $a > 1$. Consider a^y/a^x.

11. Consider $f(x) = x^2$ on $[0, M)$ for any M.

Section 5.1

3. $x = \frac{1}{n} \sum_{i=1}^{n} a_i$

7. Show that there is an interval about c where $f'(x) > 0$.

11. $\alpha > 1$

13. Use induction and the fact that $(d/dx)h^{(n-1)}(x) = h^{(n)}(x)$.

Section 5.2

5.(b) Let $f(x) = \cos x$, $g(x) = 1 - x^2/2$. Then $f(0) = g(0)$ and $f'(x) > g'(x)$.

7.(b) Consider $f(x) = x^2 + 1$.

16. Use the Mean Value Theorem.

17. Show $g'(x) > 0$.

Section 6.1

5.(a) $(1/n) \sum_{k=1}^{n} f(k/n)$ is a Riemann sum for $f(x)$ on the interval $[0,1]$ with partition $\{0, 1/n, 2/n, \ldots, n/n\}$.

5.(b) Use part (a) with $f(x) = x$.

8. Show that A_s^c is open. Do this by showing that if $t < s$ and $\text{osc}(f; x_0) = t$, then there is an interval about x_0 where $\text{osc}(f; x) < s$ if x is in the interval about x_0.

11.(b) The function $f \cdot g$ is continuous at x_0 unless either f or g fails to be continuous at x_0.

12.(b) Show that the lower Riemann sum of f for any partition is 0.

13. Use the fact that $\underline{S}(f; P_\epsilon) \leq S(f; P_\epsilon) \leq \overline{S}(f; P_\epsilon)$.

Section 6.2

1.(c) No.

2.(a) Write $\int_a^b f = \sum_{i=1}^n \int_{a_i}^{b_i} f$.

2.(b) For given $\epsilon > 0$, there is a $\delta > 0$ such that if $|x - y| < \delta$, then $|f(x) - f(y)| < \epsilon$. Choose a partition

$$P = \{x_0, x_1, \ldots, x_m\} \text{ with } ||P|| < \delta.$$

Define $g_\epsilon(x)$ by
$$g_\epsilon(x) = f(x_i) \text{ if } x \in [x_i, x_{i+1}).$$

5. First show that $||f||_n \leq \max\{|f(x)| \mid x \in [0,1]\} = L$ for every n. Then show that for a given $\epsilon > 0$, there is a point x_0 and an interval about x_0 where $|f(x)| > L - \epsilon$ for every x in the interval. Now show that, for n sufficiently large, $||f||_n > L - \epsilon$.

14.(a) No.

18. e^2

Section 6.3

2.(a) $2e^2 + 2$

5.(a) $f(x) = g(x) - h(x)$ where

$$h(x) = \begin{cases} -x^2 & \text{if } x \in [-1,0] \\ 0 & \text{if } x \in (0,1], \end{cases}$$

$$g(x) = \begin{cases} 0 & \text{if } x \in [-1,0] \\ x^2 & \text{if } x \in (0,1], \end{cases}$$

Section 7.1

3. Let $\{S_n\}$ be the sequence of partial sums for $\sum a_n$. Then $\{S_n\}$ converges if and only it is a Cauchy sequence.

8. Show that the series $\sum n^k / A^n$ converges.

13.(a) Show that $a_n \leq b_n(a_1/b_1)$.

14.(c) Use Exercise 14(b).

Section 8.1

6.(a) $f(x) = \sin x$

8.(a) Suppose there is an N such that $|x/n| < 1$ if $n < N$ for every x. Show this is impossible by choosing $x = N + 1$.

9. To show uniform convergence on $[0, 1 - \alpha]$, show that $(1 - \alpha)^n < \epsilon$ if n is sufficiently large. To show nonuniform convergence on $[0,1)$, note that if $\epsilon = 1/2$ and if N is any positive integer, $x^N > 1/2$ for $x > (1/2)^{1/N} \in (0,1)$.

10.(b) Consider $f_n(x) = x/n$, $A_i = [i, i+1]$.

13.(b) Let $f_n(x) = g_n(x) = x + 1/n$ on $[0, \infty)$.
14.(a) Let $f_n(x) = x^n$ on $[0,1]$. Choose $x_n = 1 - 1/n$.
15.(a) Suppose there are $x, y \in [a, b]$ with $x < y$ where $f(x) > f(y)$. Show this is impossible by taking $\epsilon = (1/4)[f(x) - f(y)]$.

Section 8.2

3. Let $f_n(x) = x(1-x)^n$. Each f_n is continuous, but $f(x) = \sum f_n(x)$ is not continuous.
6. Let $\{S_n(x)\}$ be the sequence of partial sums for $\sum f_n(x)$. Then $\{S_n(x)\}$ is uniformly Cauchy, so $|S_{n+1}(x) - S_n(x)| < \epsilon$ for every x if n is sufficiently large.
7. $a_k = f^{(k)}(c)/k!$
9.(a) Use $1/(1+x^2) = \sum_{n=0}^{\infty}(-x^2)^n$ for $|x| < 1$.
9.(b) $\int_0^{\sqrt{3}/3} \frac{dx}{1+x^2} = \tan^{-1}\left(\frac{\sqrt{3}}{3}\right) = \pi/6$.
11. See the proof of Theorem 8.9.
12. Use Mertens' Theorem (Theorem 7.17).
13. Use $1/(1+x) = \sum_{n=0}^{\infty}(-x)^n$ for $|x| < 1$.
14. $m^{1/n} \leq |a_n|^{1/n} \leq M^{1/n}$ and $\lim m^{1/n} = \lim M^{1/n} = 1$.

Section 8.3

3.(a) $x - x^3/3! + x^5/5! - x^7/7! + \cdots$
6.(b) Use the Maclaurin series for $\sin x$.
6.(c) Use the expansion for e^y and replace y by $-x^2$.

Section 9.1

4.(c) Use $\int x^2 \cos nx \, dx = (1/n^3)[2nx \cos nx + (n^2x^2 - 2)\sin nx]$,
$\int x^2 \sin nx \, dx = (1/n^3)[2nx \sin nx + (2 - n^2x^2)\cos nx]$.
4.(d) Use $\int x \sin nx \, dx = (1/n^2)[\sin nx - nx \cos nx]$,
$\int x \cos nx \, dx = (1/n^2)[\cos nx + nx \sin nx]$.
7. $f(x) = x^2 + x$

Section 9.2

1.(a) $\cos kt \sin(t/2) = (1/2)[\sin(k + 1/2)t - \sin(k - 1/2)t]$
1.(b) $\sin(t/2)\sin(kt/2) = (1/2)[\cos(t/2 - kt/2) - \cos(t/2 + kt/2)]$
4.(a) 2

Index

absolute value, 22
absolutely convergent, 216
accumulation point, 58, 71
additive identity, 14
alternating series test, 222
analytic, 260
Archimedean Principle, 31
Axiom of Completeness, 27

Bessel's inequality, 285
Binomial Theorem, 20
Bolzano-Weierstrass Theorem, 58
boundary of a set, 71
boundary point, 71
bounded sequence, 44
bounded variation, 193

Cantor's Theorem, 34
cartesian product, 3
Cauchy criterion for uniform
        convergence, 238
Cauchy criterion for uniform
        convergence of a series, 247
Cauchy product, 223
Cauchy sequence, 50
Chain Rule, 125
Change of Variables Theorem, 175
class C^n, 260
closed interval, 17
closed relative to a set, 70
closed set, 68
closure of a set, 72
cluster point, 58, 71
compact set, 73
comparison test, 203
complete ordered field, 13, 26
composition of functions, 8

conditionally convergent, 216
connected set, 77
continuous function, 87
convergent sequence, 40
convergent series, 199
convex function, 105
countable set, 32

De Morgan's Laws, 4
derivative of a function, 120
derivative of an inverse function, 147
differentiable, 120
Dirichlet kernel, 288
discontinuity, 113
diverge to infinity, 42
divergent sequence, 40

empty set, 1

Fejer kernel, 296
Fibonacci sequence, 58
field, 13
Fourier coefficients, 279
Fundamental Theorem of Calculus,
        174

Generalized Mean Value Theorem,
        136
geometric series, 202
greatest integer function, 109
greatest lower bound, 26

Heine-Borel Theorem, 74

Induction Principle, 19
infimum, 26
integral test, 209
interior point, 71

317

Printed in the United States
by Baker & Taylor Publisher Services